HE, LEO

He, Leo

THE LIFE AND POETRY OF LEW WELCH

Ewan Clark

Oregon State University Press   Corvallis

Library of Congress Cataloging-in-Publication Data

Names: Clark, Ewan, 1969- author.
Title: He, Leo : the life and poetry of Lew Welch / Ewan Clark.
Description: Corvallis : Oregon State University Press, 2023. | Includes
    bibliographical references and index.
Identifiers: LCCN 2023028392 | ISBN 9780870712470 (paperback) |
    ISBN 9780870712500 (ebook)
Subjects: LCSH: Welch, Lew. | Poets, American–Biography. | Beats
    (Persons)–Biography. | Beat poetry, American–History and criticism.
Classification: LCC PS3573.E45 Z55 2023 | DDC 811/.54 [B]–dc23/
    eng/20230712
LC record available at https://lccn.loc.gov/2023028392

♾This paper meets the requirements of ANSI/NISO Z39.48-1992
(Permanence of Paper).

**Oregon State University
OSU Press**

Oregon State University Press
121 The Valley Library
Corvallis OR 97331-4501
541-737-3166 • fax 541-737-3170
www.osupress.oregonstate.edu

*Oregon State University Press in Corvallis, Oregon, is located within the traditional homelands
of the Mary's River or Ampinefu Band of Kalapuya. Following the Willamette Valley Treaty of
1855, Kalapuya people were forcibly removed to reservations in Western Oregon. Today, living
descendants of these people are a part of the Confederated Tribes of Grand Ronde Community
of Oregon (grandronde.org) and the Confederated Tribes of the Siletz Indians (ctsi.nsn.us).*

For Lew. Wherever you are.

Contents

Illustrations

Abbreviations

All quotations and excerpts from Lew Welch's published works are cited parenthetically within the text by the abbreviations listed here. All other references are cited in the notes.

RB   *Ring of Bone*. Bolinas: Grey Fox Press, 1973

TT   *Trip Trap: Haiku on the Road*. Bolinas: Grey Fox Press, 1973

IL   *I, Leo: An Unfinished Novel*. Bolinas: Grey Fox Press, 1977

OBP   *On Bread & Poetry*. Bolinas: Grey Fox Press, 1977

IR   *I Remain: Letters and Correspondence*, volumes 1 and 2. Bolinas: Grey Fox Press, 1980

HWP   *How I Work as a Poet & Other Essays*. Bolinas: Grey Fox Press, 1983

HGS   *How I Read Gertrude Stein*. San Francisco: Grey Fox Press, 1996

Acknowledgments

This book would never have been written without the help, advice, and assistance of a great many people.

To begin with, those who offered advice and encouragement when it became clear that this project was going to be more than a passing fancy: Special thanks go to Clifford Burke, Andrew Hoyem, and Tom Killion for their time and thoughts on Lew. And to Trevor Carolan for his enthusiasm and encouragement.

Immense gratitude goes to Jack Shoemaker and Gary Snyder for not only taking the time to share their memories of Lew but for also offering advice and giving me the idea that this project was one worth pursuing. To Jeff Cregg and Kremas Carrigg for sharing their memories of Lew. And to Hugh Cregg for his generosity and time.

My research took me to several libraries and museums across the United States, and my thanks in particular go to Heather Smedberg and Nina Mamikunian at UC San Diego for their assistance during and after my visits to the Geisel Library. Thanks also go to Isabel Lyndon and Jim Holmes at Reed College, Alison Fraser at University of Buffalo, Vanessa Lee at Columbia University Libraries, Joel Minor at WU St. Louis, and Nancy Goldman and Jason Sanders at Berkeley Art Museum and Pacific Film Archive.

Many other people played important roles along the way, including transcribing interviews, helping to find literary executors or archival information, reading parts of the manuscript, giving me the keys to their woodland cabin so I could write in seclusion, providing permissions, copyediting, and generally offering other much-needed encouragement and assistance. Thus my thanks go to Frederick T. Courtright, Norman H. Davis, Jon Halper, Kathy Johnston, Elaine Katzenberger, the Kieft family, Cindy Leeson and the folks at the Ring of Bone Zendo, Paul E. Nelson, Sarah Serafimidis, Robert Shatzkin, Julia Smeekes, Tim and Tom at the

Utrechtse Boekenbar, Colin Tripp, Rene van der Voort, and all my family and friends. Special thanks also go to everyone at OSU Press, particularly Marty Brown and Micki Reaman. And extra special thanks to Robert Ross and Joe Lee, not only for allowing me to use works from their private collections but for their enthusiasm and generosity too.

I would like to specifically acknowledge a huge debt of gratitude to the following: Kim Hogeland at OSU Press for seeing something in this project and for giving me the chance to fly Lew's flag; Susan Campbell for going over the book with a fine-toothed comb and polishing all the rough edges; and Jaap van der Bent for his friendship and for turning me on to Lew in the first place when, as an undergrad, he suggested that Lew might be an interesting subject for my thesis.

Thanks also go to David Gremmels, Norman Fischer, Anne Longépé and Charlotte Service- Longépé, Julie Rogers, and Michael Williams for kindly granting me permission to use material from the estates of Robert Lavigne, Philip Whalen, Robert Service, David Meltzer, and Donald Allen, respectively.

And lastly, to my parents, Les and Eleanor. And to Marlou, Maya, and Finn. Without all of whom this would have been an impossible task.

Permissions

Excerpts from *Ring of Bone: Collected Poems 1950–1971*; *How I Work as a Poet & Other Essays*; *I, Remain: The Letters of Lew Welch & the Correspondence of His Friends*, volumes 1 and 2; *I, Leo: An Unfinished Novel*; *On Bread & Poetry: A Panel Discussion between Gary Snyder, Lew Welch and Philip Whalen*; and *Trip Trap: Haiku on the Road* (by Jack Kerouac, Albert Saijo and Lew Welch), Copyright © 1973, 1977, 1980, 1996, 1998 by Lew Welch. Reprinted with the permission of The Permissions Company, LLC on behalf of City Lights Books, www.citylights.com.

"My Love the Fisherman Comes Back" by Lenore Kandel. Poem originally appeared in *Genesis West* (A Supplement, August 1964). Anthologized in *Collected Poems of Lenore Kandel*, published by North Atlantic Books, copyright © 2102 by the Estate of Lenore Kandel. Reprinted by permission of the publisher.

Excerpt from "Chicago" from *The Complete Poems of Carl Sandburg*. Copyright © 1969, 1970 by Lilian Streichen Sandburg. Reprinted by permission of Mariner Books, an imprint of HarperCollins Publishers.

Excerpt from "On Teaching Without Pupils" from *The Collected Poems of Bertolt Brecht* reprinted by kind permission of W. W. Norton.

Excerpt from "On the Wire" from *Collected Poems of Robert Service* reprinted by kind permission of the Estate of Robert Service.

Material from Lew Welch's papers in the Special Collections Library at Geisel Library in University of California, San Diego, reprinted by kind permission of the Estate of Lew Welch and UCSD.

Material from Donald Allen's papers in the Special Collections Library at Geisel Library in University of California, San Diego reprinted by kind permission of the Estate of Donald M. Allen and UCSD.

Material from Philip Whalen's papers in the Special Collections Library at Bancroft Library in University of California, Berkeley and Reed College reprinted by kind permission of the Estate of Philip Whalen.

Portraits of Lew Welch and Robert Lavigne reprinted by kind permission of the Estate of Robert Lavigne.

Griffin Yearbook portrait reprinted courtesy of Special Collections and Archives, Eric V. Hauser Memorial Library, Reed College.

Author's Note

Like many poets and authors throughout the years, Lew Welch used what is now universally regarded as discriminatory and offensive language, including racist slurs, in his work. Although I have retained Welch's original terminology for the sake of accuracy, I do not condone it. Likewise, a few quoted works in this book by Welch's influences and contemporaries also contain offensive words, including the N-word.

Introduction

Once more alone, he climbs away
Into the mountains of original pain.
—Rilke, *Duino Elegies*

On November 26, 1963, six poets took part in a reading at the International Music Hall in San Francisco's North Beach district. Organized by Auerhahn Press publishers Andrew Hoyem and Dave Haselwood, the reading was a benefit showcase featuring Hoyem, Allen Ginsberg, Michael McClure, Philip Lamantia, Philip Whalen, and Lew Welch. The event was one of many such readings that had been held at venues across the city for several years, events that enabled both established and up-and-coming poets to perform for a general public ever more interested in the poetry of the Beat Generation and San Francisco Renaissance. For many of the participants, the Auerhahn reading was simply another opportunity to read their work, but for Welch it signified an important moment in his further development as a poet.

By 1963, Lew Welch was an established part of the counterculture in San Francisco. Although he had published only one work and a handful of poems in magazines and anthologies, his involvement in the poetry scene was undisputed. Welch's inclusion in Donald Allen's seminal 1960 anthology *The New American Poetry* was a watershed moment and has gone some way in maintaining Welch's poetic voice for generations of new readers in the years since. A consummate performer, Welch was confident, proud, and humorous. His performance at the Auerhahn event is among his best, fluctuating between the serious and the frivolous, the candid and the comic. However, the years immediately preceding the reading were anything but smooth. Welch had just returned from a self-imposed period of exile in the wilderness of Northern California and was

1

showcasing the poems he had written there. While his previous work had hinted at the power and eloquence he could display in his writing, the small collection of works that came from his hermitage—titled *Hermit Poems* upon its publication in 1965—showed a newfound sense of self and place he had spent years trying to find.

From his earliest works at Reed College in the late 1940s to what for many is his pièce de résistance, "Chicago Poem," written in 1957, Welch had drawn from a wealth of experience to gradually find his poetic voice. With his musical background, a keen sense of the importance of American speech, and a confident outward presence that belied the inward turmoil that would later blight his entire existence, Welch's graduate work on Gertrude Stein and his early poems suggested a surefire career in the arts. Yet by the mid-1950s his life could hardly have been more different from the one he had envisioned only a handful of years earlier. And while he would eventually become a proto-performance poet who arguably transcended the poetic frameworks of that period, his journey was fraught with obstacles, traumas, and problems that lend gravitas to the creative successes he achieved. In creating a singular and distinctive body of writing, Welch balanced and drew from the myriad derangements that helped to define, control, and, ultimately, destroy him.

In its own right, "Chicago Poem" is a career-defining work. It shows Welch at a period in his life when the notion of place in the concrete sense was becoming key to his existence. Between graduation in 1950 and the moment he sat down to write the poem in the summer of 1957, Welch had stagnated creatively, embarked on a life of relative conventionality, and struggled to come to terms with various issues he had carried since childhood. Indeed, by the time he became fully aware of the counter-culture happenings on the West Coast that he would one day become part of, Welch was married and living a traditional middle-class life. His experiences in Chicago, and in New York before that, laid a significant foundation for what he eventually achieved as a poet. It may even be argued that they were paramount. While other poets and authors were drawing widely from the artistic and cultural milieu on both the East and West Coasts, Welch was walking a tightrope of conventionality in the Midwest, teetering on the edges of sanity and domesticity, with only occasional bursts of creativity. The life he had envisioned for himself at Reed as the poetic "voice of the latter half of the 20th century" was now totally gone. He was largely oblivious to the literary gatherings taking place in San Francisco and New York, despite being tantalizingly close to both on several counts. Welch was not one of the poets who enjoyed Kenneth Rexroth's "creative coteries" in North Beach or partook in the poetry and jazz jams in Greenwich Village. Instead, for much of the 1950s, Welch navigated a path that was laden largely with contradictions, seeking answers to the personal existentialist questions that threatened to consume him. However, it was this journey—and the answers he found along the way—that would ultimately enable Welch to find the bones of an authentic poetic voice. And, when he did, the force was such that it changed the course of his life forever.

By the time he wrote "Chicago Poem," Welch had been in the city for almost six years; the poem was an outpouring of all his feelings about living there. It exemplifies a key aspect of the poet that Welch would become, including as it does his biting commentary on urban life and the blinkered insular pride of Chicago's inhabitants, as well as the carefully constructed linguistic vignettes that would become a characteristic signature of his later work. The poem rages in its condemnation of the all-consuming city, and Welch manages to establish new parameters for himself that not only incorporate his previous poetic idols and personal fragility, but also allow for the new influences he had been exposed to. It is impossible to read "Chicago Poem" without seeing similarities to Allen Ginsberg's "Howl." In the same way that Ginsberg's Moloch devours his supplicants, so Welch's poem ends with the image of a city dying if it isn't sustained by its inhabitants. And thus Welch's Chicago becomes the metaphorical equivalent of Ginsberg's angry idol. But rather than let it devour him, Welch decides to escape and find other sources to which he can offer himself. While Welch's legacy cannot be pinned to a single work, "Chicago Poem" was a significant moment not only in helping to cement his position within the canon of American poetry but also in providing him legitimacy and a sense of belonging.

In *Genesis Angels: The Saga of Lew Welch and the Beat Generation,* Aram Saroyan firmly places Welch within the framework of this literary milieu. Here was a generation of writers and artists at the forefront of a new creative America in the 1940s and 1950s, and although Welch was not in the vanguard of that movement, his place within it is undoubted. Yet where exactly did Welch fit in? By the time he had relocated from

Chicago to the West Coast in late 1957, the San Francisco Renaissance was an established fact, and the Beat Generation was arguably already waning. It might even be said that Welch is much more than a poet in the Beat or counterculture context alone. The insight and intellect he displayed in his prose pieces, the humor and concision of his plays and songs, the performative nature of his work in general, and the authenticity of his language are all aspects of Welch's writing that make him more of a proto-performer in the vaudevillian vein than a mere "Beat" poet.

Welch had initially walked a path of relative conventionality in terms of his early poetics—finding inspiration in the experimentalism of Anglo-American poets such as William Carlos Williams, W. H. Auden, and Ezra Pound—but his later poetry clearly imagines a new Eden in which humanity can erase the mistakes of the past while looking to deeply embedded belief systems for guidance. What started with "Chicago Poem" and finds its synthesis in *Hermit Poems* is the poetry in which Welch's literary narrative finds its home, in the poetics of that time, which can be easily aligned with the sensibilities of many of his peers on the West Coast. As Gary Snyder became a leading voice in propagating the negative confluence of environmental fragility and human ambition, Welch's later writing often placed the narrator in various utopias that highlighted humanity's fallibility and insignificance as well as challenging us to adapt to and preserve the world in which we live. He also highlighted the hopefulness that came with self-knowledge and the acceptance of our failings, which were often manifold and related primarily to the natural environs of the West Coast. Welch's Eden was found on the slopes of Mount Tamalpais, the gorges of the Rogue River Valley, or the Pacific beaches along the fringes of Marin County. In this sense Welch inhabited a clear space within the realms of regionalism and place. His was a work firmly rooted in California and Oregon, and although he may have long harbored the dream of venturing beyond the shores of North America, there is a poignancy and inevitability in the fact that his work highlights that he never managed this. Indeed, perhaps his poems and essays are all the more powerful as a result of his not being able to extract himself from his immediate surroundings. Yet where the landscape is quintessentially western, Welch also mixes the influences of the wider Pacific Rim into his work, and as such often extends the sense of place beyond what could be construed as "local." By envisioning the poetics and cultural histories of the Far East in some of his work, Welch uses the West Coast as a base from which to explore the wider region—and to interlink the imagined with the concrete to construct his own utopia. As such, Welch clearly intended to establish himself as a "regional" poet, albeit one more concerned with inter-regionality than anything more localized.

Yet it is the disenfranchisement of the individual that represents a fundamental aspect in establishing what "place" might have meant to the poets of the San Francisco Renaissance. Just as Kerouac actively sought to reinhabit a place on the fringes alongside the fellaheen, so poets such as Snyder, Whalen, and Robert Duncan operated on those margins already. However, their marginalization was not uniform. Indeed, as Michael Davidson writes, "For each poet, 'place' means different things, but for all . . . the fact of living in the West means living in the margins, whether this implies a radical political tradition, nontraditional religious practices, or extreme psychological states."[1]

Although Welch's place is more clearly defined by his religious practices and his psychological state, in this context it is easy to see him as being an amalgamation of all three of Davidson's categories.[2] However, the primary elements that characterized much of his work—and thus reflect place—are undoubtedly an early and serious fascination with Buddhism that was eventually usurped by the psychological traumas of his youth and, later, alcoholism. And while these aspects establish place and a sense of community, they may also offer clues to the notion that place can further be seen as an idea rather than as anything concrete. In Welch's case, his return to the West Coast into a milieu of creative freedom and poetic independence could equally be viewed as a return to what could have been, had he not left for Chicago, as well as what he intended to become once he extricated himself from the shackles of domesticity. A reconnection to his past offered him the potential artistic fulfilment he craved. In this sense, the return constitutes the rebirth of the poet and the establishment of a narrative that sought to unearth Arcadia but found America instead.

———

Since his disappearance on May 23, 1971, Lew Welch's poetic legacy has been largely ignored. While his work may never have reached the heights attained by that of his friends and peers, his writing—like the life from which it was drawn—offers a wealth of beauty, humor, hope, and despair. It provides brutal and enchanting insight into his dreams and desires as well as into his hang-ups and worries. It highlights his views on society, ecology, and language in an America emerging from the ravages of one war and entering into another, and on the ever-burgeoning rise of corporate sensibility and the strategies needed to escape it. Welch's work plumbs the depths of his consciousness and asks its readers to enter a world that abounds with the juxtapositions of urbanism and nature, corporations and communes, writing and speech, tenderness and violence. His voice speaks as loudly and as eloquently now as it always did, dealing with issues that continue to be relevant and, as such, need to be heard.

Ring of Bone, Welch's collected works of poetry published in 1973 by Donald Allen, contains a preface written by Welch before he disappeared, in which the reader is introduced to the three principal characters in Welch's work: "The Mountain, The City, and the Man who attempts to understand and live with them." Welch describes how he had envisaged the structure of the collection as "circular, or back and forth," constituting a "spiritual autobiography in more or less chronological sequence." I have taken these three characters as the vehicles with which to travel into the different facets of Welch's life and work: the circular narrative is maintained, as is the more or less chronological order. Given that "the Man changes more than The Mountain and The City, and it appears he will always need both" (RB, 3), this study similarly relies on the mutual interdependence of all three.

Part 1 focuses on "the Man," Lew Welch, weaving its way from the farmlands of Kansas through the streets of Phoenix, Portland, Chicago, and San Francisco, on to the mountains and pastures of the American West, the Pacific seaboard and Northern California, and through the corridors and hallways of universities, advertising agencies, and barrooms.

Part 2 introduces "the Mountain" and the part it played in shaping Welch and his work. It touches on his views on nature and ecology as he walks in the Trinity Alps, on Tamalpais and Shasta, and in the Sierra Nevada. The deep love that Welch felt for such landscapes is manifest alongside the hope they offered him and the eventual relief.

Finally, part 3 cements "the City" as an essential element in Welch's life. For all his disparaging commentary on the evils of urban life, without them we would never have seen his "blind, red, rhinoceros" snuffling on the shores of Lake Michigan or the "millions of terrified beings scurrying through senseless mazes." Neither would we have seen or understood the role these things played in what he felt were essential catalysts for change. Welch may have been eternally drawn to the mountain, but the city was as invaluable to the man as anything else—indeed, it was a necessity.

In my quest to write his story, I continually asked myself whether Welch's body of work justified the research. Is he worthy of a biography? Are his life and work sufficiently unique? But as he slowly revealed himself to me, so these questions faded. I was left with Lew's own maxim as my ultimate guide:

Guard the mysteries, constantly reveal them! (RB, 128)

PART I

The Man

We invent ourselves out of ingredients we didn't choose, by a process we can't control.

— Lew Welch (RB, 29)

1
Farmers, Flappers, and Fuck-Ups

Lewis Barrett Welch Jr. was born at 9:00 p.m. on August 16, 1926, in Phoenix's Arizona Deaconess Hospital, the first child of Lewis Barrett Welch Sr. and Edith Dorothy Brownfield Welch. Although Lewis was named after his father, Lewis Barrett Welch was also the name of his paternal great-grandfather, whose mother, Susannah Barrett, had married into the Welches in 1838, thus beginning a family tradition of naming that would end with the birth of Welch Jr. on that August evening.

Though this was her first child, it was not Dorothy's first pregnancy. She had delivered a stillborn sibling the previous year; given the complexity of the mother-son relationship during Welch's life, it is tempting to attribute a certain amount of extra importance to his birth, reinforced by Welch when he would later talk of being a reincarnation of both the dead brother he never knew and of his maternal grandfather, who he knew only through the stories told to him by his mother. It is also tempting to think that this experience created an extra-special bond between mother and child, a particular psychology that may have in part defined them. Indeed, Welch wrote later that "the second birth leaves residues of the first,"[1] and as events unfolded in the years following the 1928 birth of Welch's sister Dorothy Virginia, Welch would form an extremely close bond with his mother that not only contrasted sharply with the relationship he had with his father but also in many ways determined the course of Welch's future career and of his continual dependence on her throughout his life. As a result, we must start with Dorothy Brownfield to understand the rest of this family history.

———

Dorothy Brownfield met Lew Welch Sr. in the spring of 1923, and after "falling terribly in love with him,"[2] they were married in an intimate private ceremony in the Welch home at 1420 North Third Street in Phoenix on May 26, 1924. Witnesses included Dorothy's fellow socialite Dorothea Moore, who had been a close friend and confidante of Brownfield since childhood. The newspapers of the time were full of glowing praise for the bride-to-be, describing her as "a girl of rare charm and brilliancy" and being "a social favorite."[3] While the wedding was an understated affair, uncharacteristically so, given Dorothy's background, the social columns bristled with the details of her wedding attire, painting an opulent picture and reminding readers of the Brownfields' wealth and taste. The *Arizona Republican* described the bride as having

> chosen a frock of white embroidered French batiste with filet and draped over pink taffeta. A hat of white with trimming of pink ostrich will complete the costume. Her bouquet will be of Cecil Bruner [*sic*] roses and valley lilies. . . . Mrs. Harold Moore, who since early girlhood has been a close friend of Miss Brownfield, will be the only attendant. Her attractive coloring will be well emphasized by a figured chiffon gown of white background with orange traceries and a girdle of pastel shades. Her hat is of white trimmed with ostrich of gold tones.[4]

It is therefore easy to imagine Dorothy Brownfield as the archetypal "twenties flapper, pretty, and high-style,"[5] hailing as she did from a family that was on a completely different rung of the social and financial ladder than the groom. Indeed, given what we know about the respective families, it is difficult to understand how such a marriage could have come about at all, save the fact that, as Welch Jr. said, his father looked like "Tyrone Power and Cary Grant," and that this alone was apparently reason enough for Dorothy to develop "hot pants"[6] for him. However, in a deleted section from *The San Francisco Poets*, a collection of interviews conducted by David Meltzer and published in 1971, Welch elaborated that Dorothy "never saw a man like that in her life. All those fops in Arizona, and in comes Big Lew Welch. Big Lew Welch comes hopping in from Kansas. And he took her. And she was a bad ass to take! She was a skinny, frigid, horsey, rich, bad ass Arizona chick."[7]

The Brownfields were something of an institution in and around Phoenix in the years immediately after World War I. Dorothy was born in Seward, Nebraska, in May 1902, and by the time she met Welch Sr., she had enjoyed the sort of privileged upbringing that her husband-to-be could have only dreamed. She spent her summers in, among other places, upmarket Southern Californian seaside resorts such as the Coronado and La Jolla or purpose-built desert oases like the Garden of Allah Guest Ranch in Arizona. She traveled extensively throughout the United States, and upon graduation from Phoenix Union High School in 1920 (the yearbook described her as being as "irresistible as the wild rose"[8]), Dorothy sailed with her parents on the RMS *Victorian* to Europe, where her father, the eminent Phoenix-based doctor Robert Roy Brownfield, was undertaking a research trip. Among other countries, they visited Great Britain, Belgium, France, and Germany, combining Robert Brownfield's hospital visits in Edinburgh, London, and Paris with trips to the sites of the major World War I battlefields and other noteworthy historical and cultural attractions. It is easy to imagine the eighteen-year-old Dorothy on the Canadian Pacific liner sailing from Quebec to Liverpool and beyond, having the sorts of experiences her husband, and later her son, would never have.

Four months later, following their return from Southampton, England, aboard the SS *Imperator*, Dorothy commenced her studies in music at the University of Arizona in Tucson. She was part of the period's social set, as evidenced in the countless references to the events and gatherings of which she was part in the society columns of the major Phoenix-based newspapers.

The family's wealth and status were built on the success of Robert Brownfield who, having put himself through medical college, became one of the most famous and influential surgeons in the US Southwest. A graduate of the University of Nebraska, he became a fellow of the American College of Surgeons and the president of the Maricopa County Medical Society, winning praise from the US surgeon general for his development of an electrical surgical instrument that aided hearing tests. Brownfield also moved in social circles that included the department store magnate Baron M. Goldwater (father of Republican senator and 1964 presidential candidate Barry Goldwater) and the industrial design guru Warren McArthur, both of whom he knew because of their collective involvement in the Phoenix Country Club. In his interview with Meltzer, Welch admits regret at never having had the opportunity to meet Brownfield, describing him as "a man of great parts"[9] and lauding not only his medical prowess but also his toughness and tenacity. Brownfield was a man who was extremely successful in his career and who used these characteristics to great effect in the boxing ring too. Indeed, one could argue that these were characteristics his grandson would have dearly liked to have possessed in greater measure than he did.

The opportunity to meet Robert Brownfield was denied Welch because of Brownfield's untimely and tragic death in an automobile accident on May 1, 1921. Returning from a garden party in Mesa, Arizona, with his wife Edith and the Goldwaters, the car in which he was traveling swerved to avoid a horse and cart. Mrs. Goldwater, who was at the wheel, lost control of the vehicle, which hit an embankment and flipped multiple times before coming to rest on the sidewalk. Witnesses reported seeing Brownfield thrown from the car, suffering a broken neck. Despite the prompt response of emergency services and his subsequent surgery at Southside Community Hospital in Mesa, Brownfield was pronounced dead at 7:50 a.m. the following morning. He was forty-one years old.

In the years immediately following the death of her father, Dorothy lived in various cities up and down the West Coast, a precursor to the transitory life to which she would later subject her own family. Newspaper reports state that after the accident she left Phoenix with her mother and initially went to live in California. The pair spent the winter of 1921 in Seattle, where Dorothy was registered at the University of Washington.[10] However, by the following year, they had left the Pacific Northwest and returned to California. The summer of 1923 was spent in Hawai'i: they sailed from Los Angeles in early May and returned to San Francisco in late September. The passenger manifests for both the outward and homeward voyages give different home addresses, and it might be concluded that their five-month period in Hawai'i bridged a move from Pasadena to Palo Alto. Here Dorothy was about to embark on a new course of studies at Stanford, and the city would remain her on-and-off home for much of the rest of her life. By the time she returned to Phoenix in the spring of 1924, her mother had also died. Edith Brownfield's death certificate shows that she died in Los Angeles on January 2 at the age of forty-five.

It is impossible to say how this loss, following on so quickly from that of her father, affected Dorothy. While a rapid love affair and marriage to a Midwest farmer's son may not necessarily have been the best course of action, it is perhaps, under the circumstances, entirely understandable. The loss also left her as the sole heir to what was reported to be a considerable fortune. Welch spoke of his mother having "$100,000, American big ones, in 1922"[11] and of her being "super-rich." A court report published two weeks after her father's accident stated that his estate had been valued at $15,000.[12] In addition, life insurance policies paid out in the months immediately thereafter increased that amount considerably.[13] Whatever the exact sums and policies involved, Dorothy received a sizable inheritance that may have played a more prominent role in her future marriage than she could ever have imagined.

When and where she met Lew Welch Sr. is uncertain, although their engagement notice from April 1924 claims that they had met a little over a year earlier. Even their son was unaware of the exact circumstances of his parents' courtship. Dorothy had returned to Phoenix from Stanford shortly before the announcement of their engagement, which was only four weeks before the wedding. It is unclear at what stage of her studies she returned (Stanford records show that she earned her BA in psychology in 1942), but she is featured in the 1924 yearbook as a member of the Kappa Alpha Theta sorority, which she had been a member of since her days at the University of Arizona.[14] She likely suspended her studies to marry Welch and set up home in Phoenix, at 319 East Culver Street, the first of four addresses the Welches would have in Phoenix over the next five years.

For his part, Welch Sr., who had moved to Phoenix from Kansas in 1921 and was joined by his sister Rubea (who was the attending midwife at Welch Jr.'s birth) and his parents in the winter of 1922, was working at the time of the marriage for the Phoenix Savings Bank and Trust. Although there is no official record of his having studied or trained as a banker after his discharge from the army in 1919—indeed, his higher education appears to have been limited to a single semester at Kansas State Agricultural College, where, in the spring of 1920, he took courses in agriculture and dairy farming[15]—Welch ended up working at a bank in Fort Scott, and it was most likely from there that he engineered a transfer to Arizona. Lew Welch's later contention that his father was helped by Dorothy, or by her mother, or by one of the family's connections to secure a similar position in Phoenix is therefore questionable and may well be an early indication of the inconsistencies and mysteries that surrounded Welch's relationship with his father.

Yet whatever the circumstances that led to his career switch, by the time of his marriage Welch Sr. had started to make a name for himself within the business community in Phoenix, and he went on to hold important positions at the country club. He was also a regular feature on the golf circuit, playing and winning tournaments across the Southwest. Furthermore, in his capacity as the chairman of the Tournament Committee and secretary of the Arizona State Golf Association, Welch proved successful in the organization of such events as the Phoenix Open, which has since gone on to become one of the premier golf tournaments on the US PGA Tour.

What occurred in the years after their marriage is something of a mystery, but Welch Sr. has been variously accused by his son of "cheating the very people who got him placed in a distinguished banking job" and of being "an embezzler against the family," accusations that may well have played a role in the breakup of the marriage after Virginia was born in February 1928.[16]

When Dorothy left Phoenix for California in June 1929, to spend the summer in La Jolla, she was twenty-seven years old; she was never to return. Her application for a divorce was granted on August 23, 1929, and two days later Welch Sr. placed an advertisement in the *Arizona Republican* announcing a "mammoth auction" in which, having sold his house, he was selling its entire contents too.[17] Having been a bright light in the Phoenix glitterati a mere five years earlier, Dorothy was now the mother of two infant children, had been married and divorced, and had lost both her parents while they were still in their forties. Although Welch Sr. would make sporadic appearances in her life over the next few years, she was solely responsible for the upbringing of Lew and Virginia, and she never married again. She was, however, financially independent, and given her background it is hardly surprising that Dorothy would eventually go on to earn not only her BA in psychology but also a master's in social welfare from the University of California, Berkeley, and a doctoral degree in social work and home economics from Cornell. Records show her master's thesis was titled *An Investigation of the Activity for Sensory Responses of Healthy Newborn Infants*; combined with her previous studies, it is clear that Dorothy's interests were firmly focused on child development, and her future appointment as the chairperson of the Department of Child Development and Family Life at University of Nevada, Reno, further cements this impression. It is, however, somewhat ironic that her field of expertise was so focused on children and the family, an irony that could not have been lost on either of her children, given their relatively dysfunctional upbringing and the subsequent problematic nature of many of their future relationships. It might even be argued that these realities fueled her academic fire. Her single-minded independence, surely in itself a by-product of losing both her parents when she did, was therefore only one aspect of her drive. Lew Welch would later rather disparagingly go as far as to say of his mother that it had been "awful to be born to a rich, selfish shiksa," [18] a statement that now seems contrived and somewhat misplaced in light of his dependence on Dorothy throughout his life.

In marked contrast to Dorothy's upbringing, Lew Welch Sr. was born into a farming family on November 24, 1899, in Redfield, Kansas. In those days, Redfield was little more than a collection of rural buildings and farms a hundred miles south of Kansas City. When Welch Sr. was born, the townsfolk numbered fewer than two hundred; his father, Frank C. Welch, was a farmer of some stature in the town. Records show that Frank could often be found buying and selling at the rural fairs in and around Redfield, Uniontown, and Fort Scott. He was also showing prize-winning livestock as far afield as Chicago, and seems to have been, to some extent at least, relatively successful. One newspaper article from March 1911 reports that Frank bought a 360-acre working farm, which, despite being an archetypal picture of middle America that is in stark contrast to the high life enjoyed by the Brownfields, nonetheless indicates a certain measure of success.[19] It is also interesting to note the incongruities between how Welch perceived his grandparents and the apparent reality: he described them on the one hand as being "real good, straight, go-to-church-every-Sunday Kansas people"[20] and on the other hand as "fuck-ups."[21] There seems to be little middle ground, yet a farmer with a property of this size hardly seems to constitute being a "fuck-up . . . [who] had bad luck . . . and always sold his land cheap."[22] Indeed, Frank's social standing, while a far cry from that of the Brownfields, eventually afforded him the position of district judge, a tenure that clearly establishes his standing in the community and one that he held until his death at Fort Scott in 1955.

The Kansas Welches were descended from Ulster-born Irish immigrants, a fact of which Lew was particularly proud. He often spoke of himself as coming from Irish stock, a view that may have also been substantiated by the Irish ancestry on his maternal side. Dorothy's ancestors had settled in Pennsylvania in the mid-1600s, originating from both Northern Ireland and Scotland, a reflection of the political and religious situation during a time in which it was almost a foregone conclusion that families would come from both countries. Travel between the two was both frequent and inevitable, and so were marriages. And as Lew Welch developed his own sense of America and his place within it, this gritty immigrant sensibility reared its head every time he was faced with an issue that required him to either stand up and fight or turn to his mother for help. He was continually faced with the dichotomy of siding with either the Welch or the Brownfield gene in his nature. Although the intimate details of Welch's paternal forefathers are difficult to establish beyond generalizations and scraps of information from local newspapers, it is easy to see the comparisons between Welch the fisherman, the harbor worker, or the forest laborer and his rural ancestors.

In an unpublished poem from the early 1960s titled "A Poem for the Fathers of My Blood: The Druids, the Bards, of Wales and Ireland," Welch not only shows his ancestral pride but also writes of the similarities between the country in which he was born and the land of his forebears, thus underpinning this immigrant sensibility of transplanting heritage from the old to the new:

> It's no wonder I love this country which is
> so like the land of the races which made me:
>
> All the starkness and size and
>   unexpected plots of gentleness, the
>
> ferns and shamrocks and mossy stones[23]

In the various drafts of the poem he proudly writes of "all the ancient Druid-blood of my veins" and of America being a "hard land too (as they say). Just like the old countree." It is a "hard land" in which he will marry and have "fine sons," rid himself of his madness, and sing "songs so strange I cannot even write them down . . . howling in his hills of Sur!"

Welch spoke of only meeting his paternal grandparents once, when he was five, so exactly where his impression of them comes from is a mystery, given the fact that his father was not around enough to paint a picture of life on a Kansas farm. Indeed, the source of Lew's image may be more accurately attributed to his mother and her view of the Welches. It may also be more of a reflection of what he thought of his father: a paradoxical mixture of pride and disappointment.

Despite his superficial and sporadic relationship with his son, Welch Sr. nonetheless passed on a number of traits and talents that would, in one form or another, have an influence on Lew. For example, Welch Sr.'s athletic prowess was something that Lew inherited, and he also taught his son a measure of love for nature in which he could fish and shoot. Welch was particularly proud of having received his first rifle from his father, a wholeheartedly American transaction in a time when this was the very least that could be expected of fathers; indeed, when Welch met and later moved in with Magda Cregg in 1964, he bought Magda's youngest son Jeff his first rifle, too, a small 0.22 that he still owns to this day.[24]

Both Welches were fast runners and accomplished football players. Both hunted and fished while embracing the outdoors. Welch Sr. was a marksman of some note, as was his son. Both served in the US armed forces, yet neither saw active combat. Both struggled to overcome their demons, and neither succeeded. And there were other similarities that in many ways would define Welch's impassioned existence as well as his extremes and his failings in adulthood. Lew Welch, like his father before him, would finally one day also ask, "How did it get so god awful complicated?"[25]

Colonel Lewis Barrett Welch Sr. died on March 12, 1947, as a result of cirrhosis of the liver. Whether, as is often the case, this was entirely due to alcohol abuse is unknown, but it is tempting nonetheless to see some link between father and son that might help explain Welch's later problems with alcohol and the fact that he confessed to having "ruined his liver . . . by his mid-thirties."[26] Indeed, as his alcohol dependence grew, he would also suffer from the same ailment that contributed to his father's death. A case of history repeating itself? Whatever happened between father and son, whatever their relationship in life, on his death Welch Sr. returned to his roots in Kansas. His funeral service was presided over by the Reverend Benjamin Young two days after his death, and he was buried in Fort Scott

National Cemetery, close to both his parents, who outlived him. He, like both the Brownfields, died before reaching the age of fifty, an unfortunate trend that was not to end there. He was forty-seven years old.

2
Strange New Cottages in California

In 1929, La Jolla was the archetypal upmarket Southern Californian resort town. Described in resort promotions as "the Jewel of the Pacific Coast, the Capri of America," the town was famed for its beautiful coastline, high-class hotels, and stunning nature.[1] Dorothy Brownfield's choice to take her children to La Jolla was an obvious one—she was familiar with the area on account of her frequent summer visits to the coast, which had previously offered some sense of solace in the period immediately after her father's death. Their stay in La Jolla, however, was a short one. Within a year, Dorothy had made the short hop south into San Diego, and while Welch attended nursery school in the city, his two-year-old sister went to stay with their aunt Rubea and uncle Harold in Sacramento.

Although accurate accounts of his childhood are scarce, Welch describes this early period of his life in exaggerated terms, most interestingly

Figure 2.1: Lew and his little sister, Virginia. Photographer unknown. Courtesy of the Estate of Donald Allen and UCSD.

in his candid descriptions of the problems he experienced not only psychologically but also with regard to eating and to making (and keeping)
friends as a child. Describing a departure into infant "madness"—arguably his first retreat—Welch proudly claims, "I went to the loony bin
when I was fourteen months old. . . . It is the world's record. Even among
my beat generation friends. I have the world's record. I copped out, I
went crazy, split, I said 'forget it!'"[2] He attributes this breakdown to his
mother's inability to provide him with enough milk on account of her
flapper sensibilities, small breasts, and the practice of her probably "tap-
ing them down."[3] She apparently made him feel so terrible as a child
that he developed deep traumas regarding his childhood and his mother's
approach to it. He spoke of her leaving him with deep scars and of the fact
that his childhood was one he would not wish to repeat. In his interview
with Meltzer, Welch goes as far as to say "I HATE HER! BECAUSE SHE DID
SOME SHIT TO ME THAT YOU CAN'T BELIEVE, AND I NEVER GOT OVER IT."[4]
When Meltzer likens Lew's childhood to experiments on baby monkeys
that are deprived of all mothering ("no body contact. No warmth, no
holding"), Welch simply replies, "That's it."[5]

Yet there is certainly an element of romanticism in his account of his
upbringing and his infant breakdown that in some ways contradicts the
venom he directs toward his mother. He concedes that his need for Dorothy was so strong that she was unable to provide him with the attention he
desired, despite her best efforts. His self-confessed hatred of her was surely
born more out of frustration on his part than intentional neglect on hers,
and his description of Dorothy as "a terrible mother who was the absolute
form of Kali, death"[6] is thus hyperbolic and arguably unjustified.

In Welch's memories of his childhood, Dorothy is portrayed as the
destructive manifestation of the Hindu Goddess Kali rather than the
nurturing one. In providing her children with *moksha*, or liberation,
from *saṃsāra* in an eschatological sense, and with self-realization or self-
knowledge in a psychological one, Kali is actually the most compassion-
ate of goddesses, and as such the opposite of what Welch saw in his own
mother. As time passed, Dorothy gradually shed the mantle of this negative
manifestation in her son's eyes and gradually transformed into a positive
force that Welch acknowledged was invaluable to him—the need he felt
for her was such that resistance ultimately became futile. Yet irrespective
of whether his interpretation at the time was accurate or fair, Welch did

characterize his initial resistance to his mother as a major source of literary and linguistic inspiration, and of his admiration of and love for alternative goddesses in nature. In having initially rebelled against Dorothy, or her apparently destructive incarnation, Welch instead looked to nature to find the feminine influence he felt was lacking in his life and which he could worship, with Mount Tamalpais featuring in his later life as one among the greatest of these goddesses. When Welch writes in *The Song Mt. Tamalpais Sings* that "there is nowhere else to go," is he then asserting not only that Tamalpais signifies the point at which the American continent ends but also that in looking for solace, comfort, or inspiration, he turns to the mountain for guidance when many others turn to their mothers?

Indeed, at the Politics of Ecology teach-in organized at UC Berkeley in May 1969, Welch explained that he would sit on the rocky slopes of Tamalpais and ask her for advice. Before reading the poem, he posed the questions, "What is all this human movement?" and "How people come from all over the place and here we are on the last cliff of a continent and what does it mean?"[7] Her answer (and perhaps his too) was that very poem.

———

During Welch's childhood, his mother relocated her family with such frequency that Welch was always the new kid on the block. He was always the kid that attracted the class rejects because he was new and unaware of their status. He was always the kid that was subconsciously waiting for the next move to the next new cottage in the next California town. Welch remembered these feelings with a mixture of pride and melancholy, describing the sense of triumph he felt at being victorious in the playground game of pom-pom-pullaway while still being nameless to his peers. He also spoke of his very swift legs and athletic prowess, which he had inherited from his father, "Speed" Welch, the Kansas high school football star—the prowess that would later help open the door to track and field at his own high school.

After San Diego, the family moved up the coast to Santa Monica, where Welch finished the preschool education he had started at the Dale Street Nursery. Years later Welch wrote, "I spent years 3, 4 & 5 in a little stucco house [there]. My brain is still scarred by it. There was a 'tile-top' coffee table in the living room. My mother made me wear high shoes & short pants to kindergarten. The 'Gypsies' stole my trike. In helpless

frustration I used to urinate all over my clothes in my closet" (IR1, 115).
To compound this, Welch was already beginning to intentionally set himself apart from his peers in looking for solitude:

> When I was 6 yr. old I used to take my bicycle to the ocean because
> my household was filled with very nervous women and I had to
> get out of there and my friends were nowhere, just kick-the-can
> bullshit friends. . . . So I would go to the ocean and sit on an ocean
> rock and sit for hours and get all the sound of that ocean.[8]

Aside from the obvious issues that Welch retained from his time in
Santa Monica, it is also interesting to note that, in 1932 at least, Welch
Sr. was apparently back in the picture. In the annual Index to Register of
Voters for the Santa Monica City Precinct No. 31, the couple is registered
as living together at the same address, and despite their divorce three
years earlier, Dorothy is listed as Mrs. Dorothy B. Welch—one concrete
example of Welch Sr.'s sporadic presence in his family's life. But for how
long? By late 1935, after moving south again and spending the three previous years in three different houses in Coronado, California, the family
was living in La Mesa, where Welch attended yet another elementary
school before another short relocation to El Cajon. This move prompted
not only a further switch of schools but, more importantly in terms of
Welch's future career, the opportunity to meet his first real source of literary inspiration in the form of his seventh-grade teacher, Robert Rideout.
Welch later spoke of Rideout as the person who, along with his mother,
had turned him on to reading. By the time he was eleven years old Welch
had already experienced the joys of the library and the world of knowledge and information on offer to him there (courtesy of his frequent visits
to the public library with Dorothy, who would not only take him there
but also was allegedly prone to leaving Welch in San Diego bookstores,
knowing that he was going to be lost in books for a couple of hours while
she went on unknown errands).

But it was Rideout who instilled in the young Welch the wider benefits
of reading and devouring the works of particular authors. Indeed, Rideout
(under whose tutelage Welch claims to have read 160 novels in a year)
advised his charges that if they enjoyed one novel by a certain author they
would be sure to enjoy the rest. And so Welch began reading the novels

Figure 2.2: Lew dressed as a
cowboy and riding a pony, c.
1933. Photographer unknown.
Courtesy of the Estate of
Donald Allen and UCSD.

of, among others, Charles Nordhoff and James Norman Hall, Will James,
and Ernest Thompson Seton. While Nordhoff, Hall, and James provided
their readers with the typical swashbuckling and gun-toting fare for young
adolescents, Seton's influence on Welch's writing in particular may be
more telling, if Seton's outdoorsmanship and nature writing are anything
to go by. Welch's preteen focus may have been on mutiny, piracy, and
cowboys, but his later immersion in the natural worlds of Thoreau and
Jack London may have found its beginnings in Seton's tales of wildlife,
woodcraft, and Native American lore.

It was also during this period that Welch first discovered poetry. Joyce
Kilmer's 1913 poem "Trees" and Robert Service's World War I collection *Rhymes of a Red Cross Man* were among the first works to which
he was exposed. The former is a very simple and archaic ode in which a
tree presses its "hungry mouth" to the "earth's sweet flowing breast." For
the prepubescent Welch this may have denoted some semi-Oedipal con-
notation that reiterated his earlier frustrations concerning Dorothy and
his eating patterns. He even went so far as to say that Kilmer "snapped

something in my head." And while conventional in rhyme and meter, Service's poetry would nonetheless have served as an ideal introduction to the techniques and literary devices that poetic balladry such as Service's contains. It would also have provided Welch, albeit unknowingly at the time, with his first exposure to the poetic use of dialect and regional language (in this case Scottish and Irish), something he would later champion through his own treatment of pure contemporary American speech as the basis for not only his but all poetry. Indeed, many years later, when Welch was on his hermitage in the Trinity Alps, he often turned to Robert Service when reciting poetry with the locals, recounting that he felt the works were still very relevant and that Service was a poet who could write the kind of things that people can say to good friends in a mountain cabin. Perhaps some small part of Welch also had a similar desire for his own work to be revered in the same way.

In his lengthy 1969 interview with Meltzer, Welch attributes one of Service's poems in particular, "On the Wire," as the one that drove him to poetry as an eleven-year-old boy, on account of its truth: "I suddenly saw a man telling truth. I had never seen it before."[9] The image of a heavily wounded soldier tangled in barbed wire and stranded in no-man's-land offered the young Welch a view of reality such as he had never seen before. While the wartime exploits of Nordhoff and Hall provided fantastic escapism through the exploits of stereotypical heroes, Service painted heroism from a completely different perspective, that of the soldier beyond salvation taking his own life to end his suffering. The final stanza has a wonderful poignancy that, given Welch's own end, has a more than striking sense of ironic verisimilitude:

> Hark the resentful guns!
> Oh, how thankful am I
> To think my beloved ones
> Will never know how I die!
> I've suffered more than my share;
> I'm shattered beyond repair;
> I've fought like a man the fight,
> And now I demand the right
> (God! how his fingers cling!)
> To do without shame this thing.

> Good! there's a bullet still;
> Now I'm ready to fire;
> Blame me, God, if You will,
> Here on the wire . . . the wire. [10]

Although it would be some time before Welch himself was "shattered beyond repair," the signs of what awaited him in his immediate future were already visible. As "resentful guns" once again shattered peace in Europe, Welch was facing another upheaval. Dorothy had decided, after a hiatus of sixteen years, to complete her studies at Stanford, and thus she relocated the family to Palo Alto. Once again, Lew and Gig (as Virginia was called) were facing a fresh start, but this time, for Lew at least, he had newfound inspiration in the form of literature. His increasingly dysfunctional childhood, and the traumas he would later attribute to that, suddenly took on a different perspective. The novels and poetry to which Welch had been introduced by Rideout not only provided an outlet for his adolescent imagination and naïve fantasies but also gave Welch a new sense of himself. And despite the horrific events of the intervening years, it was this sense of himself that would later manifest itself into his greatest incarnation, that of the poet.

3
From 440 Yards to B-29s

By 1940, Dorothy Brownfield had moved her family yet again—this time back up the coast to Palo Alto, which over the next ten years would become by far the most stable environment the children had ever experienced. The reason for this move was most likely the one that prefigured every move Dorothy would make over the coming two decades—her desire to continue and complete the course of studies she had begun at Stanford University almost twenty years earlier—and perhaps a feeling that she had some other unfinished business to attend to.

In Palo Alto, both Welch and his sister attended junior high before going on to Palo Alto High School, which was a short walk from the family home at 1227 Fulton Street and a stone's throw away from Stanford, where Dorothy was now in the penultimate year of her bachelor's degree course. The events of Welch's high school years, like those of his earlier education, are based on his own anecdotes and memories, but what is clear about the years between his enrollment in 1940 and his graduation four years later was that the athletic talent he had shown in elementary school had not waned with age, despite the podiatric problems he is purported to have endured throughout his childhood. Welch weaves an elaborate tale of having to wear iron shoes at kindergarten, of having atrophied feet and broken arches from running races barefoot in sixth grade, and of having to walk backward for the entire time he was in ninth grade. And all on account of his mother getting "in the grips of some idiot who made a steel-trap shoe that ruined [his] feet so bad."[1] He was forced by Dorothy to wear the shoes until such time as he finally resisted and fought back. From then on there would be no more crutches, only straightforward therapeutic exercises to repair the damage—a small victory of will over his Kali mother? Or an interesting similarity between Welch's self-aggrandized lameness and that of Oedipus, who also suffered from injured feet, albeit in his case at the hands of his father?

In spite of these and other physical issues, Welch was a fullback for the Palo Alto Vikings football team, later saying that his teammates used to call him "Crunch Welch" on account of the crunching sound he made when he hit the line, due to being only 134 pounds. As he put it himself, he was "fast but light."[2] Consequently, he excelled on the track team, posting a best time of 49.7 seconds for the 440 yards, which under any circumstances is a very fine time, let alone given Lew's prior injuries.[3] By his own admission he had soft arches, and his feet broke very easily, thus making it difficult to hurdle or jump, and surely to some extent justifying Dorothy's attempt at finding a remedy. Yet following in his father's (presumably healthier) footsteps, when the time came to further his education, it was Welch's gifted legs that had made him into a track star with enough promise to result in a potential scholarship. However, before any decision could be made on that front, there was the small matter of World War II to contend with, and within months of Welch's graduation from high school he was eligible for conscription.

———

When the United States entered World War II in December 1941, Welch was still only a fifteen-year-old walking backward through ninth grade, and his involvement in the war effort would have to wait until 1944 when he was old enough to enlist. With the previous wartime exploits of Hall and Service providing him with some small sense of courage in an otherwise tumultuous time, Welch dreamed of emulating Errol Flynn and, years later, recalled that he "tried very hard to fight"[4] when the time came. Little did he know that acts of wartime heroism such as those portrayed by Flynn on the silver screen had actually occurred in his own family. War and its immediate consequences had already played a significant and very real role in his not-too-distant past.

Welch's great-grandfather and namesake Lewis Barrett Welch had been a soldier in F Company of the Eighth Kansas Regiment during the American Civil War. Based initially at Fort Leavenworth, and fighting as part of the Union Army, the elder Welch—who would eventually achieve the rank of first sergeant—was engaged in every skirmish and battle the regiment fought between his enlistment in August 1861 and his discharge as a result of being wounded during the Siege of Atlanta in October 1864.[5] This sense of military duty then made its way down the family line from father to son and so on. And far from being confined to the Welches, this sense of patriotism may also have been instilled in Lew by Dorothy, who, in late 1943, having completed the commensurate two weeks of training at the Red Cross national headquarters in Washington, DC, had been given the post of Red Cross assistant field director at the US Fleet Reserve Office in San Francisco. In her new function as a member of the Red Cross Military and Naval Welfare Service, Dorothy would use her background in psychology and family-related issues to assist and deal with problems created by the separation of servicemen from their families.[6]

At a time when American involvement in the war was proving key to the eventual success of the Allies, these acts of heroism and duty may well have rubbed off on Welch. However, his teenage visions of heroism on the battlefields of Europe were to be short-lived; by the time he was eligible to be called up again—when American forces were embroiled on the Korean Peninsula less than a decade later—he had developed a very different attitude to war and his sense of duty toward participating in it. And even if he had known about his ancestor's decorated and heroic involvement in the most American of all conflicts, it would most probably have mattered very little by then. In 1944, however, he was more than willing to be called up and so, when he got the call in January 1945, Welch set off for Amarillo Army Air Field in Texas to do his basic training.

Within a matter of weeks of his arrival, Welch had made air cadet and wrote of his anticipation at starting his preflight training. Although Welch hoped to be assigned to fighter jets, Amarillo was being largely used for B-17 and B-29 bombers, and by late February his classification had changed to B-29 remote control turret operator mechanic gunner, with duties that included repairing the turrets during combat missions if they were to jam in any way, firing on the enemy, and "certain other duties I am not allowed to mention."[7] Welch considered his classification as the best deal available to him in Amarillo while bemoaning the conditions and telling Dorothy that if he never saw the place again it would be too soon.

One of the few saving graces at Amarillo was his friendship with a local Texan cadet called Clark Gilbert. The two visited Gilbert's home in nearby Pampa as frequently as possible to escape military life and enjoy some welcome home cooking. They also spent numerous evenings in the city—something Welch did less than many of the other cadets on account of his financial situation and general discontent at what Amarillo had to

Figure 3.1: Lew in his
air force uniform, 1945.
Photographer unknown.
Courtesy of the Estate of
Donald Allen and UCSD.

offer. He often wrote of his boredom on base, and the moments of relative happiness in his correspondence coincide with his learning to play the piano and reading the novels he asked Dorothy to send him. His reading matter ranged from Hemingway's *For Whom the Bell Tolls* to the most popular novels of the day, *A Tree Grows in Brooklyn* and *Forever Amber*. Showing an early penchant for literary criticism, he also wrote to Dorothy expressing his disappointment in Margaret Mitchell's *Gone with the Wind*. Clearly, his imminent transfer to Denver could not come soon enough.

On his classification, he wrote that B-29s "remind me of flying freight trains. It takes the damn things one mile to take off," but his intent to play an active role in the war is very clear. The decision to quit pilot cadet school and become a gunner was born of a need to do his duty: "This is my war and I refuse to fight it sitting on my tail waiting to be assigned."[8]

Before leaving for Denver, Welch also showed his skill with a firearm, becoming an expert marksman. Like his father, who was now serving his country for the second time at the Naval Ammunition Depot in Indiana, Welch relished his firearms training. On the rifle range and on bivouac exercises, his accuracy and skill gained him not only a glowing reputation but also the gratitude of his fellow servicemen, who allegedly used Welch to bolster their scores in order to attain the totals necessary to proceed to the same expert level of marksmanship he had gained. He later wrote of trading kitchen duties in return for his services.

Welch had first learned how to shoot using the rifle given to him as a child by his father, foreshadowing the ironic inevitability of the similarities in their military service, and the fact that they would meet for the last time while both serving their country. In *Genesis Angels*, Aram Saroyan describes a meeting between father and son at Lowry Field:

> And one day, the old man, "Speed" Welch shows up at the base, and father and son greet and see each other, and look away and look back, for the last time. . . . Lew got a good strong look into his lifelong absent father's face before it faced back into itself, and was gone. . . . The boy had grown; the father had aged—and the two said hello and goodbye for the first and last times under the flat American sky, each as grown men.[9]

This meeting must surely have been as poignant as it seems to have been definitive, given the situation the country was in at the time. Unlike his father, who was too old by this time, Welch could easily have been called up to fight at any moment, and whatever their differences, both must surely have understood the potential finality of this meeting. Was it possibly an attempt by a father to reconcile with his son before it was too late? Or was it a merely a wartime coincidence? Two servicemen on an air force base in the course of their duties? Either way, it appears to have been a moment when both men understood the nature of their relationship and their respective places in the world. In a letter to Dorothy shortly afterward, Welch wrote that his father was "quite a guy" and that he was "looking forward to the time when he came through again."[10]

Lew Welch Sr. had enlisted at Camp Pendleton, the US Marine Corps recruit depot in San Diego County. His recruitment records from July 1942 state his rank as private first class, with the remark that he should be a reservist. He had been a corporal in World War I and was stationed at Parris Island, South Carolina, where he was attached to the rifle range detachment. He had excelled as a marksman (local newspapers covering his discharge and return to Redfield reported that he been awarded "the expert rifleman's badge"), and it was here that his love of shooting was nurtured and further developed—a love he would eventually pass on to his son, and that would remain with Welch Jr. for the rest of his life (Welch later wrote that "this is the one thing Lew [Sr.] gave me. . . . I came to guns not as a stranger.") Welch Sr. remained at Parris Island for a year until his discharge in 1919, never seeing active service. This was also the case during his second military stint, which was divided between the naval depots in Burns City and Crane, Indiana, and the Naval Supply Depot in Norfolk, Virginia, where he worked as a clerk in the quartermaster's office. On his discharge in April 1945, his commanding officer, Captain Frederic E. Fowler, signed off his certificate of service with honorable conditions— a commendation that his son would also receive before the year's end.

Interestingly, the discharge certificate gives Welch Sr.'s address as 1227 Fulton Street in Palo Alto—the address Dorothy Brownfield had been living at since the early 1940s with Lew Jr. and Virginia. As in Santa Monica in 1932, Welch Sr. seems to have been reunited with Dorothy, but again it was relatively short-lived, and as quickly as he had reappeared in their lives, he was gone again. While Saroyan writes that Welch knew his father, any reconciliation was temporary. In late 1945, Lew wrote consolingly to Dorothy that he had been "really blue to think that your home life is not working out better." The reasons for this may have been more than simply marital, as records show that Welch Sr. had undergone a serious operation in San Francisco's Fort Miley VA Hospital in October 1945, and he may have been forced to stay with Dorothy to recuperate.

Indeed, it is clear from earlier letters that Lew's father went to stay with Dorothy immediately after his discharge and that he was in need of some sort of rest. In fact, there is even a suggestion that their relationship had taken a positive turn, with Welch writing that he was "glad to hear that all has been well lately & that you & Dad had a nice rest together."[11] In another letter from the same period, Welch writes that he is "glad to hear that Lew is out of the hospital & feeling all right,"[12] thereby indicating that his father's stay was more than an attempt at resurrecting his marriage. As already mentioned, before long Welch Sr. had returned to Kansas, first on account of his sister Rubea's funeral (another forty-something death in the family) and then for what became a long-term stay in the Winter General Army Hospital in Topeka—presumably a consequence of his earlier illness and almost certainly the precursor to his own death in March 1947.

———

When Welch arrived at Lowry Field (now Lowry Air Force Base) in Colorado on April 15, 1945, to do his technical training, the war in Europe was in its final few weeks, as German forces capitulated on all fronts, but Welch was still hopeful that he would see some action in the Pacific. He was also more than happy to have been transferred there, writing to Dorothy that "Denver is about the prettiest & best town I have ever seen"[13] — and certainly a better place to see out the war than Amarillo. Indeed, he enthusiastically wrote of finding "a little hangout you'd really like. It's the only place in Denver with any sophistication. . . . They have a small Negro band that is really wonderful. I spend every dime of my pay in there & listen to that wonderful jazz & can forget all about this lousy Army."[14]

It was solace in the form of music, the like of which he only otherwise found in the little chapel on the base in which he played piano to escape the dullness of his comrades and the stagnancy of a nonactive conscription. Indeed, in the end he would sit out the next six months confined to base, ending his military career as a "firing furnacer," tending to the fires in the barracks and making sure the coals were always hot so that the airmen's quarters were kept warm—hardly the swashbuckling Errol Flynnesque last stand he had once envisioned. Welch's war ended in November 1945 after what seemed like months of waiting. By the time he left the air force he had served less than a year, and the ambition and intent he had shown only six or seven months before had mutated into a weary acceptance of his role and a longing for it to come to an end. Years later—perhaps in a rush of nostalgia—he spoke of his disappointment that, like his father, he never saw any active service, a fact he later unceremoniously described as a bummer.

———

Arriving back in Palo Alto just before Christmas 1945, Welch was now a clean-cut nineteen-year-old ex-serviceman with a high school diploma and a love of piano, reading, and sports. His next step was not to immediately study English literature or develop his athletic ability but instead to reenroll for the civil engineering course he had started briefly at UC Berkeley after high school.

Yet, on returning to the Bay Area, his previous enthusiasm for the course quickly waned, and Welch dropped out in early 1946 to do odd jobs for a few months until he decided what course of study he really wanted to embark on. When the new school year started in September—having worked as, among other things, a car mechanic and a haberdasher—he seems to have found his muse, and he enrolled at Stockton Junior College to study English, music, and painting. And although his time there was relatively short-lived, it was to be the first and, in many ways, the most influential seat of learning he would ever attend.

4
Kamm's Brick Block

Having enrolled at Stockton, Welch, now twenty, moved into a small apartment at the Riverview Housing Project, which had been opened in 1943 to provide temporary accommodation to workers aiding the war effort at the local Pollock Shipyards and which was now being used to accommodate students. However, his stay there was to be relatively short, as the course of his education took an unexpected twist in the spring of 1947. Indeed, this twist not only altered the course of his education but also cemented his entire purpose of being. In a matter of hours, he went from being a bachelor student to a would-be poet on the cusp of a journey into language and literature. In his interview with Meltzer in 1969, Welch recalls the moment when, suffering from some forgotten student crack-up, he went to visit his teacher and mentor at the college, James Wilson. Finding Wilson out of office, Welch picked up one of the many open books littering the teacher's desk and began reading, determined to "wait until Wilson got back, no matter how long it takes."[1] The book was *Three Lives* by Gertrude Stein. As Welch further explained with unemphatic clarity, "Suddenly reading 'Melanctha' I felt as though I had been invited to a very distinguished party, a weekend party in the country, and at this party there was Shakespeare, Poe, Stein, Joyce, Dickens, Chaucer, all the other people that I admired."[2] This metaphorical party would be an invitation for him to serve a literary apprenticeship in which he listened to and learned the language, rhythms, and music of his fellow revelers. Welch also spoke to Meltzer about his "humbleness," and of remaining silent until his craft was such that he felt able to hold his own at such a party, stating "I didn't say a word. I just listened. I listened for a long time, and it was good long party."[3] Indeed, it lasted the best part of ten years and grew to include not only Thoreau, Hawthorne, and Sherwood Anderson but also their successors, Welch's friends and peers: Allen Ginsberg, Gary Snyder, and Jack Kerouac to name but three. Yet Wilson's initial impression of his student was that he was destined to become a great painter rather than a poet. Indeed, Wilson remembered Welch more for his artistic ability than his writing prowess, recalling that Welch "was the most talented young painter I have known. He was my wife's student at the college and he was the best she had ever had."[4] Welch's unexpected yet fateful discovery of Gertrude Stein on Wilson's desk was the moment at which, as Wilson puts it, "when I wasn't there, when none of us was there, [Welch] turned into a writer."[5] It may have been a surprise to Wilson (and a shock to his wife), but it was a circumstance that prompted him to encourage Welch to take his talents elsewhere. Why Wilson suggested Reed College is anyone's guess, but it is tempting to think that Wilson knew about the progressive nature of the faculty and student body in Portland and saw something in Welch that would flourish there. So, at his mentor's behest, Welch enrolled at Reed and moved to Oregon.

Tall and thin, red hair still cut short, a sea bag on his shoulder and a $15 Comoy pipe in his mouth (his latest kick: great pipes), his dark excited eyes (hunting or hunted you cannot say) are taking everything in. He walks down a sunlit street still sparkling with that morning's gentle rain and busy with warehouse-bustling men . . . along the damp pavement pigeons waddle along before his advancing foot . . . walking by warehouse storefronts . . . then above a narrow doorway . . . a single milk-glass globe (which is never lit) — a sign and standard for the Kamm Building, his newly rented home. (IL, 15–16)

Then, like now, Reed was a college renowned for its liberal arts program and high level of scholarship, creativity, and social engagement. Founded in 1908 and named after Oregon pioneers Simeon Gannet Reed and his wife Amanda, the Reed Institute, as it is officially known, was established as a seat of learning with the purpose of providing a "more flexible, individualized approach to a rigorous liberal arts education" and was not, as is often assumed, a response to student protests and cries for unconventional forms of experimental education as evidenced in the United States in the 1960s.[6] Indeed, Reed College offered a legitimate West Coast alternative to the East Coast Ivy League model, which perpetuated a tradition of the highest quality academia coupled with the elitist substrata of varsity sport, fraternities, and exclusive old-boys' clubs. Reed's refusal to buy into this tradition and to offer an intellectually and socially viable alternative made it a college whose purpose was to devote itself to "the life of the mind"—with that life being understood primarily as the academic life that was not only coeducational and nonsectarian but also entirely egalitarian.[7]

Set amid woods and wetlands, Reed is an eclectic mix of architecture and nature. While many original buildings echo their Ivy League counterparts in Tudor Gothic opulence, later additions are firmly rooted in the contemporary modernist style. The campus is bisected by the Reed College Canyon, a national wildlife preserve that includes Crystal Creek Spring, and it is tempting to imagine Welch, who had only just turned twenty-two at the time of matriculation, strolling through the campus and further developing what was becoming an ardent love of and fascination with nature. The seeds of this passion had been partially sown on the all-too-seldom fishing trips during which his father had given him a first fleeting glimpse of what Welch later termed "the World that is not Man" (RB, 172). Indeed, his early poetry not only reflects this developing fascination but also sketches images of the flora and fauna he encountered daily at Reed. By this time he was endeavoring to develop a writing style that would help him control the composition of his poetry as his consciousness of language grew. He writes of taking concrete steps to counteract the "linguistic constipation" that this increasing consciousness created, and of assigning himself the task of very simply describing, each day, one scene that came to his attention while walking to school. One such example of this is Welch's vivid description of the skunk cabbages in the Reed pond. He recalls writing what he remembers as one of his very first poems:

> I began to be a poet at Reed College and one day I was walking around the lake they have there . . . a very lovely lake. . . and there is also a swamp behind it. . . . I saw this thing and it was really weird. It was a skunk cabbage.
>
> Slowly in the swamps unfold
> great yellow petals of a

savage thing, a
tropic thing—

While no stilt-legged birds watch,
no monkey screams,
those great yellow petals
unfold.

Rank Plant.[8]

Although written by the hand of a poet still finding his true identity, "Skunk Cabbage" displays the accuracy and consiseness that he learnt from Stein and that subsequently became a key feature of his oeuvre. The poem varies in pace and balance. There is also a veiled exoticism in the description of the plant as it unfolds, and a final comment that serves to immediately deconstruct the simplicity and beauty of the primary image. He also shows an economy of language that would characterize much of his subsequent work. In an essay titled "Poems and Remarks," written to accompany a small collection of what Welch called his "minimal poems," he highlights the thought that went into the creation of "Skunk Cabbage" and the precision he employed: "The line 'no stilt-legged birds watch' has always pleased me—not for the image which is almost ubiquitous—but for the stuttering consonants and the resulting necessity to read slowly which in turn gives an intensity to the words, and insistence of tone, which is here useful" (IR1, 35).

Welch was writing with form and language in mind, with meaning relegated to a secondary tier of importance, underlining the fact that he was "more interested in the manner of the saying than . . . in what is said" (IR1, 39–40)—an introduction to his interpretation of the anecdotal Chinese artistic concepts of Sh'n and Yee that underpinned much of his work at that time and the basis for what would become his primary philosophy as a poet.[9]

———

For Welch, Reed College provided him not only with a natural and intellectual environment in which his talents could burgeon but also with inspirational courses and classmates. Among his closest friends were

Gary Snyder (whom Welch referred to as "one of the finest people I have ever known" [IR1, 1]), Philip Whalen (lovingly dubbed "the portly poet laureate of the school" on account of his being "by far the best writer" [IL, 53]), and Edwin Danielsen ("painter and cocksman" [IL, 20]), with whom he roomed in the Kamm Building in downtown Portland—an imposing, stately block designed as an office building for the celebrated Swiss-born entrepreneur Jacob Kamm and haunted by the ghosts of its nineteenth-century past. Bordering skid row, the building had been ear-marked for demolition by late 1948. Welch's disdain at the prospect of such a regal building being flattened is clear in a letter he wrote to his mother: "The fine old building we moved out of is almost demolished. There is a rumor that this one must give way to the fire dept's demand for wider streets. Dr. Faustus paves the world!" (IR1, 7). This is one of the earliest indications of Welch's growing discontent with the role of govern-ment institutions and their effect on the city. Gradually this discontent would spread until the city itself was the Faustus figure and nature the victim of his destructive ways.

The story goes that Welch met Whalen when the two were sitting in adjoining booths in the Reed College coffee shop. In order to see who was commanding so much attention behind him, Whalen looked over to see Welch talking in a characteristically imposing voice and singing a country-style tune about a murderer attacking his victim's "thin breast-bone." Whalen saw a kindred spirit, and a brotherhood was born. Welch, like Whalen, had a particular way with words, concocting unconventional combinations that focused primarily on intonation and American authen-ticity rather than explicit meaning. And he often did so in song. It is easy to imagine Whalen sitting listening to his future friend expounding in the manner of William Carlos Williams. Indeed, it was a friendship that prompted a keen mutual appreciation, and one that provided Welch with a great deal of encouragement when faced with the many demons that he would later encounter. In an interview with David Meltzer, Whalen describes his first meeting with Welch:

> I was sitting with somebody or another, and in the next booth there was this guy sort of spouting off all sorts of wonderful nonsense. I kept listening to it. I heard him say, "Red glass birds! Thin brass dome!"

I said, "Wait a minute." So I got up and went over to the next booth where this redheaded guy was, talking to people. And I said, "What was all that about red glass birds?"

"Oh," he says, "no, no." It was a song that he had been working on that had to do about how, "She hollered and roared and tore all of her hair. And I carved my initials on her breastbone." "Thin breastbone" is what I heard as "bright brass dome" or something like that. And, ahh, "Told her don't cry little darlin' that the mark of the man." And so we continued talking and telling, "You ought to write things down, for god sake." And he said, "Well that's no good."

And so we went on from there talking about Gertrude Stein and about Williams and all sorts of things. He was very funny.[10]

It may have been a chance meeting, but it was one that would have a huge effect on Welch, not only in terms of poetry but also with regard to language, philosophy, and the meaning of friendship. If Welch had always been something of an outsider, Whalen had been a popular child who was "one of the guys" at school and who valued friendships ever after. Clearly Welch correctly intuited early on that in Whalen he would have a friend for life, regardless of circumstance.

———

As the 1940s drew to a close, Welch became increasingly unhappy about the state of the nation. Tensions on the Korean Peninsula were steadily increasing, while at home Senator Joseph McCarthy was ramping up his anti-Communist rhetoric, with the subsequent witch hunts just around the corner. Indeed, Welch himself would later be a subject of the Red Scare on account of his leftist leanings and fleeting involvement in the Portland branch of the John Reed Club, a Communist group active throughout the country during those years that was later investigated by the House Un-American Activities Committee (HUAC). In the written transcript of the "Investigation of Communist Activities in the Pacific Northwest Area," Welch is mentioned as having been "in the party a very short time" in the testimony of Homer Owen, who had been called to give evidence before the committee in Portland on June 18, 1954.[11] Owen's was one of many testimonies in which more than two dozen former Reed students and faculty members were named as having been members of the party. Indeed, the resulting recriminations ended with Snyder's former professor Stanley Moore being fired and fellow professors Lloyd Reynolds and Leonard Marsak being forced to choose between their careers at the college and testifying.[12] One of those who provided names and information was the dean of students and the chairman of the Portland Chapter of the American Veteran's Committee, Robert Canon, who later said that naming Moore and Reynolds was the worst thing he had ever done in his life.[13] Ironically, it was later suggested by Snyder that Welch's membership was actually far more likely to have been due to his need for social inclusion in a tribe, a brotherhood, than any strong political sentiments he may have harbored.[14] Welch might also have been aware of the tendency for such clubs to organize events at which poetry and music were integral. In San Francisco, for example, Kenneth Rexroth had used these cultural aspects to create a branch of the John Reed Club as a stage for Communist propaganda, and it is reasonable to think that Welch likely saw a poetic opportunity rather than a political platform.[15] Either way, his involvement was short-lived and never investigated any further.

So, in the face of such discontent (and with the recent dissolution of his love affair with fellow Reed student and girlfriend of Gary Snyder, Robin Collins), Welch hatched the ill-conceived plan of leaving the United States and its ever-more-destructive ways altogether and moving to Santiago, Chile. In a series of highly critical letters to Dorothy, Welch vents this discontent and anger at a country that he feels is through and in which he cannot live and flourish as an artist. He wrote of the United States being a "goddamned, time-ridden money-hungry decadent country" lacking vitality in which he is "looked upon as a freak simply because I want to be creative" (IR1, 23). This critique is the first concrete indication of what would become another central theme in his work, a theme that, in a letter to Charles Olson many years later, he simply and honestly called the "vulgar and dangerous din" of his American life. A din he was now on the verge of extricating himself from.

Welch was now almost twenty-four years old, and his idea was simple. Escape the United States, avoid being drafted into a war that he felt was "utterly without sense" and that would kill him "physically or psychologically" (IR1, 18).[16] Instead he would live a life with simple needs and easily attained desires, in Chile.

This was all a far cry from his previous disappointment at not being able to see any combat during World War II and displays his increasing political sensibilities. He talks matter-of-factly about preparations to leave and of the geography, culture, and climate of Chile, and of the opportunity to have enough freedom there to create art. His description of Santiago as an "urban community, which I must live in . . . since I do not care for primitive existence, although a simple one is to my taste" (IR1, 16) is revealing of his need for relative comfort in a city—which is in sharp contrast with his feelings about urban living in later years. He is apologetic at the prospect of not seeing Dorothy for many years but deems this a necessary consequence for the achievement of his artistic liberty (and desire to avoid military service). On paper, it all sounds very plausible and even noble (if highly idealistic). However, the plan depended largely on one key aspect, namely Dorothy Brownfield—the "rich, selfish shiksa." As neither Welch nor his intended traveling companion, Ed Danielsen, had the necessary finances to fund such a venture, Welch was perfectly happy to ask his mother for the money. In addition, he had no qualms about canceling earlier plans he had made with her (much to her disappointment) or of taking advantage of her eternally generous nature in regularly sending him money already, a trend that continued until the end of his life. His idealism and impetuous nature take precedence, and Welch reveals a paradoxical aspect of his personality: the desire for independence and the inability to extricate himself from what he called the aristocracy of family and money into which he was born.

However, with the wheels seemingly in irreversible motion ("Today we got the wished for news. We are definitely booked for passage on the *Nootka*"), the plan fell through.[17] Not because Dorothy refused to fund it, but because in the end the captain of said ship refused to take them. While no explanation was given for the captain's abrupt change of mind, Welch wrote to Dorothy that both he and Danielsen were relieved by this turn of events. Danielsen developed a sudden unwillingness to leave his girlfriend, and Welch was increasingly unsure about whether his partner would be much of a companion on the trip anyway. After four months of planning, researching, and variously substantiating his reasons for going, Welch found himself stuck in Portland with a thoroughly unsatisfying feeling that it had been a "half-cocked" experience from which he was sure, in time, to learn some valuable lessons. His conviction about leaving had been founded on political truths and emotional sentiments that were slowly undermined by the reality of red tape and indecision, but while unsuccessful in execution, this idea of escape was one that would recur again and again; over time, Chile would be replaced by Big Sur, Forks of Salmon, Mount Tamalpais, and, ultimately, the mountains of the Sierra Nevada.

In this case, however, it turned out to be a blessing in disguise. Only a matter of weeks after Welch's proposed departure, William Carlos Williams visited Reed on a lecture tour—an event that not only offered Welch some sense of solace but also provided yet more motivation to set aside far-flung plans to leave the country and instead to refocus his attention on poetry, Gertrude Stein, and learning to live in the United States.

———

Like Robert Rideout in El Cajon and James Wilson in Stockton, it would be Lloyd Reynolds that provided some of the greatest inspiration to Welch among the Reed College teaching staff. Despite Welch's keen interest in the medieval and Renaissance literature courses taught by Professor Ralph Berringer, Reynolds's classes provided him with an outlet for both his literary and his creative talents. Alongside Snyder and Whalen, Welch took Reynolds's classes on creative writing and was introduced to the possibilities created by applying calligraphic techniques to such writing. During Reynolds's calligraphy classes, the students were encouraged to investigate the myriad of possibilities and opportunities open to them through calligraphy, and which they in turn were required to open up to. Reynolds encouraged associative thought and laid bare his thoughts on the human condition in the hope that by elevating the "simple, utilitarian practice of handwriting" from a daily activity to an art form, a student would begin to look differently at other seemingly simple aspects of their life and thus develop themselves accordingly. Reynolds approached his classes with a willingness to discuss anything that would benefit his students, and many found a connection in the various elements his teachings encompassed, including "Zen Buddhist ideas, the celebration of nature, the merging of beauty and function, and a belief in the individual and what one person can do."[18] Reynolds instilled in his students William Morris's "'craftsman ideal,' that natural beauty, simplicity, and utility should infuse everyday life and everyday objects, and that goods should be made by artisans rather than by machine."[19] And this ideal is clearly one that Welch and company took with them on their respective poetic journeys. As the influence of Eastern culture grew in the United States, in a period when Chinese and Japanese immigrants were coming to North America in ever more significant numbers, the cultural climate was highly "conducive to introducing Buddhist influences."[20] Reynolds facilitated and encouraged this at a time when Welch and his fellow housemates were first exposed to the haiku tradition through the writings and translations of R. H. Blythe, among others. In discussing Reynolds's philosophy as "autobiography," former student Jackie Svaren describes it in terms that would equally apply to meditation—a simplicity of thought and an allowance of complete openness to nothing but the core essence of living: "He rolled this thing out: I inhale, I exhale, hallelujah! Just break it all down to the basic, breathe out, breathe in, and stop fretting about all the rest of this stuff."[21]

In the preface to *Festschrift for Lloyd Reynolds* (1966), a collaborative ode to Reynolds's life and work, Philip Whalen writes that "the profundity and scope of [his] influence, [his] inspiration"[22] was too great to be done justice to in a single volume of works by his former colleagues and students, regardless of their status and renown at the time of publication. The collection included works by, among others, Gary Snyder, James Dickey, Jonathan Williams, John Cage, Whalen, and Welch. Indeed, such was the regard in which Reynolds was held by his students that Whalen went as far as to say that had everyone who was influenced by Reynolds contributed, the project would have run to multiple volumes. In even more glowing terms, Whalen had earlier written that Reynolds had "succeeded in changing my mind about the hopelessness of writing poems or anything else in this late and decadent period of the world; his encouragement and advice and friendship cut through all the fogs and megrims which I had contracted from reading the 'New Criticism' and The Partisan Review."[23]

By the time Welch entered his classes, Lloyd Reynolds had been teaching at Reed for almost twenty years and had become one of the foremost authorities on calligraphy in the United States. His influence on Welch is visible not only in the calligraphic works that Welch produced—the poem "Step Out onto the Planet" from *Hermit Poems* is accompanied by a Japanese *enso* drawn by Welch that provides a perfect example of how a single fluid brushstroke can encompass a profound symbolic meaning that in turn deepens the interpretation of the poem—but also in how Welch approached language and its practice in general. Reynolds encouraged his students to write their essays, letters, and other texts in the *italic* hand. He asserted that this was both an excellent mental exercise to absorb the text as well as an opportunity for their fingers, hand, and arm to learn "the motion of calligraphy."[24] Welch, like many of his fellow students, never forgot this practice, and in *Hermit Poems*, which he dedicated to Reynolds, his choice to have his handwriting reproduced adds a uniqueness to the collection that underpins its identity and reinforces its strength. The simplicity of the brushstrokes and openness of the lettering are a metaphorical breathing in and out. The symbols may be complex, the underlying motivation equally so, but the aesthetic effect offers a calmness and a quietude that can be easily associated with Reynolds's vision of calligraphy as "a link between one's own simple, utilitarian practice of handwriting and the accumulation of knowledge and scholarship."[25]

Also taking classes under Reynolds was another master calligrapher from whom the students could take inspiration, Charles Leong. A Chinese American who, like Welch and Whalen, was attending Reed on the GI Bill, Leong was studying Occidental humanities, having previously trained as a seal (stamp) carver in China. Despite the difference in age and experience (Leong, who was in his early forties when he enrolled at Reed, had served in the US Army Medical Corps in Italy during the war), Leong became friends with many of the younger poets, who were intrigued and inspired not only by his calligraphic skill but also by his heritage, education, and wisdom. Indeed, Gary Snyder later said that he had learned his calligraphic skills not from Reynolds but from Leong ("with Charlie as my guide, I learned to hold the brush as well as the pen").[26] Reynolds himself would eventually hire Leong to work for Champoeg Press, the small publishing house he cofounded at Reed in 1951.

Leong's stature within the group is further captured in anecdotal form in Welch's unfinished autobiographical work *I Leo*, in which he describes one meeting with Leong at the house into which Welch and Danielsen had recently moved after the death of its previous inhabitant. Prompted by the imminent demolition of the Kamm Building and their enforced eviction, Welch and Danielsen resided briefly in the now-famous Lambert Street house with Snyder and the rest but quickly stumbled on this new apartment and immediately recognized its potential despite the particularly unpleasant and somewhat eerie nature of the prior tenant's death. However, any doubts that the new residents of what became known as "Anna's House" (an ode to the prior tenant) may have had about the possibility that the house was somehow haunted were immediately dispelled upon Leong's entry: "Charlie, then, was the first guest. And what he did was walk around the place with his small cup of tea in his hand and say: 'Ah, spider webs.' He explained that spiders didn't spin their webs in places where ghosts were—ghosts broke the webs and the spiders left. Then he said: 'This will be a lucky house'" (IL, 53).

———

Much has been written about the friendship between Snyder, Whalen, and Welch at Reed. It was there that each honed their writing skills and took their first steps toward the development of the philosophies and ideologies that would later define their respective poetic lives. It was in the house at 1414 SE Lambert Street, which they shared with fellow students such as the poet William Dickey and the novelist Don Berry, that the trio was immersed in the kind of creative milieu that David Schneider describes, in his biography of Philip Whalen, *Crowded by Beauty*, as being "a forerunner for the many urban communes of a decade later; politically left-wing . . . intellectually serious and stimulating, and religiously open."[27] Schneider writes of the house being located on the intersection of some magical axis where "if the constellation is right, [it] may attract certain inexplicable, temporary 'blessings' so that the group and its members stand out as memorable, influential,"[28] something which is certainly true of the poets the house harbored. Berry later spoke of the discussions and arguments among the inhabitants of the house as being "probably the birth canal for the Beat Generation—classic postwar Bohemianism [in which] the quality of the minds was extraordinary."[29]

Many of the Lambert Street students, including Welch, who lived there briefly between his stays at the Kamm Building and Anna's House, were also involved in the establishment of *Janus*, a literary magazine that afforded its founders an opportunity to contribute their poetry and enabled Welch to hone his analytical and critical skills as its coeditor. While *Janus* was certainly not the first literary magazine at Reed, its intentions were clear from the start, with Welch later stating that the focus of the magazine "was to treat literature as a craft which can be learned and talked about—the mystic, mysterious soul-rending bullshit which we have inherited from that least of centuries, the 19th, have confused the 20th until student writing is either derivative or hysterical" (IR1, 9). Snyder and Whalen were also part of the magazine's editorial staff, and *Janus* can boast some of their earliest published works. In addition to Welch's first poem, "Monologue of One Whom the Spring Found Unaccompanied" in the May 1950 edition, Whalen's "November First" and Snyder's "Walking Lonely on a Fall Day" were featured in the January 1950 edition. Welch's contributions as a literary critic highlight the practical application of his work on Stein as well as his own personal poetic preferences. His first critical commentaries are peer reviews of a poem by John Rogers and of a short story by Joan Baker, both of which introduce a critic firmly focused on the accuracy of simple language to convey a clear and interesting message as well as the need for the narrative to be uncluttered. In describing Rogers's use of language in his poem "The Prophet Stood Atop his Minaret" as "not simple, but plain, uninteresting" Welch establishes a standard that he would attempt to maintain in everything he wrote—language in and of itself must be of far more value than merely the image it creates or the meaning it imparts.[30] This ideology would become only more evident as he developed his philosophy on language through a deepening interest in Gertrude Stein and in how he incorporated his analysis of her work into his own poetry. Indeed, his discovery of Stein at Stockton was the catalyst for what would become the most comprehensive study of language he ever wrote, his thesis at Reed, "The Writing of Gertrude Stein: Its Nature and Principles," which he completed in April 1950.

Janus, however, was not the only creative outlet open to the budding poets. Welch and his housemates also founded the Adelaide Crapsey–Oswald Spengler Mutual Admiration Poetasters Society, a playful yet relatively serious endeavor in which the group played a word game using words chosen at random from either various works of literature or from the dictionary to then write poems. The choice of Adelaide Crapsey and Oswald Spengler as the writers the group was mutually appreciative of was based on their admiration of the former as an experimental poet famed for creating the cinquain, which it has been suggested is the only "serious poetic form to emerge from the United States,"[31] and of the latter as the historian whose influential work, *The Decline of the West*,

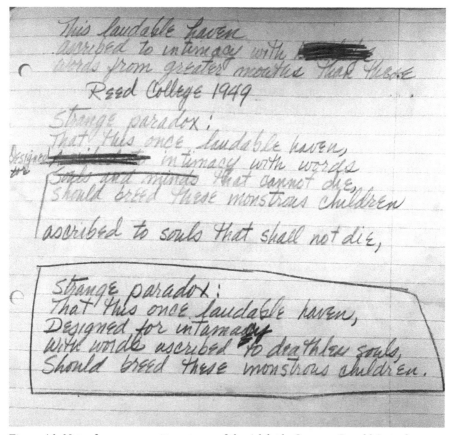

Figure 4.1: Notes for a poem written at one of the Adelaide Crapsey–Oswald Spengler Mutual Admiration Poetasters Society evenings at Reed College, 1949. Courtesy of the Estate of Lew Welch and Reed College.

would become a key influence on Beat writers such as Jack Kerouac and William S. Burroughs.

Crapsey's poetry, although limited in scope because of her untimely death in 1914, aged only thirty-six, would certainly have appealed to Welch for its brevity and accuracy. Her cinquain form employed concision and formality similar to Japanese forms such as the tanka, albeit with a slightly less expansive syllabic structure. Indeed, like both Pound and Williams, Crapsey appears to have admired the aesthetics of Japanese poetics, and it is easy to understand how the young Reed poets would have been influenced by this too. Like Crapsey, Welch also sought the "perception and expression" manifest in Japanese poetry.[32] Indeed, these were attributes on which he built much of his later poetic identity.

For his part, Spengler's work appealed to many writers and poets at the time because his theory, as they understood it, was less about the total annihilation of Western culture than about the need for people in America to look further afield in search of alternative cultural offerings. For Welch and company, living on the Pacific Rim and having a more than passing interest in Eastern philosophy and culture, *The Decline of the West* read more like a manifesto than a warning.

The idea for the game itself came after a discussion in which one of Welch's girlfriends and fellow English students, Kate Ware, had claimed that some words were simply unusable in any form of creative writing. Welch and Whalen vehemently disagreed, saying that poetry was receptive to all forms of language and that they would prove this to be the case. The idea was to select five words and work them seamlessly into a poem. The only criterion was that none of them would know what the words they chose meant (thus underpinning Welch's idea of Sh'n and his focus on the "manner of saying [rather than] what is said"). Welch later claimed that in a dictionary of 360,000 words, the poets in attendance on the first night could find only three that they were unfamiliar with: *flamen, liripipionated,* and *bema,* and thus allowed two others to be included that they knew but had never used before, namely *geode* and *propolis.* They then had fifteen minutes to write a poem. Welch explained the outcome with uncanny clarity many years later in what turned out to be his final lecture at Reed College in March 1971.

Whalen comes up with:

> Sated flamen
> Liripipionated
> Paces, chanting, in the bema
> While demented hillside bees
> Pack geodes w/propolis and honey
> Sweet rock eggs
> Tribute to no walking god.

Isn't that incredible? Wow. Boy that guy really has it. That Whalen is so good you just can't believe it. (HWP, 74–76)

Welch's effort with the same words was equally interesting. And although he was never one to shy away from using his own work as an example of how poetry should be written, it is surely indicative of his own feelings of insecurity at that time that he used Whalen's poem rather than his own during the Reed lecture. Yet despite any misgivings he may have had, the poem displays a clear sense of Welch's poetic ear, if not yet his increasing interest in concision:

> as if receiving solace from geode
> or visions viewed through propolis
> he stumbles
> causing Mr Montague to titter.
> he rose however un-liripipionated
> as a flamen would rise
> from his bima [*sic*] to be crowned.[33]

The young poets met regularly and would spend entire evenings playing the game. And, far from being confined to their rooms, the society organized and carried out a benefit reading in the Anna Mann Building at Reed, which was proudly advertised in the school newspaper *The Quest* under the headline "Crapsey Society Plans First Public Meeting." In addition to being the society's first reading, it was also Whalen's first public performance. As he remembers it, the idea came about as a means of generating some money for him to pay his way to San Francisco. The young poets would provide refreshments to a select audience and would read their poems with and without musical accompaniment.[34]

Welch's role in the event is clear. Aside from reading his own works, he was also charged with explaining the history of the society since its inception. The musical entertainment was provided by Roy Stillwell, who was also a resident of Lambert Street and an old army friend of Whalen. Stillwell was an accomplished violinist who went on to join the Oregon Symphony upon graduation in 1951 and as such was the perfect choice to provide the accompaniment to the poems. The Crapsey Society archive contains a rough draft of how the evening was divided. Part one included Welch's introduction and explanation, followed by a reading of "original poems" by Welch, Whalen, and Ware in part two. The third part was "Four Poems with Violin," including readings by Welch, Whalen, and Les

Thompson, the origins of which were words taken from Henry James's *The Wings of a Dove*. The event concluded with more original poems.

Among the most accomplished of Welch's works that night, and one that may foreshadow a key image in his first published work, "Wobbly Rock," is a poem titled "Voice from the Tomb of Virginia Woolf."[35] Of the many drafts that Welch wrote (and that have survived), this was one of a handful that he chose to type up for the performance. The words to be used on this occasion were *jaw, silence, engagement, steady, and inert*—none of which can be considered difficult or in any way obscure, as prescribed by the initial rules of the game—and Welch's poem underwent various phases and amendments before he settled on the final version.

> There would be silence here;
> But beneath three gulls
> Is that steady sound of engagement
> Of inert and mobile powers
> The sea
> And this bleached jaw of land.

In the way that "Wobbly Rock" envisions the confluence of sea and land, so too does this poem. There is also the necessary relationship between the inert and powerful, the sea being described simply so and the land being a "bleached jaw," thus implying a certain subservience in terms of the relationship. Indeed, while the word "engagement" suggests parity, one cannot help but *hear* the sea as having the upper hand. It is a vision that dominates "Wobbly Rock" but that is suggestive here of an early focus on not only nature but also the mysteries therein.

The same might be said of a second poem from that reading, "Vision on the Bus," in which Welch writes that

> A clear enamel hide
> Divides
> The empty air and the pond
> If you thrust your hand
> Far into the coolness,
> You will find, among three stones
> A quiet secret.

Again, there is the juxtaposition of two primary elements, in this case water and air, and the mystery of what lies within. Where "three gulls" are complicit in the first image, so are three stones in the second.

The reception the poets received that night is unknown, but Whalen later said that the earnings were spent on "food and books" and thus he was no nearer to amassing his fare to San Francisco.[36] However, he had instead experienced his first public reading and, one assumes, gained more praise and stature among his many friends.

The praise that Welch heaped on his college friend was not just confined to his Crapsey Society poems. Whalen's poetry (and his way of approaching it) would become a continual source of inspiration for Welch and a means to defend poetry in the face of what he felt was undue criticism for many years to come. During his years in Chicago, when he was cut adrift from his Reed housemates, Welch often turned to this and other experiments with language and poetry to remind him of the path he had once walked and from which he seemed to have partly strayed. He wrote to Whalen that his memories of Reed and the Crapsey Society helped to fill the void that had developed since then and point to a poet who, despite the problems that beset him throughout the early to mid-1950s, had retained a certain faith in poetry and in the methods he and his friends had created to write it. In a despondent show of how his life in Chicago was a far cry from his Reed years, Welch wrote to Whalen of having tried to start a Crapsey Society there too. However, it turned out to be a dismal failure; all that Welch managed from it were a few lines of poetry that had failed to buoy the assembled group. Gone were words such as *bema* and *propolis*, replaced by *somnolent* and *down-baked*, *myriad* and *splintered*. The poem he produced, which was far simpler and more downbeat than anything they had written at Reed, contains vocabulary that is positively mundane when compared to the earlier fare. And it is hard to imagine such words being used in those heady days at Reed. Indeed, Welch admits to having forgotten what five words were chosen prior to writing the poem, and although it is difficult to imagine that he did not know what any of them meant, it is another reminder of how far removed he had become from his poetic roots.

> Crag in the wind, and on it
> Somnolent.
> Eyes watery, from the wind's cut.

Speak old man, thy wisdom's meant for me

Tell us how those myriad spears did glint
Those thousand points grown into a hedge
Whereon the splintered sun down-baked. (IR1, 92)

———

By early 1949, Welch was already considering a potential thesis topic
and how he would go about graduating within the next year. Although
Gertrude Stein had very much been the catalyst for his decision to go to
Reed, she was not yet in his immediate thoughts as a final subject for his
capstone work. Indeed, his initial idea was far from linguistic; in a letter
to Dorothy from January 1949, Welch writes about how he had been captivated by the work of the Austrian psychologist Alfred Adler, especially
notions relating to childhood trauma and its relationship with neurosis.
It is hard to read these passages without seeing a clear and obvious link to
Welch's own relationship with Dorothy and how, albeit subconsciously,
he is speaking of this when he writes:

> It is, for example, not that a child has a trauma concerning wetting
> the bed, but that he was made to feel <u>inferior</u> by being <u>scolded</u>
> for wetting the bed. In other words, what does the trauma, as
> trauma, mean to the child? Then, according to Adler, the child
> develops a symptom of neurosis in an attempt to establish or regain
> a superiority—a protective or "safety mechanism."[37]

Welch continues that, on the basis of Adler's hypothesis, he would
like to further investigate the link between "Literary Genius and Neurosis," taking Byron, Goethe, and others as his examples. His a priori
conclusion is that only through the early childhood formation of neu-
roses and other complexes can true genius be realized, and that the
reason some artists "ended up with only an unrealized potential rather
than a gigantic contribution in arts & science . . . is simply that they
weren't neurotic enough."[38] In some respects, this correlation between
literary genius and childhood trauma introduces an interesting personal

link in itself, especially considering Welch's assertion that "perhaps a stupid mother (such as Byron's) while building a psychological mess is also building greatness."[39] The plan, whether in a thesis or a paper, was simple. Read Adler, Freud, and Jung, investigate Lewis Terman's *Genetic Studies of Genius*, and debunk or substantiate Welch's own conclusions. The question he was never able to answer, however, was whether his own mother's role in his development resulted in any modicum of greatness. Or, whether his traumatic childhood simply ended in a failure to channel the complexities that she helped to build.

Although there is a strong conviction in this letter that psychoanalysis and Adlerian theory were what was exciting him most at that time, in the end he turned (inevitably?) back to Stein and his love of language. And when he finally sat down to write his thesis, it was finished in a matter of weeks, and with the aid of his supervisor Frank Jones, the young poet and critic was given the opportunity to truly express his ideas.

If Stein was the inspiration, Jones provided the validation. A teacher of comparative literature at Reed, Jones focused on French literature despite teaching "all literature for his province."[40] Indeed, Jones was supposedly responsible for introducing the poetry of Jacques Prévert to an American audience long before Lawrence Ferlinghetti's translation of poems from *Paroles* for the City Lights Pocket Poets Series in 1958, often seen as seminal in the dissemination of French poetry to a wider English-speaking audience. Jones was a respected critic and widely published poet, and as such seemed like the perfect choice for Welch's chosen study. Openly admitting to his student that he knew next to nothing about Stein, Jones was nevertheless prepared to take on the project purely on the strength of the conviction Welch showed when canvassing the Reed staff for a supervisor. No one else seemed prepared to take a risk on the student or his intended subject matter, and Welch was quick to point out possible reasons why. In a letter to his mother from September 1949, he not only explains the situation regarding his thesis but also expresses his admiration for Jones, an admiration that was quickly reciprocated when Jones realized his student's talent and ability:

> That no one would be in a position to advise me regarding Stein, I knew. . . . No one on the faculty has ever read her, much less studied her writings or her approach to them. . . . [However,] after

outlining my possible approach to the subject my adviser, a Mr. Frank Jones who I now admire for reasons obvious to you in a moment, said: "I have never heard a thesis subject outlined in a manner which better convinced me that the student was upon sure ground, and more capable of producing a worthwhile dissertation." (IR1, 7)

At the time, Welch's thesis was a unique piece of writing on a subject that had never before been tackled to such an extent. The magnitude of his work is highlighted in Eric Paul Shaffer's introduction to *How I Read Gertrude Stein*, the version of Welch's thesis published by Grey Fox Press in 1996, in which Shaffer writes of Welch working in a "critical void" and underlining the fact that almost all comprehensive critical studies into Stein were conducted after Welch's. Indeed, Welch cites only one critical study in his bibliography; as such, he was working in a state of total autonomy. Despite her standing in the literary world, Gertrude Stein remained something of an enigma among academics, most of whom were apparently reluctant to take on the academic challenge of analyzing her work and its place in the canon of American literature. So it was with some courage and no little amount of confidence that Welch set about his task. Knowing that he was on virgin ground, however, only served to bolster his ambition. Jones had applauded his decision to venture on an uncharted course, and Welch felt a certain sense of power in the knowledge that Jones was in a precarious position due to his ignorance of Stein and the perceived fear and insecurity Jones had regarding her work. Welch wrote of Jones,

> He is intelligent, educated in the field of literature, receptive but doubtful of Miss Stein's worth. But most important of all, he has not read Stein because he has not been sure that she is sincere or indeed anything other than a literary curiosity, also I feel that he is somewhat doubtful of his security with regards to literature, that is his security is one that is maintained by great labor and which is an important part of his security as a whole, and therefore I feel he is really afraid of Stein. (IR1, 8)

Lewis Welch

An Examination of the
Writings of Gertrude Stein

Figure 4.2: Lew's graduation
photo in the Reed College
yearbook, *Griffin*, from 1950.
Courtesy of the Estate of Lew
Welch and Reed College.

This apparent bravado appeared to be justified. Jones not only accepted his young student's topic but also realized and understood Welch's sense that he could educate his supervisor in the ways of Stein. On completion, his thesis brought high praise from the examination committee and, more important, drew particular interest from William Carlos Williams, who "was very excited by it [and] had learned a lot from it" (IR1, 40). Welch described his oral defense of his work as being "conducted from a great overstuffed armchair, tolerating all questions, answering magnanimously" (IR1, 165–166)—a stance he later admitted showed, justifiably or otherwise, the arrogance of youth. And surely all the more so, given that the oral board contained scholars such as Professor Donald MacRae, who was not only a significant presence within the English Department at Reed but was also, it was later claimed by Whalen, familiar with Stein's work. For his part, many years afterward, Frank Jones said that what Welch wrote in his thesis on Stein "still delights me, more than twenty years after" containing as it did "the freshness of personal discovery."[41]

The thesis itself is a detailed and illuminating investigation of Stein's approach to language using many of her works. It is an essential tool to understanding Welch's own work because it provides a framework within which many of his early poems, in particular, can be analyzed. At the time, however, it was the catalyst for what turned out to be a decade of momentous change, prompting a short-lived poetic adventure on the East Coast that mutated into a period spent in a barren wilderness of creative inactivity and psychological turmoil. It was a period that also saw Welch settle down and tread the conventional path of what he later called "the American Homemaking Bit" (IR, 145).

5
You Can't Fix It. You Can't Make It Go Away

After graduation, Welch turned his immediate attention to the impending visit of William Carlos Williams to Reed. Having written an essay on poetics in the meantime, Welch waited to see if he would get the opportunity not only to present it to Williams but also, more importantly, to discuss his thesis and the possibilities of having it published. Williams arrived in Portland in early November 1950 and was scheduled to spend one week of a three-week reading tour of the Pacific Northwest there, which would also include a five-day stint at both the University of Washington and the University of Oregon.

By this time Williams was in his late sixties, and the thought of such a demanding tour was apparently daunting. In a letter to Louis Zukofsky dated less than a week before his departure from his home in Rutherford, New Jersey, Williams wrote "pray for me."[1] Around the same time, Welch excitedly wrote to Dorothy about the imminent arrival of such a great American literary voice and what this might mean for him. Riding high on the praise his thesis had received, Welch hoped that Williams would see his works in the same vein or, at the very least, "find them interesting . . . [or] at least be able to know who I am." He also specified his underlying motive, writing, "I will explain my plans for the book—at the same time hinting that I will send him the completed thing and ask his help re: publishers" (IR1, 26).

When Williams arrived, Welch and the other young poets were immediately under his spell. Prior to his planned reading at the college, Lloyd Reynolds had taken Williams to visit them at SE Lambert Street,[2] and Philip Whalen wrote that "he was interested in what we had to say. He made us feel like poets, not students anymore; he talked to us as if we were his equals."[3] If, however, Williams was daunted by such a workload, it certainly did not show. Welch wrote that "Dr Williams himself is the sweetest old guy I have ever met . . . very lively. He is so exciting to be around. . . . He was reading his poetry last night and got very worked up himself and very intense and simple" (IR1, 41).

These were qualities that may have appealed to the twenty-four-year-old Welch, but it was not so much how Williams delivered his poetry that appealed to Welch but the language that Williams used to do it. Like Stein, Welch could hear elements of language in Williams's work that underlined Welch's growing assertion that the only real American poetry was poetry written in "American." When he wrote to Dorothy that "we do not speak the language of England and our poetry should not have the English form. We talk American, and the poet's job is to intensify this dialect, sharpen into poetry, keep the words clean and sharp" (IR1, 41), he was not only repeating Williams's advice but also very much making a statement of intent, a manifesto. His words, excited as they were, reflect the confidence Welch felt in himself and his writing. And this grew still further when Williams announced his delight in Welch's thesis, going so far as to say that he had actually learned from it. In his autobiography, Williams writes that "I found another chap with a first-rate essay on Gertrude Stein, one of the most lucid I have yet encountered."[4]

Where Welch had displayed a certain amount of arrogance during the defense of his thesis, Williams's praise took Welch's idea that he was to become "the voice of the latter half of the 20th century" to new heights. He gushed to his mother that he had found a friend in Williams and that, after quickly polishing his thesis and erasing any last vestiges of a collegiate youthfulness, Welch would be able to simply "sit back and autograph books." This was a far cry from the way his work had been received the last time a literary luminary had visited Reed. Two years earlier, in October 1948, the poet Stephen Spender had briefly lectured at the college and read a selection of Welch's poems during a class. Welch later described the reading to Williams, commenting not only on Spender's attitude toward his poems but also on the Englishman's reading skills, which he called abominable: "Stephen Spender read some of them to a class at Reed and remarked 'I suppose one has to call it poetry, but it's rather minimal don't you think'" (IR1, 29).

Welch also came to the conclusion that there was nothing for him on the West Coast anymore and that his future lay in the east—more specifically, in New York City. However, reading between the lines, Welch's self-confident bravado is very closely connected to his need for validation and support. His decision to move east, while obviously linked to a desire to increase his proximity to Williams—who had invited Welch to visit him at his home in Rutherford, New Jersey—and the literary doors he could open, may also have been because of his deep-seated connection to Dorothy. His mother had recently enrolled at Cornell University to study child psychology, and this move may well have represented something of a fracture for Welch. While in Palo Alto, Dorothy had been relatively close, but Ithaca, New York, was a different matter. His sister, who had married the wealthy New York businessman Maitland Lee Griggs in 1949, was also living on the East Coast (in Ardsley-on-Hudson), thereby leaving Welch essentially cut off from his family. Going east was now, in a sense, almost like going back home.

———

Welch moved to New York City in late November 1950 and within weeks had not only secured himself a job as a clerk at Sterns on Fifth Avenue but also found a small apartment at 33 West 82nd Street in a rooming house run by a Scottish-Irish couple called Mr. and Mrs. Angus. Welch's apartment was on the Upper West Side, which was then a largely Jewish neighborhood and was, as such, immersed in the cultural idiosyncrasies of Jewish life. While the reason for Welch's choosing to live in that neighborhood is unknown, it is possible that he found it through his former Reed College roommate and friend, Marshall Kolin, whose parents lived in the building opposite. Whatever the case, it was neither a neighborhood nor a city in which Welch flourished. Within five months, he had left again, having cultivated an enduring and largely negative impression of the city in that time. Indeed, some years later, he wrote to Snyder that his time in New York was one "without friends or means," adding that it had been "pointless and fruitless—I became only bitter and afraid" (IR1, 80–81). Yet, this bitterness and fear could have so easily been avoided if he had only known what was happening in other parts of the city and found a way to get involved.

When Welch arrived in New York, it was very much the "noisy rock" he later described it as. But while that term was a negative one for Welch, the "noise" that the city was creating was anything but negative. In the years immediately after World War II, New York had unashamedly established itself as an economic and cultural center unrivaled on a global scale. The war had seen to it that the city was now the dominant force in manufacturing and finance, and the influx of artists and writers who had escaped Europe in the preceding decennia and settled in the city also meant that New York had taken over from Paris as the center of the art world. Indeed, many of the foremost European artists of the 1920s and 1930s, including Arshile Gorky, Marcel Duchamp, and Fernand Léger, did much of their best work in the city and laid the foundations for some of the major postwar American art movements. Given Welch's interest in and aptitude for art, it is a wonder that he wasn't more inspired by New York's art scene. In the five months he stayed there, major exhibitions were staged, from *Abstract Painting and Sculpture in America* at the Museum of Modern Art to the *Annual Exhibition of Contemporary American Painting* at the Whitney. Yet the only mention he makes of such events is his intended visit to the Modigliani show, which had opened at MOMA in April 1951, and which he planned to go to with his sister Gig.

In addition to the art scene, New York had firmly reestablished itself as the center of the jazz world too. The city was already home to legendary clubs such as the Hickory House, Minton's Playhouse, and Café Society, and as clubs opened and reopened throughout the city, bebop began to emerge as the new expression in jazz. The impact of bebop—a spontaneous and free-form style that would innovate not only music and dance, but also literature—on Beat writers is well-documented, and it is interesting to note that when Welch was living in New York, John Clellon Holmes was finishing his novel *Go* only a few blocks to the southeast. A fairly straightforward tale of disaffected American youth "torn between the negativity and despair that were prevalent . . . after World War Two and the new, more positive attitude towards life" that the end of the war had engendered,[5] *Go* didn't include the spontaneous prose and syncopated rhythms of later Beat texts (including Holmes's jazz novel, *The Horn*), but it is celebrated as being the novel that first coined the term "beat generation."

At the time, Holmes was living with his first wife on Lexington Avenue between Fifty-Sixth and Fifty-Seventh Streets, and while Welch was apparently stagnating on the Upper West Side, Holmes was writing, entertaining, and recording music and poetry readings in his apartment. These recordings, many of which have been preserved, feature among others Ginsberg, Kerouac, and, on one occasion during that time, Neal Cassady, and provide a snapshot of the amalgamation of music and literature in a new Bohemia. And just as Holmes was finishing *Go*, Kerouac was about to start on the first draft of *On the Road* across Manhattan on West Twentieth Street. Any number of notable artists and writers were scattered throughout the city and could easily be found in Greenwich Village haunts such as the Cedar Street Tavern, the White Horse Tavern, or the Stable Pony Inn (where Ginsberg and Corso first met), or Times Square cafés such as Chase's, Grant's, and Bickford's. It is tempting to imagine what this literary and artistic milieu would have meant to Welch had he been able to connect with it.

As it was, however, within days of Kerouac finishing the first scroll draft of *On the Road*, Welch had left the city altogether, seemingly oblivious to its myriad possibilities and bohemian spirit. In essence, Welch was cut off from it all. His contact with Williams was the only literary one he had, and even though Williams was a huge presence in that world and had a tremendous influence on Ginsberg during that time, he never really signified anything more to Welch than a friendly and encouraging ear at a time when Welch was more in need of a helping hand. Indeed, the irony of the situation becomes even more apparent when one considers that Welch's visits to Williams coincided with those of Ginsberg, yet for some reason Williams never thought to introduce them. This is even more interesting if one considers the similarities between the two at that time: both were twenty-five, both were struggling to find an authentic poetic voice, and both were experiencing emotional and psychological issues—Welch was the verge of his first serious "crack-up," and Ginsberg had recently been released from an eight-month stint in the psychiatric facility at Columbia Presbyterian Medical Center. Moreover, shortly after his release, Ginsberg attended a poetry reading by Williams at the Guggenheim, an event that "would dramatically alter the direction of both his poetry and his life" in much that same way that Williams' reading at Reed had done for Welch.[6]

But for Welch, New York had become a "noisy rock," the first of a number of such rocks to engulf him in the "din" of America. Even regular visits from Virginia and Dorothy, the arrival of his old friend Dave Brubeck for a series of shows at the Hickory House, and a serious attempt at studying work by Stein in the New York Public Library were not enough to lessen his growing disquiet. What was supposed to be a journey to literary stardom ended in tatters as a result of Welch's first significant psychological and literary crisis.

Before this crisis fully materialized however, Welch took up the offer extended by Williams at Reed to visit the poet's home in Rutherford. To the young graduate, arriving in New York with both an invitation and a commendation from one of America's foremost poets must have felt like a slowly opening oyster shell, gradually revealing the pearl within. And although his visits to Rutherford were few, they would become invaluable to Welch. If the initial effect was small, the long-term impact that Williams had on Welch was immense. Indeed, if he was indebted to anyone other than his former teachers Rideout, Wilson, and Jones, it was to Williams, of whom he said, "I really became a poet only because of Williams. Williams and Gertrude Stein."[7]

These visits to Williams's household had a profound effect on Welch's poetic sensibility. He speaks of them resulting in "total mind transmission" and of the precision and diligence in Williams's methods, as well as of the purity of his speech and the humility of his actions. Years later, after Welch published *Wobbly Rock* and sent it as a thank-you to a then-frail Williams, the latter replied that though he had seen that Welch was on the verge of a breakdown in New York, the intervening time had not been wasted if such poetry was the result.

With such inspiration, it is hard to understand how Welch was able to write poems only sporadically. Instead of polishing his work on Stein as he had predicted, Welch found that he was increasingly at odds with himself as to how to actually tackle it. He spent time in the Special Collections Room at the New York Public Library poring over rare Stein manuscripts that might help him complete his work for potential publication, admitting all the while that he was struggling to work and that his research was merely a means of killing time and absorbing information that may not necessarily ever surface again. He wrote to Williams that "this has been a very bad couple of weeks. Just can't work. Best thing is to take a lot in if you can't make a lot come out. So I do research on Gerty and watch the people do the things they do and listen to 'em talk" (IR1, 46). The bigger issue however was not even his thesis but the fact that one of his friends had criticized his poems, saying they were incomprehensible and that Welch was "obscure, antisocial, and 'like all modern artists'" (IR1, 44). This created a crisis of confidence the like of which Welch had yet to experience, and his work naturally suffered.

With the exception of Williams, Welch had ever shown his poems only to his friends and peers. By 1950, he had had nothing published save "Monologue of One Whom the Spring Found Unaccompanied" in *Janus*. At Stockton College he had written the sports pages and helped others with their writing assignments. At Reed he was principally an editor and critic who rarely had to expose himself to the very criticism he handed out to others. Within the safe confines of the Crapsey Society, criticism may have been sharp, but it was also more easily accepted. Now in New York, he had shown his work to a friend who then more or less condemned it.

To a poet who strove to deliver concise, uncluttered images that surpass meaning on account of their apparent simplicity, this criticism came as a complete shock. In a letter to Williams, Welch summarizes his ideology perfectly when he says,

> I insist that if a thing seen, just a simple god damned thing seen, seen intently, and for no other purpose than just seeing, beginning and ending with it being a seen thing, a thing with completeness to it for me, a god damed [*sic*] THING, I, to repeat, insist that if this thing is something to remember with pleasure . . . although I may not be able to USE the god damned thing . . . it is worthy of being made into a poem. (IR1, 44)

Part of Welch's approach to poetry was to capture not only the things he saw but arguably, even more so, the things he heard. He was very conscious of listening carefully to people around him when they were talking—a practice that remained central to his work and his idea that language is speech and that speech can, by extension, be shaped to create perfect and legitimate poetry. Useless perhaps, but poetry nonetheless. In the preface to *Ring of Bone*, Welch offers an anecdotal example of what this sense of legitimacy means even as he contradicts notions of useful/lessness in the process:

> I once took a guided tour through a California winery and the guide . . . droned away with his memorized speech of facts and figures, chanting them perfectly in that guide-chant all of us have heard, and suddenly he stopped and shouted "Whose kid is that!" A small child was determined to fall into a 500 gallon vat of wine.

The force of real speech slammed right against false speech was startling as a thunderclap, and not because he called out loudly. I vowed never to release a poem of mine which couldn't at least equal the force of that guide's "Whose kid is that!" Pound said that poetry ought to be at least as well written as prose. I say poetry ought to be at least as vigorous and useful as natural speech. (RB, 5–6)

For Welch, it was of critical importance that poets "know what their tribe is speaking."[8] "Tribe" was a term that Williams had previously used in relation to what he called "the theme of much I had to say of the arts,"[9] and which was no longer used to refer only to Native Americans and Indigenous groups but to define what Gary Snyder described as a "new type of society now emerging within the industrial nations."[10] In his essay "Why Tribe?," Snyder further elaborates that the tribe "proposes a totally different style" and is made up of, among others, "alienated intellectuals, creative types and general social misfits" who had come to recognize each other's kindred spirits in the years after the world wars in which consumerism, corporations, centralized government, and capitalism began to take hold.[11] As these tribes looked to create a greater sense of community and embrace notions of syndicalism in order to counteract the evils of contemporary society, so there was a need for language that transcended this. This was the language that Welch would listen for in his search for particular ways of speaking that were pleasurable to his ear and that he deemed to be American in origin. For Welch, the language of the tribe was key. It was all part of his plan to distance American poetry from its English roots and have it represent a new poetry that took none of its idiosyncrasies from old influences but instead collected from the natural dialects and vernaculars of a country now establishing itself more than ever as its own—even in its present state of fragmentation. He expressed amazement at how Whitman could have produced such works of genuine greatness using a language that was ostensibly "the rubble of Europe," and proposed that Mark Twain was the first author to use American sentences. To write such sentences, authors and poets needed to

go out into the street and listen to the way people talk. You really have to listen to the kind of things that people say. You have to listen to the birds that are in the air, the helicopters, the big rush of jets. . . . You have to have your ears open. You have to have your goddamn ears open or you're not going to be a poet. Or you're not going to be a writer of any importance whatsoever.[12]

Listening with accuracy was a skill he had developed all his life, and which he was constantly seeking to improve on and embed into his work. Indeed, Welch once claimed that he could decipher structure from meaning in everyday speech when he was as young as four, remembering an incident in which Milton, the family gardener, stated that "I ain't got nothing"[13]—a clear departure in Welch's young mind from what the ear would normally hear and process as correct. Even at this young age, Welch contends that his ear was attuned to the American idiosyncrasies of such colloquialisms and their apparently grammatical deficiencies and social prejudices. It must surely have been with some interest that he later followed the furor surrounding the publication of the 1961 *Webster's Dictionary* in which *ain't* was listed as having received unqualified approval by renowned linguists (thus eradicating the contention that *ain't* was confined to the uneducated lower classes). James Sledd, Welch's linguistics teacher in Chicago, had coedited the dictionary and, like Welch, apparently understood the importance of such colloquialisms. He could not, however, have foreseen the barrage of criticism the inclusion received from media and academic sources alike, and he famously responded in 1964 by sardonically stating in *Language* magazine that "the agonizing deappraisals of Webster's Third International show that any red-blooded American would prefer incest to ain't."[14] However, Welch's increasingly unique approach to poetry was not enough to rebuff the (self-)criticism he felt, and as a result, the next six or seven years were to be among the most barren of his entire writing career. While he enclosed various poems in his letters to Williams, Snyder, and Dorothy, the sum of his work between 1950 and 1957 amounts to no more than two or three dozen poems. He shelved his work on Stein for good, and instead started to dig about in what he called the "garbage" of his mind in an attempt to find some peace and comprehension concerning his childhood and its complexities.

———

In early 1951, Welch had the first of a series of breakdowns. While Welch felt that the reasons behind it were related to his childhood, there is a

strong suggestion that Welch may have suffered from a more serious disorder, namely, as it was known at the time, manic depression. In 1951,
manic depressive illness—or bipolar disorder, as it later became known—
was as yet unclassified in the United States. Indeed, the Diagnostic and
Statistical Manual of Mental Disorders, or DSM, the manual published
by the American Psychiatric Association that classifies mental health disorders, was first published in 1952; when Welch had his first breakdown,
a clinical diagnosis would have been difficult to come by. The term manic
depression, first coined by influential German psychiatrist Emil Kraepelin, was widely used to refer to mental illness that had an emotional
or mood-based trigger rather than a cognitive one. Kraepelin purported
that the degenerative nature of a disorder such as schizophrenia should
be differentiated because it was rooted in cognition and thus resulted in
an irreversible loss of cognitive functions. This, in turn, made it more
difficult to treat. On the face of it, manic depression was more easily treat-
able by using barbiturates and other drugs such as promethazine, but not
until after Welch had this breakdown did more concerted clinical trials
result in lithium being determined the most effective drug to treat bipolar
disorder. Where barbiturates had been employed to induce "sleep states"
in manic or depressive patients, lithium acted as stabilizer. Yet, in the
United States especially, groundbreaking research into lithium use that
had been spearheaded by John Cade in the 1940s was largely discredited,
and as a result the drug was not classified for prescription by the FDA
until April 1970, a little over a year before Welch disappeared.

For Welch, this potential disorder, coupled with his severe and crip-
pling alcohol abuse, seems to have gone unchecked. Throughout his life,
we find many references to the ebb and flow of his moods, or the highs
and lows of his productivity. Yet never is there any mention of a significant, and diagnosed, mental problem. Indeed, much of Welch's erratic
behavior is often put down to his alcoholism rather than to any condition
that might have been exacerbated by it. So when, in 1951, he first decided
to face his demons, it was by means of an altogether different, and highly
controversial, approach.

In spring 1951, Welch began attending meetings and therapy sessions
conducted by a group he called the "Dianetics people"—the first incarnation of what would later become L. Ron Hubbard's Scientology movement. By the early 1950s, Dianetics had become a controversial alternative

to more conventional means of psychiatry/psychology. L. Ron Hubbard's "bible," *Dianetics: The Modern Science of Mental Health*, had been published in May 1950; by the time Welch became involved in the process of elevating himself to the status of what Hubbard called the "clear" or, dianetically speaking, "optimum individual," the book had already sold very well and created some considerable commotion in medical circles. Hubbard's philosophy was simple: this new science of the mind was about eradicating "all psycho-somatic ills and human aberration."[15] He was of the opinion that all humans are essentially good and that negative aspects in a person's life are a result of the "reactive mind," or as he describes it, "that portion of the mind which files and retains physical pain and painful emotion and seeks to direct the organism solely on a stimulus-response basis."[16] The role of Dianetics in curing the reactive mind was to identify and unlock so-called engrams that are the moments of unconsciousness wherein such pain or trauma occurs. As such, the reactive mind has a reservoir of engrams that are beyond conscious thought and, when they are unlocked and discharged, all manner of suffering is alleviated—physical and psychological. Hubbard claimed that this process was not only helpful for ailments such as arthritis, myopia, and heart disease but was also helpful in dealing with schizophrenia, manic-depression, neurosis, and dipsomania.

The treatment was similarly straightforward: each patient, or "preclear," was assigned an "auditor" to facilitate a therapy in which the "target is the engram."[17] The patient was encouraged to regress—or "return" in Dianetics terminology, "regress" having too many associations with hypnotism—into their memories such that they relived that memory not as a memory but as a concrete moment. Upon consciously experiencing that moment, the pain or trauma unconsciously attributed to it is then removed, and the memory is discharged from the reactive mind bank and refiled in the standard memory banks of the analytical mind.

For Welch, who spoke at length in later life about the dysfunctionality of his childhood, this process offered him a chance to unlock certain traumatic aspects of his early life that he felt were aberrations, and that were still apparently causing him some anguish. His correspondence from that period fails to mention any catalyst for his sudden involvement in Dianetics, and it may even be that his breakdown was a *result of* it rather than a treatment for it. Given the attention that Welch paid to these issues in the years that followed, mostly by means of conventional psychology, it may be that Dianetics forced him to confront aspects of his childhood that he had long buried. However, regardless of chronology, he is clear about the fact that Dianetics was a "therapy"—he even successfully persuaded Gig to get involved so she could unlock some of her own childhood traumas—and that, despite increasing skepticism and warnings being issued by the medical world, Welch saw nothing dangerous in it.[18] Indeed, in a letter to Dorothy from April 1951, he reiterated that "there is no danger involved" and "I don't think one need worry about dangers" (IR1, 48). That said, before long he was beginning to question the efficacy and uniqueness of Dianetics, writing to Dorothy that, while the experience of returning to, for example, a prenatal state was an enlightening one, he was unsure how this theory differed from the theories of psychiatrists and psychoanalysts such as Karen Horner, Erich Fromm, and Harry Sullivan. Yet, he was sure that the new *goal* of therapy established by Dianetics would significantly alter the methodology in getting there.

During these sessions, Welch was encouraged to relive his childhood through what he termed an "intensification of memory" that allowed him to refocus certain aspects of his current life through arriving at a deeper understanding of the fact that

> a whole set of my attitudes and opinions has been resting upon things that happened [in my childhood], and that when I went through them again, and cried them "off," so to speak, I had, and have, noticeable increase in liveliness during everyday actions, and a strange feeling of a structure, or structures, being torn down. (IR1, 47)

Welch writes of "judgment structures" that, when encountered and undone, release memories of experiences that had never before been realized or known and that, upon release, provide greater clarity and a sense of enlightenment about the self that was until that moment blocked and "bottled up in the past." He also writes of "the prenatal memory business," in which he suggests that Dorothy's behavior while he was still in the womb had a detrimental impact on his life. He contends that his tendency to procrastinate was the result of Dorothy saying "not now," "let me go," or "leave me alone." In fact, he even attributes severe psychosomatic pain to a conversation he "overheard" as a fetus in which abortion is discussed.

His correspondence with Dorothy on this subject may have also unlocked certain misgivings about her role in his childhood, and, given the fact that she was studying child psychology at that time, it must surely have made for uncomfortable reading. Perhaps not unsurprisingly, little of how she felt about this was ever mentioned in her correspondence with Welch. Contrary to sharing his thoughts on being afflicted with a "generalized terror," Welch only occasionally mentions his writing in these letters, and the ones that follow his departure from New York are extremely revealing of his inner turmoil, foreshadowing much of the trouble he would subsequently go through to give his life the same balance and clarity he looked for in his poetry.

Upon leaving New York, Welch initially went to Florida, where he joined other followers of the Dianetics movement who were spending the summer in Sarasota. In another letter to Dorothy, Welch explained what the trip had in store for him and justified this choice as being the result of his having "become rather dislocated" and experiencing a feeling of "general uprootedness, or unrootedness, about my present condition that I shall yield to—since after all what else?" (IR1, 49). What transpired in Florida was a further investigation of his psyche through the practices he had initiated in New York, but by the summer's end Welch wrote that his work was done and that he felt he now had the necessary tools to continue the task alone elsewhere. It seems that he had achieved—to a certain extent at least—his intended goal of becoming "cured . . . or in full possession of nature faculties for unhampered use in present situations."[19] He also writes of having now recognized the inflexibility of the mind where ideas are concerned "('ideas' means not only the concepts, opinions, attitudes, etc., . . . but also the objects [of these ideas] themselves)" (IR1, 55). Humans are used *by* their ideas rather than making use *of* them. The inability to "properly relate objects" results in the most extreme of outcomes, "in (or un) sanity" and only with the help of a mediator (such as, to quote Welch, psychiatrists, auditors, or mediums) can the patient come to the realization that, to relieve such mental states, the "little defenses and stabilizing mechanisms" in the mind must be trained to relax so that these mental citadels can more easily be examined and self-knowledge increased. Having made this discovery, Welch specifies a very clear intent to continue his self-analysis by delving into Raja Yoga and Maharishi Patanjali's *Yoga Sutras*—another nod of recognition toward the Eastern philosophies he had first encountered at Reed and a course of thought that would recur to varying degrees intermittently throughout the rest of his life and writing.[20]

Having undergone the course in Dianetics therapy, Welch most likely understood that even more rigorous discipline was required for the focused application of such yoga practice, but in gathering together a number of different mediators to aid his quest, he may have overestimated his own ability to maintain such discipline. Although his intentions were noble, his self-knowledge would require significantly more development before such a practice could be successful. While it is possible that Dianetics had unmasked certain blockages in his psyche, others were still very rigidly in place and, more important, still entirely invisible to him.

His summer in Florida also provided a chance meeting with two former Reed classmates, Peter Oser and Dan Drew. Oser, the great-grandson of John D. Rockefeller and the grandson of the industrialist Harold McCormick, had flown from his family home in Geneva, Switzerland, to Sarasota, where his latest girlfriend had a house. The couple met up with Oser's friends there, Welch being one among them. Oser was an eager follower of the Dianetics movement, having been introduced to it by his former wife Mary, who had gone to New York in 1950 in the hope that she could find a Dianetics practitioner to treat her husband for what she called his "money sickness." Oser had been finding it increasingly difficult to function normally on account of the immense wealth he had inherited, or, as it was also called, his "'havingness,' the affliction of having it all."[21] In New York, Mary met John Starr Cooke, a young Dianetics auditor who became Oser's teacher and traveled extensively with the couple to Europe and Africa. Oser lavished money on Cooke and the movement, such was his initial fanaticism for it. However, rather than cure Oser of his burden of inheritance, Cooke saw to it that Mary and her husband grew further apart, with the result that the couple divorced the following year (Cooke and Mary would marry in 1952).

When Oser joined Welch in Sarasota, he was not only still burdened by his "affliction" but also more than happy to fund Welch's trip back to Chicago, footing the bill for the whole group's travel and hotel expenses. On arrival, they stayed in the Stevens Hotel, which at one time was the largest hotel in the world and a landmark property that reflected Oser's wealth and status. It was a gesture not lost on Welch, and he would profit from Oser's generosity for the better part of a month.

By August, when the time came to leave Chicago again, Welch went to New York with Oser and Drew to collect some of his remaining things from the Anguses. Again, Oser took care of the tab for their stay—this time in the Park Sheraton Hotel, before he returned to Switzerland. He would never feature in Welch's life again, being merely a passing, and convenient, coincidence.

Though his relationship with Welch was fleeting, Oser's legacy is perhaps most interestingly portrayed in a novel by Brion Gysin titled *The Process*. A scathing parody of the Dianetics and Scientology philosophy he had encountered when the Cookes visited and befriended him in Tangiers, the novel sheds light on what Gysin felt were financial malpractices that bordered on embezzlement. While Gysin admired and was enchanted by the couple, he was far from impressed by their "religion." The story is recounted in Peter Manseau's *One Nation, Under Gods*:

> In allusion to Scientology's auditing obsession, Gysin used his novel
> . . . to recount the way [Cooke and his wife Mary] had deployed
> a spiritual practice called "Grammatology" to bilk [Mary's] first
> husband, "Peter Paul Strangeblood, the richest little boy in the
> world," out of a cardboard box full of cash they had arranged to
> have parachuted into the desert. "I'd been giving Strangeblood
> various occult exercises for his havingness," Gysin wrote in the
> voice of John. . . . Poor PP, there really wasn't much anyone could
> really do for him except take all that money away from him and he
> knew it.[22]

Welch, like Cooke, also knew it. His final words on Oser are an echo of those spoken by Cooke and captured by Gysin. He rather flippantly wrote to Dorothy that "Pete . . . is responsible for making this trip a possibility—he having a staggering income. . . . This trip is at his expense. One person more or less will only add a few hundred to his prolly tremendous htl. bill, which makes no difference when one's income (annual) is in six figures. . . . Also—and most important—Pete understands it so" (IR1, 57).

———

By this time Welch had made up his mind to return to university in Chicago. Having first mentioned the idea when he was still in New York, the final decision to actually enroll in graduate school to study history and philosophy was made in Florida. Welch wrote that the year off had done him good, and he was now ready to commence the next phase of his "synthesis." After a brief visit to New York, where he also saw Dorothy in Ithaca, he settled into his new Chicago dwelling at 4713 South Woodlawn Avenue, ready to start his studies. His excitement is clear ("I am all rarin'") and his plans are once again voluminous and ambitious; first a master's, followed by a PhD, and then settling down to fulfil his previous intention to open up "the minds of the American youth" (IR1, 50) as a future professor. Having removed (some of) the blocks in his own mind, Welch suddenly felt in harmony with himself, and his former exuberance returned. In early November, he wrote to Dorothy of taking "a truly first-rate" course on Plato's *Republic*, given by "one of the authorities on Plato & Aristotle," the eminent historian and philosopher Professor Richard McKeon.[23] Welch was also registered for a course in Greek history and one in logic. He wrote of the requirements and teaching as being on a par to those at Reed and that he felt very much at home at the University of Chicago. One might, therefore, expect that this would also have resulted in the reestablishment of his literary inspiration and provided him with the motivation he needed to rekindle his work. On the contrary, however, rather than helping him refocus his energies, life in Chicago seemed to do little other than detach him still further from the grand literary and academic plans he had had not six months before. And, true to form, within a matter of weeks circumstance would dictate that he make yet another decision that once again turned his life on its head.

In late November, Welch made the short journey south to spend some time in Bloomington, Indiana, where Gary Snyder had enrolled at graduate school in the fall and was studying anthropology and linguistics. Snyder was living in Bloomington with another of their old Reed friends, Dell Hymes, in an apartment above a Chinese restaurant on Kirkwood Avenue.[24] Welch had joined them on their journey back after both had attended the American Anthropological Association's annual conference in Chicago with Indiana University associate professor of anthropology David Bidney, from whom Snyder took a course in Western philosophy.

Given the almost constant fluctuations in mood that Welch had been experiencing, the visit provided him with a much-needed, if short-lived, boost of clarity and self-confidence. Snyder apparently helped him appreciate that "the important thing is to become whatever it is that man can be" (IR1, 61), thereby prompting Welch, who always took heed of what Snyder had to say, to take immediate stock of his position. He planned to enroll at the Kenyon School of Letters on account of its being "the heart of the excellent [literary] criticism in America," and if all went well and Kenyon hired the right people, Welch even saw himself transferring into the Letters Department there to "be in on the ground floor of a very revolutionary, and very worthwhile development in Education."[25]

Upon returning to Chicago, Welch thus decided it would be in his best interests to immediately switch from philosophy to English — another decision that might be expected to have added impetus to his literary motivation. He described his decision to take philosophy as "an adult way of staying a little boy" and noted that the English course was not particularly challenging. In a letter from December 8, 1951, Welch indirectly thanks Snyder for allowing him to see that "the acceptance of oneself is a great relief" — which in many ways was a continuation of the self-analysis he had begun with Dianetics and that would continue still further during his next five years in Chicago.

Although he describes his visit to "Blummerberg" as the most significant event of the year, the desired result would be anything but forthcoming. He subsequently described the University of Chicago's English Department as stifling, considered leaving the university altogether and joining Snyder in Indiana permanently (Snyder himself left Bloomington after only one semester around the same time), and wrote that he was "lonely, depressed, and bored" — all of which are indicative of his transition from what he described to Williams as "a condition as near to a mental breakdown as one can come" (IR1, 60) to the full-fledged psychoanalysis he would undergo within a year. Here, perhaps for the first time in writing, Welch exposes himself as someone who is susceptible to manic swings in mood and motivation: the intense high of graduation, followed by an unwarranted loss of confidence and the first manifestations of serious doubt. Then therapeutic analysis and the gradual unearthing of his real self, leading to more grandiose statements of intent, only for reality to bite back and leave him further adrift than he had been before. It is hardly surprising then that, while his path to Chicago seemed to be a positive one, the mental breakdown he wrote of, and which he later called his "big freeze," was firmly on its way. Indeed, Philip Whalen described him many years later: "Lew was sort of bipolar. He was fine sometimes, and other times he was down. . . . It was terrible. He'd grouse around about how he was broke and was having a hard time, and then, you know, presently, a couple of hours later, he'd be sparkling all over everything and having a grand time."[26]

In contrast to the various problems he seemed to be facing, Welch spent the Christmas period working in the relatively mundane surroundings of the US Post Office. He had secured the job on account of his landlord being the assistant postmaster, and whatever encouragement he had brought back with him from Indiana seemed to have disappeared with the furious ebb and flow of his moods, never more apparent than at this time. In one letter to Snyder, he writes of his frustration at the literature courses on offer and talks of accompanying Snyder to the West Coast because "these people hate pomes" (IR1, 65). A fortnight later, however, in a second letter he admits to doing well in his studies, having stopped writing poetry altogether and having become "a scholar." These letters are also conspicuous for mention of explicit references to his increasingly excessive use of alcohol and its influence on his writing. If drunkenness blighted his creative output, it also awoke a less attractive side of Welch's nature that would often rear its ugly head, first, in one letter, only for him to apologize for having said something out of turn or inappropriate in the next. His letters sometimes contain language that has a wicked undertone and is as revelatory as anything positive he wrote. Yet this tone was often fleeting, and within a paragraph he was upbeat and optimistic again, writing of new ventures or ideas.

When Snyder returned Welch's visit in late January 1952, he was welcomed with open arms and, for want of money, entertainment in the form of good conversation only. In turn, Snyder wrote of spending his time in Chicago in one of Welch's "dark and sunless rooms where I read five days"[27] before boarding a Greyhound bus for Portland, where he was planning to visit the pair's former philosophy teacher at Reed, Stanley Moore. This brief get-together appears to have been one of only a few occasions in which the pair saw each other while Welch lived in Chicago. Their correspondence would similarly dwindle, and while Snyder's path would take him to Japan via Berkeley and the famous Six Gallery reading in San Francisco, Welch stayed in the Midwest for another five years, a period that was to be defined as one largely comprising frustration, psychoanalysis, and domesticity.

———

Of course, it would be remiss to suggest that the time Welch spent at the University of Chicago was entirely fruitless and, similarly so, to suggest that his time there can be defined only by difficulties and dramas. On the contrary, during a poetry reading in San Francisco in November 1959, Welch, despite seemingly telling his audience to avoid going to Chicago at all costs, nonetheless speaks with a real sense of candor about how his studies in linguistics, and more specifically dialects, enabled him to further develop his ability to listen and to hear speech accurately. This in turn was essential learning for him.

He was fortunate enough to have followed classes by the renowned structural linguist James R. Sledd, who employed his own southern accent—which Welch once described as the broadest and flattest Georgia accent imaginable—in class as a prime example of how a particularly distinct American dialect can construct sentences of both beauty and balance without being seen as inferior and without compromising on authenticity. Welch listened and absorbed, later saying that Sledd "really altered my mind—really showed me the wonders and mysteries of this thing we call language. First of all, he forever removed my prejudices regarding dialect" (HWP, 34).

Years later, perhaps in part because of these lessons, Welch said that until the early 1950s he had failed to fully identify "real" American speech and that, consequently, his pre-Chicago poetry retained a refinedness that he equated to more traditional English than to the speech of the tribe he was part of. He also wrote that these early poems were merely "a handbook. A schoolboy's tablet. The record of my hammering at the language. But they are now too small, too crabbed" (IR1, 60). Welch goes on to say that he suddenly realized that "the vigor of the language you hear on the streets is greater than the vigor of the language in poetry"[28] and that he began to focus on the aim of combining the two into a form that would allow him to "perform it in American diction."[29] For Welch, Sledd's use of language was one example of such diction. Here again was a teacher exerting his influence. And part of that influence was to make Welch see that to truly hear American speech, he needed to go back and understand what distinguished it from English in the first place.

In addition to Sledd, another key influence on Welch's development at the time was the Danish linguist, Otto Jespersen. Welch's graduate studies in Chicago would almost certainly have included a reading of Jespersen's *A Modern English Grammar on Historical Principles* as well as *Essentials of English Grammar*, and it is clear from Welch's later essays and readings such as "Language Is Speech" that Jespersen's ideology concerning the true nature and use of language helped Welch to justify and vindicate his own thoughts on the subject. It is difficult to differentiate between Jespersen's assertions that "language is *primarily speech*, i.e. chiefly conversation (dialogue), while the written (and printed) word is only a kind of substitute,"[30] and Welch's own statement, "Language is speech. Any other form, the printed one or the taped one, is a translation of language" (HWP, 30).

So, along with his studies into the nature of language, Welch also found himself taking classes on Old English and Chaucer in order to learn something about the history of the English language. And in combining linguistics with history, Welch seems to have found a legitimate means with which to concretize his own learning. If Sledd and Jespersen had given him the theory, his finding the differences between British and American speech in literature provided the basis for his own poetic development. The enthusiasm that ensued—the flow preceding the inevitable ebb, as it were—even resulted in his first published work since Reed College, when the *Chicago Review* published a short poem titled "Aubade" in its Winter 1952 edition. Later renamed "Lines to an Urban Dawn" in an attempt to distance it from more traditional poetic forms (an "aubade" or dawn serenade, is a song or poem that refers to lovers parting at dawn) and to emphasize the "lines" of the poem in the same way as Sh'n focused on the lines of the drawing or painting, the poem harks back to Welch's experiences in Portland and is another clear example of his ability to take everyday observations and mold them into intimate vignettes—albeit in a language with which he may no longer have been content. Despite this, Welch's willingness to submit the poem to such an important literary review highlights his positive frame of mind at that time, and its publication was the reward.

However, his enthusiasm was again short-lived. Circumstances conspired to change the course of his life once again. The ebb and the flow. It would indeed be many more years, and the renewal of old acquaintances and a now-legendary poetry reading on the West Coast, before Welch's literary life would finally get back on track.

———

Welch quit his studies for good in March 1953, and although the Korean War was nearing its end by then, he must have quit in the knowledge that he was now safe from conscription. By summer he was working for the advertising department of Montgomery Ward.[31] Letters to Dorothy from 1952 and 1953 deal almost exclusively with financial issues relating to his impending psychoanalysis, his work, and various declarations of love for his girlfriend at the time, Barbara, with whom he was having an intense and passionate relationship that bordered on marriage. Entries in Gary Snyder's journals from late 1952 mention Welch spending time with Barbara Wuest, who was the best friend of Enid "Tommy" Thompson, a Reed alumnus and wife of Welch's friend, Grover Sales. In one entry, Snyder writes, "Lew sitting up until 5 a.m. at Barbara Wuest's, why?"[32]

Either way, Welch's relationship with Barbara was volatile as well as passionate. In August 1953, he writes of marriage "inside of a year" and the "idea of permanence" being difficult but ultimately what they both want. Yet, by November, they had broken up, due to what Welch called "growing pains expressing themselves in a resisting of added responsibilities" (IR1, 76). In a period in which Welch was looking ever more inward for answers about his past, such a volatile relationship was obviously something that added to his sense of, in his words, being "very high strung." Yet, it was also through Barbara that Welch met a group of people who would offer him a view on urban living that would later underpin much of his thoughts on not only Chicago, but on city life in general.

At the time, Barbara was living in a building alongside several designers and architects who were enrolled at the Institute of Design (ID), then housed on the city's North Dearborn Street. Established in Chicago in 1937 by László Moholy-Nagy, the "New Bauhaus" was an attempt by Moholy-Nagy to take the same ideas from his teachings at the Bauhaus in Germany and develop them further in the United States. Moholy-Nagy, like other intellectuals and artists, had fled Germany in 1933, settling first in the Netherlands, then in London. In Chicago, Moholy-Nagy firmly established a clear notion of how he could develop the traditional Bauhaus theory of segregated, craft-based distinctions into a more multidisciplinary approach that focused on three departments: architecture, product design, and light workshop (advertising arts). It was the institute's approach to functionalism that most appealed to Welch. He wrote of urban inhabitants having a "right to light and air" and of how the ID had made him radically rethink his opinions on architecture, furnishing, and decoration. His apartment on North Sedgwick Street, in a predominantly white Mid-North neighborhood, became a personal statement of New Bauhaus design: minimalist, functional, clean, and, perhaps most importantly, cheap. In fact, such was the importance of this new design ideology to him, that Welch wrote to Dorothy that he was considering "taking some of the night courses at the Institute of Design just to learn the handling of materials and the basis theory of this approach" (IR1, 74).

In time, however, it would be less the theoretical approach to decoration and furnishing that would remain with Welch than the right to light and air that the ID was championing. Except, in his case, before too long Welch would develop the idea such that this right eradicated the need for urban living at all.

Aside from his on-off love affair with Barbara, during this period Welch underwent his most significant therapy to date, entering into a concerted period of psychoanalysis that lasted four years and that required him to find a decent job to cover the costs. Although he would initially profit from a trust fund set up for Dorothy after her father's death, Welch eventually took on the responsibility for the payment of his treatment himself. Indeed, he argued that the terms of him using the trust fund undermined this very responsibility and that, rather than letting him take control over the financing of his treatment, the trustees were depriving Welch of what he felt was necessary to take control of his future. To some extent this heightened sense of looking after his own affairs may also have contributed to the cessation of his full-time studies.

Welch started attending individual therapy sessions in September 1952. Although he had hoped to begin earlier, his psychiatrist, Dr. Joseph G. Kepecs, had suggested they wait until after the summer so that the holiday would not interrupt the continuity of the work—a decision that surely offered Welch a certain sense of comfort, given that Kepecs apparently saw no need for urgency in beginning the sessions—this despite the fact that Welch had apparently recently experienced another severe break-down and clearly wanted to get professional help.[33]

Welch most likely came into contact with Kepecs through the renowned psychologist Dr. Samuel J. Beck, who had performed a Rorschach test on Welch in July. At the time, Beck was one of the foremost practitioners of the test and worked alongside Kepecs at the Michael Reese Hospital in Chicago. The Rorschach test—which Dorothy was keen to point out had cost her $50—was routinely carried out in the United States as the first stage in studying and determining a person's character and personality before embarking on a specific course of analysis.[34] The results of Welch's test were likely destroyed when the hospital closed, but given Kepecs's lack of urgency, there seems to have been little to suggest that Welch's character, in his doctor's opinion at least, was deeply flawed.

Welch's relationship with Kepecs lasted for the better part of four years and seems to have been a more than satisfactory one. Time has told us that the effects of these sessions were more temporary that either would have liked, or that maybe Welch's issues were far more deep-seated for analysis of this type to deal with in the first place. However, not only would Welch dedicate a future poem to Kepecs, but he also described his analysis as "the best and easiest way of growing into a person mature enough to enjoy what is available to enjoy & to make available what is presently lacking in one's world" (IR1, 77).

While it may have initially been unclear what Welch felt was lacking at that time, there is a certain sense of maturity in how he now seemed to be approaching his life, and before too long he would develop an intimate understanding of what analysis was doing for him.

Kepecs, who at the time was an associate psychiatrist at Michael Reese, listened, counseled, and paved the way for Welch to gain a more gradual understanding of himself. Welch was able to "see himself more clearly in Joseph Kepecs,' and psychiatry's mirror: a friendly, and attentive silence which allows one to speak and listen at the same time."[35] He later described the analysis:

It is easily the most important thing I've ever done in my life, and the thing from which the most satisfaction has come. I don't think you [Dorothy] realize (because no one does who hasn't been through analysis) that one doesn't wait, breathlessly, until he is secure enough to quit. Instead you go because you badly need affection, warmth, and council—and when you begin getting it, it itself is a deep pleasure. It is not a patching job that attempts to send you off on your own—quite the reverse, it is a process by which desperately lonely people learn that they are not really alone, . . . that they can trust people to supply them with friendliness and love without which there is no point to living.[36]

In answer to his mother asking why he did not feel "secure enough" to see Kepecs less often, Welch bluntly answered her question with one of his own: "Why should I deprive myself of seeing the person I am more fond of than anyone I have ever known?" Indeed, in an unpublished letter to Dorothy from January 1954, Welch gives the fullest and most heartfelt indication of not only what Kepecs meant to him at that time but also of his thoughts on the role his late father had played in his life:

Kepecs has become a real father to me—one resists the relationship, of course, but it finally happens. A boy just has to have a father to become a man—it's that simple. If you don't have one or you had one that wasn't capable, then you have to buy one later on.[37]

This notion of "purchasing" a replacement father seems to bolster the sentiment of need that Welch conveys here, and there is no doubting the sincerity with which he writes. His commitment not only to analysis but also to Kepecs is clear, and he even went as far as to express his disappointment that his sister did not seem to have a similar desire to enter wholeheartedly into analysis—a statement that must have had its foundations in some sort of truth and again seems like a reproof of Dorothy (and their father). Was Virginia as miserable as Lew? She had apparently "withdrawn into her unhappiness"[38] and, not having "any devices left for combating it,"[39] should she not also try to benefit from such analysis? And in so doing find a surrogate father? Like brother, like sister?

————

In his work, Welch was now a nine-to-five junior advertising executive earning close to $4,000 annually, charged with planning, writing, and setting adverts for anything from electrical devices to home appliances. In the late summer of 1953, Welch proudly informed Dorothy that the ads for "television sets and radios which will appear next September in newspapers in 580 cities and towns in this country were written by me" (IR1, 74). He seems to have been genuinely content. His sessions with Kepecs were apparently paying dividends and his ideas of publication and PhDs were, for the time being, a thing of the past. Letters that once contained the plans and poems of his dreams were now filled with elaborate descriptions of his work, of how the company was structured, and of his place was within it. He wrote of promotions and merchandising, and admitted that "I'm much more fitted for this than I was for a teaching career—I'm not scholarly enough for that. . . . I'm in as good a spot as there is at Wards and I'm doing well" (IR1, 74–75).

In the three years it would ordinarily have taken him to get his master's degree, the self-styled "voice of the latter half of the 20th century" would now rather make a decent salary and was in a position to become qualified for a big-money job. Yet although Welch had entered the mainstream, the literary candle still flickered dimly among the statistics, targets, and projections. He was no longer the adolescent poet-critic but the mature realist, aware of the need for his own self-censorship to play a passive role and allow his creativity, however stifled and intermittent he felt it to be, to blossom on its own terms. From time to time he saw that with increased maturity came the possibility of better writing. Where he had largely shied away from submitting poems to magazines previously, he now sent one to the *New Yorker*. The criticism once given him by a mere student would surely pale in comparison to what he might have received from one of the country's foremost literary magazines. Yet Welch sent it off regardless. The flickering candle.

6
Panegyric for Illinois

> Laughing the stormy, husky, brawling laughter of Youth, half-naked,
> sweating, proud to be Hog / Butcher, Tool Maker, Stacker of Wheat,
> Player with Railroads and Freight Handler to the / Nation.
>
> — Carl Sandberg, "Chicago," *Chicago Poems*

For the next few years Welch tread a path that combined apparent domesticity and commercial humdrum with occasional bursts of creativity. His correspondence (at least that which has survived) appears limited to the odd letter to and from Gary Snyder and Philip Whalen. Welch also wrote to Dorothy, who was now living and working in Vermont, as regularly as he could, even apologizing at one point for not writing enough ("I seem never to be able to find time to write anyone" [IR1, 68]). He continued his sessions with Kepecs, and finally decided to break up with Barbara for good sometime during early to mid-1955. In a letter to Snyder that summer, Welch included part of a poem that he had written in response to having been sent a copy of Paul Verlaine's poem "Colloque Sentimental" by "an estranged beloved." Whoever the potential sender—and it's difficult not to see it as Barbara—Welch's reply is telling in its repeated use of Verlaine's line "Pourquoi voulez-vous donc qu'il m'en souvienne?"[1] Welch also writes of beauty being treasured only by "vain, deluded men" and of his narrator being "betrayed by the hope that we might love again" (IR1, 84). Indeed, he ends his letter by outlining how the poem will also include a description of Venus returning to the waves—once again, a veiled reference to beauty lost? To Barbara? If so, it was a sentiment that seems to have faded fairly quickly. No sooner had he broken off one relationship than another began.

Around the same time, Welch met a young Chicagoan named Mary Garber. Within a matter of months, the two were married in a whirlwind romance, not dissimilar to that of his parents thirty years before and true to what Welch had once written to Dorothy about his marrying someone he had known for a relatively short length of time.

Born in 1924 and raised in Cook County, Illinois, Mary was the only child of Emil and Loretta Garber. Emil Garber was a well-respected newspaperman, having worked his way up from humble beginnings as a copyboy at William Randolph Hearst's *Herald & Examiner*. Emil later worked for the *Chicago Daily News* and the *Sun-Times* before becoming promotions manager at the *Chicago American*.[2] In addition to his newspaper work, Garber was also involved in the Chicago dancehall scene, acting for a time as the promotions man and media representative for two of the city's most prominent ballrooms, the Trianon and the Aragon. He had also been heavily involved in the *Herald & Examiner*'s sponsorship of competitions for talented young pianists, which may have provided Welch with a faint echo of not only his own interest in music but that of his mother's adolescent musical world as well.

It is possible that Welch, now twenty-nine years old, ran into Mary at one of these ballrooms. As a college student, Mary had harbored dreams of becoming a nightclub singer and after graduation she sang with different bands on the circuit. Welch's cultured background, eloquence, and musical heritage must surely have made him attractive to women, and Mary was no doubt as captivated by him as he allegedly was by both her and her father. Aram Saroyan suggests that perhaps at a certain point Emil Garber began to gradually replace Joseph Kepecs as the primary male presence in Welch's life, and that Welch "really loved this man" to such an extent that "this might also have been a primary motive for marrying Mary."[3] Indeed, what Welch most likely saw in Garber was a surrogate father: a man of courage, genuineness, trustworthiness, and tenacity.

Many years later, at a poetry reading at San Francisco's Glide Memorial Church, Welch described Garber as a "gentle, good, Jewish man," and he must have been thrilled by the fact that his future father-in-law was also a man of words and music.[4] In the only published anecdote Welch ever told about his father-in-law, he rather comically depicts Garber as, first and foremost, a reporter who would go to any length to get a story. Recounting an incident when Garber was reputed to have spent an inordinately long amount of time in a phone booth (Welch hyperbolically suggests his siege lasted three days) during a Chicago race riot in order to see the action up close and phone in his stories, Welch inadvertently underpins an important element of Garber's character: his willingness to take chances in situations fraught with danger. Is it possible that his daughter was tarred with the same brush? That she was soon to find herself with a very similar dilemma: taking a risk when the chances of success were surely against her?

Welch also portrays a streetwise reporter who was toughened by his urban surroundings to such an extent that he could not see for himself the potential dangers that lurked around him. In the same aforementioned anecdote, Welch uses Garber to highlight his own growing disaffection with city life and his incredulity at the nature of urbanites who find themselves fearful of what can be seen and found outside of the city. On one of their weekend visits to Lake Wisconsin, Garber was apparently afraid to go for a stroll in the evening in case they stumbled upon a deer which, he remarked, "will put [your] eyes out with their horns." Here was the same man who had also saved the life of his wife when she "fell between two cars of an elevated train while trying to cross from one to the other."[5] Who had also broken the steelworkers picket line in the embattled city of Gary, Indiana, during the infamous 1919 steel strike; who had lived through and witnessed the regular violent confrontations on the streets of Chicago. And now he stood trembling in a forest, afraid of nature's intimacy and unable to see the purity of the experience. Only later, when finally confronted with "a trio of deer only twenty feet away" was Garber able to appreciate their beauty, after which he was said to have wept. Much to the amusement of the audience, Welch's metaphorical headline read "City Boy Makes Good. Aged 56." Yet although Welch may have made fun of Garber to make a point, he was nevertheless full of admiration and respect for his father-in-law to the extent that perhaps it was indeed more for Garber than for love that he married Mary. Whatever the truth, the couple were married at the Joseph Bond Chapel on the University of Chicago campus on October 15, 1955, and *her* perseverance and tenacity would indeed be put to the test in more ways than one during the next three years.

―――――

In Chicago, the Welches seemingly fulfilled their respective roles of provider and housekeeper, living the "ritualized dream of American life."[6] Having given up on her dream of becoming a singer, Mary gradually began to inhabit the position occupied by so many young women in 1950s America—that of housewife. However, Mary was far from being the subordinate wife that was typical of the era. She also worked in advertising and was encouraged to do so by Welch, who later wrote to Dorothy that Mary would be the "breadwinner while I get all my PhD. courses finished."[7] Indeed, Dorothy's example as a single working mother was a positive one in terms of bucking the trend regarding contemporary notions of the roles of husband and wife, so much so that Welch may have sought to replicate it in his own marriage. He writes of them both working hard and having to "work out some kind of a meaningful daily pattern out of a pretty chaotic schedule" of sometimes working different shifts and seeing one another only occasionally.[8] That said, after her son's disappearance, Dorothy wrote to Donald Allen that Mary's longing for domesticity and a husband who would do the things that "millions of other business men and husbands were doing" may well have contributed not only to Welch's increasing dissatisfaction about his life in Chicago but to his alcohol intake as well.[9] She wrote that

> Lew tried to be part of the establishment but it became increasingly difficult for him. He seldom, if ever, wrote any poems or even talked to anyone about the things that interested him. To make life bearable he started drinking heavily. The alcoholic habit that was to cost him his life was established at that time.[10]

So, even though Welch continued to work at Montgomery Ward and was quietly making his way upward within the company, he was becoming slowly discontented by his new life, with all the consequences that that seemed to bring. Yet contrary to Dorothy's assertions about his seldom writing poetry, his letters to Snyder included various poems from that time. While he wrote of being "very troubled about all this working for a living and the unreal goals I sometimes feel to be mine" (IR1, 80), Welch also talks of his increasing poetic independence since shedding Yeats's influence and finding his own voice. He writes of having increased control in the poetic process and of how he has found a way to capture the conscious/unconscious in his poetry. His inclusion of poems such "Epithalamion" and "Compleynt of His Mistresse" is an admission of this progress and is the first time in a long time that Welch talks of his work in positive terms, stripped of all but the barest self-criticism that had, and would again later, plague his creative output. Despite everything, one feels a growing sense that his working life is being definitively usurped by his creative life, and that Welch now has a keen sense of himself as a poet gaining originality but not yet entirely free from influence.

————

By late 1955, Welch and Mary had moved into an apartment block at 5433 South Dorchester Avenue. The building, which was situated in the Hyde Park neighborhood, had been designed by the renowned architects Keck & Keck[11] in 1949 for the Hyde Park Cooperative and, moving away from the more conventional dwellings in the city at that time, represented "a new departure in urban living."[12] The building, known as the Pioneer Co-op, provided a range of affordable housing in varying sizes in an area that had been in gradual decay, and was a venture in which individuals would make a down payment in exchange for shares in proportion to the size of their property, thus creating a co-op that would circumvent the need for external financiers. In total, twenty-three units were built, featuring state-of-the-art appliances and housing families and couples from various backgrounds and ethnicities—a situation that, in a city becoming ever more racially segregated, was also a departure from the norm. Welch later described living in the co-op to Dorothy as being

> the answer to all the scratching and stabbing in America that drives all sensitive people wild. You simply get together and cooperate. As far as I can see there is no other way to live in large cities. The people in Co-ops are of all nationalities, groups, religions, and trades—they are together only because they really have faith and respect in people. (IR1, 87)

Welch was clearly happy in his new abode, if perhaps less so in his work. He wrote to Dorothy that there was nothing partisan about the co-op but "only people" and that any man in America can "live with dignity and pride if he will only make the small effort to overcome his fear and his hate and walk over to his neighbor and be known to him." He goes on to say that people should show humility, delight in who they are, and not endeavor to be something they are not. The irony of this is that, within a matter of months, Welch would no longer see anything positive in Chicago, and any sense of pride he had in the city would be gone. The answer Welch thought he had found to all the "scratching and stabbing" was apparently gone too.

———

While James Sledd had been a key influence in Chicago in terms of language, and the ID in terms of creativity and a new philosophy on urban living, Welch also looked to the city's fervent music scene for examples of rhythm and melody that could help him further hone his poetic output. As the home of jazz, Chicago was as good a place as any for a poet so focused on structure and speech patterns. The free-form and improvisational style of jazz offered another perspective from which to view poetry. In the same way that Kerouac and others on the literary circuit lauded jazz pioneers such as Charlie Parker, Miles Davis, and Dexter Gordon in their work, so Welch also made a point of immersing himself in the jazz scene in Chicago, further developing his deep sense that the American voice could be heard in different mediums.

Living in the city at that time gave Welch the opportunity to see some of the greatest jazz musicians of the day, who would regularly play any number of the clubs dotted throughout the city on an almost continual basis. In addition to Thelonious Monk—who Welch often referred to as an influence, once writing that on seeing him play in San Francisco he had had a "tremendous affirmation" on account of Monk's "willingness to pause and wait [and his] absolute disdaining of transitions and developments" (IR1, 181)—one of the most prominent figures on the circuit at the time was Charlie Parker, who Kerouac later said was "musically as important as Beethoven" and who Welch had the privilege of seeing play in various clubs in Chicago.[13] Although by the early to mid-1950s, Parker was suffering from alcohol-induced ulcers, chronic self-doubt, and systematic self-abuse at the hands of drink and drugs, it is clear from Welch's oral descriptions of Parker that he was still a figure to be revered.[14]

Much has been written about the link between jazz and poetry during the 1950s and 1960s, with Kerouac even going as far as to say, in the *Essentials of Spontaneous Prose*, that his writing technique resembled that of a jazz musician "blowing . . . on subject of image."[15] Yet some have argued that Kerouac never fully defined that link between the two mediums (more than merely suggesting a similarity in delivery), and Welch is no more pronounced in explaining this influence on his works. In the Meltzer interviews, Welch goes into some detail about Parker's presence on and around the stage, but never describes Parker's music as a specific catalyst for his poetry. Describing Parker to Meltzer as "big and strong . . . with hands like a fucking farmer" and with his "horn hanging from his neck like a big necklace," Parker's enormous presence is obvious, but other than constituting an almost awe-inspiring figure for the "high school kids" who were backing him, Parker seems to be little more than another anecdotal footnote in Welch's life—albeit an impressive one.

In addition to his gradually increasing poetic output, Welch also began to long for his former friends from Reed, asking Gary Snyder if he knew the whereabouts of Whalen, Danielsen, or Grover and Tommy Sales. There is a sense in his letters that, despite his apparent marital bliss and weekends in the country with Mary and her parents and friends, Welch was becoming increasingly conscious of his detachment, and he describes not only his desire not to prolong his current lifestyle beyond a year or two but also his fears at what the consequences might be: "I have to make a change with a year or two at the latest. . . . My big fear is to discover myself without friends or means if I make the break and I can't stand being with-out either. I tried that in New York. . . . It was pointless and fruitless—I only became bitter and afraid" (IR1, 80). Indeed, by the time Welch fully reestablished contact with Whalen and company, the Beat Generation was already making headline news. In October 1955, while he was taking his first steps into married life in Chicago, across the country in San Fran-cisco one of the most important events in postwar American literature had just taken place in a small art gallery on Fillmore Street—an event that would not only impact American letters for the foreseeable future but in time would constitute a watershed moment for Welch too.

The now-famous Six Gallery reading took place on a cool October night, initiated by Kenneth Rexroth as a means of showcasing a group of young up-and-coming poets in a joint reading of their work. Snyder and Whalen were both asked to contribute, joining Philip Lamantia, Michael McClure, and Allen Ginsberg on the bill. Although Snyder was selected to end the evening with a reading of "Berry Feast" (Whalen read third after Philip Lamantia and Michael McClure), it was neither of Welch's former housemates who made the evening the success it was. Rather it was Ginsberg, who read the first part of "Howl" for the first time in public and was immediately catapulted onto an unremitting wave of fame and notoriety, the importance of which Welch would eventually feel thousands of miles away in Illinois.

Welch had temporarily lost touch with both Snyder and Whalen, and he was oblivious to their reading at the Six Gallery. Had he known, Welch would surely have looked on in awe and envy, taking stock of his own development in the American corporate machine as his poetic sensibilities stuttered and his friends realized their (and perhaps even his) dreams. He had stagnated. The poem he had sent to the *New Yorker* in 1953 had been rejected. Seemingly happily married and prosperous, Welch was the antithesis of an impoverished poet. When he finally read about the Six Gallery reading and the literary developments on the West Coast in newspaper clippings in early 1957, it acted as a wake-up call. In the same way it has been seen as a catalyst for its participants, so it also acted as a rallying cry to poets who had lost their way, even if it was now more than a year since the event itself. In much the same way that San Francisco had attracted hordes of young people looking for a creative outlet in the years immediately after World War II, so this reading seemed to give them some sense of legitimacy. And no little sense of inspiration.

It was Philip Whalen—with whom Welch had reestablished contact in late 1956—who first informed Welch about what was happening in San Francisco. Enclosing articles by Richard Eberhart and Lawrence Lipton in a letter from New Year's Day 1957, Whalen provided Welch's first opportunity to read firsthand about what his friends had been doing since he lost contact with them.[16] Although Snyder was in Japan by this time, Whalen promised to forward Welch's address so that he and Snyder could also get back in touch. In addition to the articles by Eberhart and Lipton, Welch was given another insight into the burgeoning West Coast poetry scene when a friend gave him a copy of an article by Louise Bogan from her column in the *New Yorker* in early April. Like Eberhart, who had written that the poets of the scene were "finely alive . . . hostile to gloomy critics" and ready to "kick down the doors of older consciousness and established practice,"[17] Bogan wrote that these poets were a "new generation of 'experimentalists'"—a description that must surely have appealed to Welch—and he made a point of mentioning the column to both

Whalen and Snyder in his correspondence. It is difficult to say whether any of these articles unsettled him to such an extent that they contributed to his feelings about life in the city, but by the time he sat down to write "Chicago Poem," in June 1957, he was clearly dissatisfied about living there, and it is easy to imagine that any longing he had harbored about returning to the West Coast was only increased by what he had read.

Either way, although Welch never actually heard the declarations of glorious defiance that Ginsberg uttered that night, or shared in the gallon jugs of California red that Jack Kerouac collected money for and passed around to the willing audience, their reverberations nonetheless rippled the waters of Lake Michigan such that Welch woke from a deep slumber, surveyed the scene outside the window of his "poorman's penthouse" on South Dorchester Avenue, and put pen to paper in a way he had not yet managed thus far. What eventually emerged was Welch's own desperate howl, a counterpunch to the proud and muscular city that Carl Sandburg celebrated in his ode to Chicago. Welch wrote his own apocalyptic vision of urban America, a panegyric for Illinois.

———

Welch completed the first draft of what was to eventually become "Chicago Poem" in June 1957. After more than five years in Sandburg's "Stormy, husky, brawling, / City of the Big Shoulders," Welch had had enough of it.[18] The discontent that had been slowly brewing now had an outlet, and Welch was finally on his way to truly rediscovering himself as a poet. He suddenly realized that he had lived in Chicago for years and had never written about it—despite the fact he hated it. He had tried to fulfil his role as the typical middle-class businessman and husband doing all the things that millions of other businessmen and husbands were doing in what he saw as a large and unattractive middle-Western metropolis, but it was becoming increasingly difficult for him. He sent a draft of the poem to Snyder and Whalen, with the former applauding the "sharp satires & images" while bemoaning both the lack of structural tightness and the flat ending. Although the poem was one of many new works Welch was writing at the time, and would not be published for another three years, in many ways it was his personal "Howl." Indeed, years later he compared it to Ginsberg's poem, saying that he wrote it in thirty minutes and that its creation was similar to vomiting.[19]

While the poem itself is a pointed and carefully constructed critique of Chicago, in essence its message could just as equally be applied to any American city. The original epigraph Welch chose for the poem, but that was never used in its eventual publication, highlights both a reflection of his own predicament and his own feelings: "For a little while I was back in the world, with my heart set on its music, on revels of midnight. But now the hate is rising within me."[20]

Choosing a fourteenth-century text from Japanese Noh theatre called "Tsunemasa," Welch aligns his poem with what was a more gentle and melancholy tale. To emphasize his point, however, he focuses on an aspect of its dramatic intensity—the "rising hate" he too feels for the city. The epigraph also has a certain amount of poignancy, given that this hatred is preceded by a sense of hope and enjoyment—things that Welch too had experienced, albeit briefly, in Chicago.

The language in "Chicago Poem" is immediate and conversational, provocative yet resigned. Welch's speaker (and it impossible not see this speaker as Welch himself) does not profess to being capable of affecting any change—the urban machine around him is too great for that. He is merely making a confession and announcing his desire to extricate himself from the city. The imagery is fierce and confrontational, portraying Chicago in bleak vignettes likened, on the one hand, to the American pre-war industrial propaganda of steel mill documentaries and, on the other hand, to the murderous experiments of Nazi scientists, a comparison that immediately negates any sense of positivity. The strength of his imagery is found partly in its provocation. Not only did he challenge Sandburg and call into question the pride that the citizens of Chicago felt toward their home, but his mention of Nazis in connection to the city is surely an attempt to answer Sandburg's challenge to "show me another city with lifted head singing so proud to be alive and coarse and strong and cunning."[21] Welch implies that his Chicago is the same city. But rather than "laughing" like the fearsome canine protector of Sandburg's people, Welch's speaker is suffocating and needs an escape into the countryside, into nature, into "the imagined dream of the American wilderness."[22] Indeed, referring to this misguided sense of civic pride, Welch later wrote that "proud people cannot live" in Chicago. And as Sandburg's "dog with tongue lapping for action, cunning as a savage pitted against the wilderness"[23] mutates into "a blind, red rhinoceros . . . running us down" as it "snuffles on the beach of its Great Lake" (RB, 11), it is very tempting to imagine the ferocious dog facing off against the blind rhinoceros, both representing the same thing but fighting for completely different corners.

"Chicago Poem" was the first of a number of critiques of urban living that Welch would write, very much foreshadowing his later essays and treatises on such cities as New York, Baltimore, and even San Francisco. His sensibilities were developing such that escaping Chicago and turning his back on it was now an inevitability. In a long, unpublished poem titled "Brought" that he wrote just after leaving Chicago, Welch disparagingly describes driving "through more / Miles of concrete and people than / Have any right to exist" and that the inhabitants are "worthless bastards" living in "dull silence" in a "huge disappointing NOTHING!"[24] It is arguably more in resignation than in hope that he therefore posited the final theory in "Chicago Poem" that "maybe a small part of it will die if I'm not around feeding it anymore" (RB, 11). But it is also a theory that in many ways challenges the reader to question their own part in the urban machine and that would later develop into Welch's rallying cry for everyone to abandon the cities and take to nature en masse.

――――――

Welch prided himself on the economy and accuracy of his language use — in both his poetry and his prose work. And if his careful choice of words is a reflection of a personal linguistic philosophy, then we can be content that this choice was singular in its intent. Yet, such a philosophy cannot be the vehicle with which to excuse certain language choices that are clearly racist or homophobic. And "Chicago Poem" provides examples of both. On the one hand, there may be an argument to say that such language was a generational or cultural norm that was "accepted" as such — after all, it was not unusual for writers and poets to use such language routinely in their work. Indeed, many of them can be accused of using inappropriate language on any number of occasions. Is Welch any different?

Yet, on the other hand, as Welch knew as well as anyone, language in any form is often an outward manifestation of an inward ideology and sense of self, as well as of a certain complacent privilege that white poets like Welch enjoyed. In the same way that Kerouac (and the Beat movement in general) has been criticized for an adoration of Black/African American culture that might be construed as cultural appropriation, so

Welch enjoyed a position of relative power that enabled him to use offensive language without fear of immediate chastisement. The fact that he uses some of his racial slurs in correspondence with his friends should not distract from the fact that he used it—in fact, under the guise of "humor" or "friendly banter," it may be even worse.

In her introduction to the book *Gertrude Stein: Selections*, Joan Retallack writes that, in spite of current attitudes about the use of racial slurs, it remains almost impossible to reconstruct what these words meant to the writers at the time. Apparently, they were freely used in "romantic-exotic conjunctions and alliterative phrases," and, as such, is it now the case that, as Retallack suggests, they were "comparatively innocent gestures of lyrical 'color' that simply fell victim to retrospective misfortune?"[25] It is certainly true that in Beat texts, this celebration and appropriation of Black culture includes widespread epithets in a romantic-exotic context. Kerouac refers to Mardou Fox, the heroine of *The Subterraneans*, as a "big buck nigger Turkish bath attendant," seeming to do so in naïve celebration of her racial identity as something he covets.[26] Kerouac made no secret of his desire to be African American, once writing that "the best the 'white world' had to offer was not enough ecstasy for me, not enough life, joy, kicks, darkness, music, not enough night."[27]

While the racist language in Welch's poems may be limited, it is present nonetheless, and is used in different ways to different effects. At the Magic Lantern reading in April 1967, Welch opens with a detailed description of acoustics in concert venues and how poets should have the same command of a room as opera singers by virtue of their voices. He argues that "if you can't fill the room with your voice, you don't belong in that room," and that problems of pitch and key while performing are down to a lack of understanding or overworking the space. While explaining this, Welch somewhat ambiguously states that "this terrible mind, this whitey mind—I'm actually a nigger—has the idea to set up the hall so it's in complete control, every minute."[28] The ambiguity of this remark as a throwaway generalization is clearly problematic. What does Welch mean? Is this a reference to his heightened awareness of spatial acoustics? Is it a backhanded compliment, similar in intent to Kerouac's? Whatever the case may be, the delivery as a by-the-way is clearly intentional. But is it done in innocence or malice?

The same is true for his poems. "Chicago Poem" not only includes a racist reference but a homophobic one too. Yet there is a subtle difference between them. In using "fairy, thrilled as a girl" to describe the filmmakers who captured images of laboring steelworkers, Welch fulfils a trite homoerotic image. Yet when using the N-word, he employs quotation marks to emphasize that it was someone else's word rather than his own. A reported image, and thus a secondhand term. Is this Welch appropriating language to make a poetic point? Is he using language to mask an offensive indiscretion? Or, as Garret Caples suggests in his discussion of the two different Chicagos as described by Welch and Sandburg, is the word used to "foreground the intensity of racial prejudice against African Americans in Chicago"[29] at that time? By way of comparison, Sandburg uses this and other racial slurs on multiple occasions in various poems, as does William Carlos Williams in his earlier work. Yet, in one, titled "Nigger," Sandburg's use of the word is seen by one critic as "a product of deep thought and serious treatment."[30] Indeed, in his autobiography Langston Hughes called Sandburg his "guiding star" and describes him as "a lover of all the living."[31] Clearly Hughes never saw Sandburg as a racist despite the latter's repeated use of derogatory language. And if Sandburg's work is the product of "deep thought and serious treatment," then arguably—in certain situations at least—so is Welch's.

However, in "Din Poem," Welch's more gratuitous use of such language immediately calls that careful aforethought into question. In his creation of a collage of contemporary America, Welch expresses the polyvocality of American speech by depicting the many ways in which language is used. He does this by employing a host of racist slurs alongside advertising slogans, jingles, news items, and snippets of everyday dialogue. But, again, is it acceptable to use such language to make a point? Where "Din Poem" markedly differs from "Chicago Poem" is in the range of words that Welch includes. While the latter poem features a single, albeit highly offensive, racial slur, the former has a myriad of epithets that relate to multiple races and ethnicities. In addition, it also uses homophobic and political barbs that overload and accentuate the poem's message. However, are words such as "wop," "spic," or "faggot" more or less offensive that "commie" or "gypsy" to the recipient of the slur? Are some words more socially acceptable than others, and where can the line be drawn? Welch appears to use them wantonly and the inclusion of musical notes in "Din

Poem" to indicate that the section in question should be sung rather than narrated, implies a frivolity that only serves to accentuate the offensiveness.

And this tendency to overlook, ignore, or indeed intentionally employ racially insensitive language was not confined to singular poems or anecdotes at readings. When Welch was compiling the poems that would eventually make up *Hermit Poems*, his intended title for the collection was "After Chinks." As a crude salutation to the Chinese poets who had influenced him during his time at Forks of Salmon, Welch later noted in a letter to Donald Allen that the title was "perhaps too disrespectful" and that Allen should "consider it a private joke" (IR2, 119–120). Indeed, when Welch submitted the collection for Allen's comments, Allen had written "I honestly don't understand the title, but I like the poems very much."[32] It seems that this was enough to change Welch's mind and in the very next letter, he apologizes and takes Allen's advice to call the collection "Early Summer poems."

Welch also used racist slurs in his correspondence and they often relate to his friends. While it is arguable that the language he employed in his poetry can be more easily contextualized, the ease with which he uses anti-Japanese slurs in particular to refer to anyone who had an association with Japan, is clearly different. In his correspondence with Snyder for example, Welch uses derogatory phrases to refer not only to people of Japanese heritage but to Snyder himself. A comment such as "Trust the local carriers will ferret you out and put this in good order before your worthy, and no doubt by now slanted, eyes" (IR1, 79) is not only racially offensive but childishly implies that Snyder, who was then in Berkeley studying Oriental languages, will somehow be physically as well as spiritually altered by his immersion in Asian culture. And this clichéd reference to appearance was not limited to passing comments about his friends. Despite his own interest in Buddhism and poetry from East Asia, Welch, often in moments of drunkenness, was capable of issuing similarly cutting insults directed at monks, poets, and philosophers too.

Yet, was Welch racist and homophobic? Or was his use of such language merely a reflection of white society in the 1950s? Whatever the case may be, it makes for uncomfortable reading and adds a grimy veneer of casual prejudice to an otherwise careful and democratic poet.

7
Westward the Course of Empire Goeth

When Welch finally packed up and left Chicago in mid-October 1957, he did so by negotiating a transfer to Montgomery Ward's Oakland, California, office. While his increasing dissatisfaction with the city and the growing literary success of his friends were certainly mitigating factors in the move, he was still married and still very much part of what he called the "'Murcan machine"—indeed he was now the chief copyeditor in charge of overseeing the output of forty-five copywriters.

There may be some small sense of frontier romance in Welch heading westward like so many before him, but, unlike Jack Kerouac's heroic protagonists on the freight trains and flatbed trucks of those early cross-country adventures, Welch and Mary flew to the coast on a lavish expense account, with the company taking care of their furniture removal and other assorted expenses until such time as they were settled into their new home. Yet this transfer was very much the beginning of a number of ends, and even the expense account could not alter Welch's growing sense that "I can't do it anymore"—with "it" being variously interpretable as work, marriage, or life itself. Whatever the case may have been, change was coming, and not a moment too soon.

A poem written in late 1957 called "I Fly to Los Angeles" describes Welch's thoughts on one of his monthly business trips from San Francisco to LA to attend various meetings and trade conferences. The poem encapsulates his early feelings of jealousy and longing for a life other than the one in which he found himself, waiting in undersized coats for public transport to low-paying jobs in the freezing Chicago winters, while high above he "saw the silver glint of 4 motor airplane" and wondered "Why? Why me way down here. . . . Why me in Chicago Christmas Post Office 9.90 a day ain't got no place to go" (RB, 18). Yet before long, Welch was in those airplanes himself at company expense, and despite his initial desire to be part of this corporate machine and play the happy husband, he becomes aware that this is not what he wants after all and instead is hiding from his true self.

The poem outlines how, while waiting for a flight in San Francisco, Welch bought a copy of Gregory Corso's *Gasoline* and wept at the revelatory messages he saw in it. He then denounces his "sinecure at 31 with 7600 a year" and thanks "all the patient people! Mary most of all and Whalen and Snyder and Kepecs and Williams and Wilson and all who loved me once" (RB, 20). Read as such, Welch's relocation to the West Coast was far more of a homecoming than a journey into the unknown. Indeed, in a reflective mood, he wrote of his incredulity of having lived for so long in Chicago and his feeling that he had never left the West Coast, realizing "that nothing has happened in all that time except I cracked up and finally got back together again. Chicago is uninhabitable"[1]

His joy at finally having arrived in San Francisco is clear in a letter he sent to Whalen from his temporary home in one of the city's finest hotels, the Alexander Hamilton:

> When I first sniffed the specific ocean I honest to god choked way up and then those huge pelicans kept breaking off the rocks and far out there the evening fog started to build up and the sun dropped through like a marble. (IR1, 115)

Within a month, the couple had found their own house on Leavenworth Street and the next chapter of the American homemaking manual could begin in earnest—or so the fairy-tale ending would have us believe. Mary made plans for their first Christmas in San Francisco and wrote to Dorothy:

> We are terribly happy here out. Our little Japanese Palace is so perfect for us. . . . Lew is happier than I've ever known him to be. And he's writing, Mother B! In such an outpour . . . and like an angel. He's so very, very good. We've much to be proud of, you and I.[2]

Welch himself also seemed, temporarily at least, to be living under the illusion that he was blissfully happy and had grand plans to allow "the big frantic mindless machine" of work and the establishment to grind away without his help. In another of his carefully constructed schedules to escape corporate

America, he outlines his plan to quit his job at Montgomery Ward once he is free of debt, in order to focus on his other aspirations. This plan, he estimated, would be accomplished within six months. Indeed, debt became something of a fixation for him in later years, with not only his suicide note specifying that he owed nothing to anyone, but in an unpublished essay titled "How to Survive in the United States," he wrote "MAXIM NUMBER ONE: Stay out of debt. Thoreau used to say we should always measure the cost of anything in terms of time."[3] He then included a calculation of money borrowed or owed versus the time wasted paying it back, which is why he was so able to frequently estimate how long he would need to become debt-free. This in turn would allow him to devote himself entirely to writing, reading, fishing, and hunting, while Mary continued to work. Any plans Welch had to finish his MA at night school—as he had attempted to do in Chicago—were also quickly forgotten. His desire to become a writer no longer involved academia. Indeed, Welch wrote that "universities are now fine for many disciplines, but if you want to be an artist you have to find your school outside of museums and even the best of Universities must, so far as art is concerned, act mainly as a museum."[4]

In one of the most revealing works from late 1957, "Summer Goes Down the Other Side of the Mountain," Welch outlined the many questions he had about his recent move and described a situation in which his perceived contentment at moving back to the coast is actually somewhat tempered.[5] Despite his obvious joy at having returned ("After wasting nearly a fourth of my life in Eastern cities") he realizes that, as long as he is part of the machine, life will never be as satisfactory as he had hoped—irrespective of location. His working day is described in terms of repetitiveness and purposelessness, and he asks his poet friends directly for answers to his questions of whether what he has is enough and, indirectly, whether he should be content with it:

> Allen, is your brilliant, upturned world of wild, wild, flowers and a
>   stilled sun beating on that glorious receptivity of yours
>   enough? Is it enough? I hope so. Mine is not always
>   enough.
> Philip, is your cottage everything that we hoped it might be?
> Gary, how did the lifetime war with the terrible responsibility of
>   all your cleverness go? Did we make it? Did we

> do much more than save ourselves from our special
> kinds of ever-imminent destructions?
> Did ever anybody really think he got enough?[6]

Yet the poem is not sad or despondent. There is a genuine sense of relief and a feeling that these existentialist questions are not a burden in the way they had been previously. The end of summer is a metaphor for the life he has left behind—and is in no hurry to repeat. Or to face its "many indignities again" or be forced to turn away from it in terror. Indeed, this new life is one filled with love, with Mary who "opens the door to warmth, our cat Jerome, good whiskey, and long long pointless evenings of sheer indulgence."[7] And as summer fades, so Welch will follow it, knowing that there

> will be new things to see. There will be love so complete and simple only I, the lover, and those I love will ever even dream that it existed. Summer goes down the other side of the mountain and I will follow it.
> It will be good.
> It will be good.[8]

Although Mary was still an obvious part of his life, his depiction of her in the poem is an archetypal image of the busy housewife, sewing, scouring, and waiting for her husband's return from the office. Whether this was Mary's idea of a happy marriage is anyone's guess, but by early 1958 both were openly having doubts about their compatibility. For his part, Welch had already considered filing for divorce but decided against it, most likely on account of his belief that the move west may add some impetus not only to his creativity and literary ambitions but also to his marriage. Only six months previously, he had written a poem titled "For Mary" that clearly indicated how he still felt about her. However, this feeling that she was "all the women of the world . . . all the paintings of all the women of the world" seems to have quickly dissipated.[9]

Mary too was beginning to have serious doubts about whether this was what she had envisioned when she signed the wedding register only three years earlier. It was becoming increasingly clear to her that she was unable to provide her husband with the things that he wanted—no matter how hard she tried. And, importantly, she was starting to understand that their individual wants and needs were, and perhaps always had been, completely different. Mary had filled a void in Welch's life at a time when he had been cut adrift from anything resembling constancy and regular emotional warmth, but now that his need had crystallized into something beyond what she was able to provide, Mary saw that their relocation to the coast was a means to an end, and that he had gone some considerable way to replacing her in this sense. Back amid his friends, Welch appeared to be in his element, and the void was now filled with the inspirational familiarities of the past and the literary opportunities of the immediate future. While the snapshots of wedding bliss captured in Welch's letters appear to portray a couple devoted to each other—the previous Christmas, for example, Mary had bought Welch a rifle as a surprise gift, prompting him to write "she keeps doing things like this, which is why we have a wonderful marriage" (IR1, 86)—the ebb and flow of Welch's moods, increasing use of alcohol, and growing comfort in this new literary milieu were obviously also taking their toll on what seem to have been, on the face of it at least, precarious foundations from the start. Moving west only served to disrupt these foundations even more, as Welch began to reacquaint himself with his former friends and revel in the burgeoning literary scene. Within weeks of their arrival, he had been to poetry readings by Kenneth Rexroth, Kenneth Patchen (whose work he was particularly fond of reciting from memory), and Lawrence Ferlinghetti; had reestablished contact with Grover and Tommy Sales; and had seen his old buddies Dave Brubeck and Paul Desmond perform at the Blackhawk. His friendships with Whalen and Snyder also enabled him to quickly join this growing American poetry renaissance and become an accepted member of what was now well on its way to becoming a nationwide phenomenon. As such, the weeks and months that went by in their new Californian home only served to highlight the already glaring discrepancies between husband and wife.

———

The impact of Viking Press' publication of Jack Kerouac's *On the Road* in September 1957 was yet to be truly felt, but nonetheless Welch's return to San Francisco could not have come at a better time. With the national media already firmly focused on the "Beat" movement, due in part to City

Lights' publication of and subsequent obscenity trial surrounding Allen Ginsberg's "Howl," San Francisco had become the center of a literary and countercultural explosion that was to leave its mark on not only the United States but also much of the Western literary world. When Judge Clayton W. Horn ruled in favor of the defendants—Ferlinghetti and his clerk, Shig Murao—his decision gave the poets and artists of this emerging generation carte blanche to experiment and publish in any and all forms of expression. Immersed in a milieu of such creativity and apparent opportunity, Welch's sense of himself as a poet became ever clearer, and so his marriage began to irreparably unravel.

Yet, despite the newfound inspiration and ever-increasing output, Welch still wrestled with the issues of self-doubt and perfectionism that lurked at the end of every stroke of his pen or click of his typewriter. These issues were accompanied by a now almost unbearable dislike of his daily work and the seemingly unobtainable dollars associated with living such a life. The only apparent saving grace was that his position as chief copyeditor afforded him an opportunity to work with the structures and intricacies of language again—albeit at a more mundane level. At the same time, it became ever clearer that Mary—not unreasonably—longed for a more conventional existence, with the securities and trimmings that went with it. Welch's original plan to work six months to pay all bills, then kick the "Business Habit," seemed not to include any of her wants or needs. His personality appears to have been unaccustomed to the conventionalities of homemaking and the accompanying need for "shelter and babies and love" (IR1, 146), having himself grown up in an environment that was, for the time, unconventional at best. Could the peculiarities of his upbringing and his relationship with Dorothy now be coming back to affect his relationships? Was Welch's former interpretation of Dorothy as the destructive incarnation of Kali still exerting its influence? He admits in a letter to Whalen of having little concept of Mary's needs, saying that it is "all soap and machinery ritualized to a point beyond my understanding" and, further, that such is the disparity in their views on marriage that, as she "does the many sweet things she loves to do, . . . I bungle around like a guy trying to play shortstop for the Giants while wearing the equipment for, and observing the rules of, badminton" (IR1, 146).

As such, their eventual separation in the summer of 1958 surely relieved them both of an increasingly unnecessary burden of responsibility and freed them to pursue their own personal notions of happiness and fulfilment.

In addition, Welch had recently been given a timely reminder of his own mortality, something that may have acted as an equally timely catalyst. His hospitalization at Mount Zion for a serious allergic reaction to an inoculation containing horse serum left him close to death ("I damn near got erased"), and he was apparently lucky to have escaped the seizure without permanent "stroke symptoms. Paralyzed face and all that" (IR1, 138). His brush with death profoundly affected him. He wrote to Dorothy:

> The emotional shock of my ailment lasted quite a while. . . . It was all so sudden and weird—it still doesn't seem at all real. The Cortisone which I had to take all the time in the hospital and for 10 days afterward is, I discover, a dreadful thing for the nerves of some people. I kept having awful dreams and hallucinations upon first waking and just before going to sleep. Thank goodness it's all over![10]

Welch later wrote of his hospitalization as having also prompted an awakening of sorts ("I am beginning to understand something. It is what you say about no separation") and a rebirth that helped him focus on certain aspects of himself and his life. In a draft letter to Philip Whalen from his hospital bed, he outlined not only the frightening visions he had had of "Bosch drawings" under the influence of hydrocortisone but also how he now saw the relationship between himself and the wider world, as well as how this affected him. It may well have been born of medication, but it was nonetheless an apparent attempt to make sense of his life thus far and to locate his position/role/existence within the "world : self" narrative.

> It is gradually dawning on my poor old brain that what is wrong is this
> WORLD turned on : SELF darkened
> turned on SELF : WORLD darkened
> darkened WORLD : darkened SELF
> and that my recent troubles have been:
> (1) now live in turned-on WORLD
> (2) if turn on SELF : (3) ENLIGHTENMENT!
> Obviously all three propositions in error.[11]

The question seems to have been how was he going to reconcile this "world : self" relationship such that he could find contentment? He had been saved from death and now had to take back control of his life.[12]

In the weeks immediately after his hospitalization, Welch went to Gary Snyder's cabin in Mill Valley regularly, from where they would make the short journey to Muir Beach to fish, talk, gather firewood, and watch the waves break onto the rocks. It is a place that would later become hugely symbolic for Welch—the importance of which may well have found its beginnings during those moments when he was recuperating and finally transitioning out of marriage and into what would become his new life as a poet.

Therefore, it was something of an inevitability that, when the separation finally came, he made the short trip to Snyder's again, where he would spend his first weeks as a free man and poet unburdened, possessing only "a few suitcases and fishing rods"[13] and learning "all about Coleman stoves and trimming the wick of [his] lamp" (IR1, 146). Joanne Kyger remembers him entering the cabin somewhat emphatically by announcing, "'That's over,' as he hung his wedding ring on a nail sticking from the wall."[14] In a letter to Whalen written on July 7, 1958, Welch said equally emphatically "I'm not sad. I'm not even tired" (IR1, 146).

———

By now, Welch had also parted company with Montgomery Ward, finally having been fired in spring 1958. In a pointed condemnation of his employer and his fellow colleagues, he had previously written to Snyder (in response to the disciplined and formal nature of his friend's days in Japan) about what happens after his morning rituals:

> I get into the car and onto the speedway with a thousand idiots who can't drive. Arrive at work, invariably late, invariably discovering that some supposed crisis has occurred. Work consists of doing the urgent. A friend describes it: "pissing on small fires." All day long inferior people demand that I do things in far less time than it takes. I do it. Then they change it—it is not "better" it is just different. Then I do it all over again in even less time. It shouldn't have been done in the first place. Then onto the speedway where

same idiots are now furious as I am. We try to kill each other for a half hour. (IR1, 108)

If this was indeed his daily routine, it is unsurprising that, like many people in the same boat, he felt such derision for the apparent rat race in which he found himself. Within a month of leaving, however, he was far more philosophical about it, writing to Whalen that he simply didn't fit in any longer. At the time, Mary had a job, and his vision of having to work for only one more year—then their debts would be cleared and he would be able to get a teaching job and have more fun—was as clear as ever. However, this particular letter—as revealing as any he wrote during that period—offers a far greater insight into Welch's life before and after his commute to work. And it is more likely here that the root of his feelings about life on the corporate merry-go-round lies.

Although occasional references to his increased alcohol use can be found in earlier letters, it is clear that this use was gradually morphing into abuse by the mid-to-late 1950s. Despite his correspondence being frequently dotted with candid and often brutal descriptions of his drinking habits, he was often less revelatory about the impact this may have been having on his poetic output and his marriage. It is easy to read between the lines and see that his life—on all fronts—was gradually fragmenting into an ever more complex mix of external influences and internal struggles. And his employment status was only a tiny part of that mix.

The next few months of Welch's life were filled by bartending and driving cabs for the Yellow Cab Company in San Francisco, the jobs a means to a necessary financial end; his focus was now firmly fixed on the development of his poetry and the making of plans for a future in which all his time was his own. As the San Francisco Renaissance continued unabated around him, Welch had finally found a milieu in which to further nurture his talent. He met the artist Robert Lavigne, read at a poetry reading organized by Jack Spicer, and, most important, moved into one of San Francisco's key cultural communes, the East-West House, where he lived variously with, among others, Claude Dalenberg, Joanne Kyger, and Albert Saijo. His correspondence radiated a definite sense of optimism, and he talked of embarking on what he called "Buddha Studies" and selling his shotgun too. Clearly, his liberation from the shackles of work and marriage was already beginning to bear fruit.

Figure 7.1: A portrait that Robert Lavigne did of Welch using magic markers, 1958. Courtesy of Joe Lee and the Estate of Robert Lavigne.

Figure 7.2: Oil painting of Robert Lavigne that Welch painted in Lavigne's studio at the Wentley Hotel in San Francisco in 1958. Courtesy of Joe Lee and the Estate of Robert Lavigne.

8
East-West, Home Is Best

When Claude Dalenberg arrived in San Francisco in the early 1950s to study at the American Academy of Asian Studies, he moved into the "splendidly rambling old mansion on Broadway Street" in which the academy was at that time established.[1] Along with six or seven other students-in-residence, Dalenberg helped run the place, acting as house manager for a time. The house was very much a communal living space, where the residents split the costs, ate together, and shared a common interest in Asian arts and culture. However, the building was owned by the academy, and when, in autumn 1956, it became apparent that it was no longer receiving the backing of either the major financiers or the trustees, the then-dean and driving force behind it, Alan Watts, felt compelled to pull the plug on his involvement in the venture. As a result, Dalenberg and several of the others decided that, in order to continue living in the "kind of communal style" to which they had become accustomed, they would have to start their own place instead. And thus the East-West House was born.[2]

Located at 2273 California Street, on the edge of what would later become Japantown, the house was established to create an environment in which the inhabitants could continue to revel in what Dalenberg later called the "ideal of East meeting West."[3] Given that Watts was the primary reason that many of the students were studying at the academy in the first place, they were hoping that he would continue to give lectures in the house on the various aspects of Buddhist theory and practice that prepared them for the journey that many longed for, and would subsequently make, to Japan. The house "thrived on the study and discussion of philosophy, literature, language, and religion, held in every room but primarily in the communal kitchen."[4] In addition to the "very nice library of the Buddhist books that were then available in English," there was a strange juxtaposition of lofty intellectual objectives and a "vigorous social life" that prompted hard drinking, partying, and various sexual liaisons.[5] Succinctly describing his many visits to the house, Jack Kerouac wrote in *Big Sur* of it being a "regular nuthouse," with wild parties and a seemingly endless stream of girls and visitors carrying all manner of liquid refreshments. So, when Welch moved into one of the rooms in 1958, it could not have been a more welcome (and potentially dangerous) cocktail of work and play after his marriage and self-proclaimed Chicagoan stagnation.

———

After hanging up his wedding ring, Welch stayed with Snyder in Mill Valley through July, but by early August had moved into a room in the building adjacent to the East-West House. Although he describes his accommodation as being more than adequate, with cleaning ladies and a regular supply of clean linen, the situation was a temporary one, and within a month he had moved into a room in the Queen Anne–style house next door. For much of the next three years, 2273 California Street was the place Welch called home.

The move also coincided with an increase in his creative output and his first serious poetry readings — indeed, he admitted that until his arrival on the West Coast he had never read out loud to strangers. The intervening years had seen a profound change in his poetic style, and San Francisco now provided an opportunity to test the waters and see how his work would be received. Welch wrote that he was finally getting to work and typing up his poems for possible publication. One potential recipient was the Grove Press West Coast editor, Donald Allen.

Over the course of his life, Welch had assembled a significant number of important and influential people: his mother, his teachers and mentors, and various fellow poets and artists. However, in terms of always believing in Welch, and always being prepared to promote and publish his work — both while Welch was alive and posthumously — one of the most important was Donald Allen. From their initial encounter in 1959 to his championing of much of Welch's hitherto unpublished works through his Grey Fox Press in the decades after Welch's disappearance, Allen was an ever-present counsel to which Welch could turn. If he needed to test new works or make contact with small magazines or editors at publishing houses, Allen was often on hand to help. And by the time the two started working together more closely on *Hermit Poems*, Allen had an expansive network in the publishing industry and was as well positioned as any to lend his expertise to the fledgling poet.

Donald Allen's route into the publishing business was a circuitous one. Having graduated from the University of Iowa with an MA in English, Allen worked as a college professor in Davenport, Iowa, before a period of traveling eventually took him to work in China. He arrived in Canton during a Japanese bombardment in September 1938, as the Second Sino-Japanese War began to escalate. It was a situation in which he initially helped to resettle and assist Chinese refugees in Hong Kong rather than carry out the teaching job he had gone there to do. It also exposed Allen to the perils and tragedies of war. Indeed, within a matter of only three years the Japanese would expand the Pacific theatre of war by bombing Pearl Harbor; after returning to Iowa, Allen volunteered to join the US Navy. He was sent to University of California, Berkeley to learn Japanese and within a year was working as a translator on, among other things, the interrogation of Japanese POWs. By the end of the war, Allen had served in the South Pacific, been awarded a Purple Heart, and worked with British intelligence in London. However, his military service also exposed him to the work that would later become his career: publishing. While working for naval intelligence, Allen was involved in relaying vital information between military units as well as publishing a weekly magazine. While a far cry from the type of literature he would later be involved with, the work nonetheless afforded him the opportunity to experience how publishing and editing work was done. And by the time he met Barney Rosset in New York in 1951, this experience, along with the freelance editing he had been doing, proved invaluable.

Allen's military service had ensured that he developed a keen interest in not only the Japanese language but also the wider influence and impact of East Asian culture in the United States. He had seen close-hand how fellow students of Japanese origin had been relocated out of Berkeley as part of Roosevelt's Executive Order 9066, and having traveled and worked in China and Japan, Allen had already developed a love of literature and culture from the region. Indeed, years later during his first stint as an editor at Grove Press, it was Allen's intimate knowledge of East Asian poetry that helped persuade Barney Rosset to publish an anthology of Japanese poetry. He also helped Grove Press reprint the works of renowned translator and sinologist Arthur Waley, without whose books it is "unlikely that the classics of the Far East would have become such an important part of [American] heritage."[6]

However, it was during his preparatory work for the anthology *The New American Poetry* that Allen first came across Welch's work. When, in 1958, Allen started to compile the poems and essays that would feature in the anthology, he was still working with Rosset at Grove Press in New York. The success of *Evergreen Review*—especially the second edition dedicated to the "San Francisco Scene," which had been Allen's idea— had been one of the catalysts for an idea to create a more expansive work that would not only showcase more established poets he had befriended, such as Charles Olson, Robert Creeley, and Frank O'Hara, but also give those younger and lesser known the opportunity to have works published, in some cases for the first time in anything other than "fugitive pamphlets and little magazines."[7] As a result, Allen moved out to San Francisco. He had had enough of New York by that time anyway and negotiated a position for himself as West Coast editor for Grove Press. In addition to being able to reacquaint himself with many of the poets he had met after the war, he was now in a better position to assess the younger talent at readings across the city. Indeed, Allen often arranged readings himself that were attended by a vast array of people. Although when published, in 1960, *The New American Poetry* was seen as an arbitrary collection of poets who Allen divided into equally arbitrary groups, in time it became the definitive work to which people turned for a legitimate overview of American poetry during that period—despite the savage criticism the book received from critics such as James Dickey ("fairly low-grade whale-fat"[8]) and Cecil Hemley ("This volume has neither historical nor aesthetic coherence"[9]). But how did Welch find himself included when other more established poets and writers such as Joanne Kyger, George Stanley, and Richard Brautigan were not? And what was it about his work that appealed to Allen so much?

Welch's still relatively obscure position in the literary milieu at the time made it both exciting and frightening to approach Allen with a view to being included in the anthology. In his letters to Philip Whalen, he asks whether he should send particular poems to Allen for consideration, displaying the same self-doubt that had been a constant in his work. Indeed, Welch even went as far as to ask Whalen to soften Allen up for him in advance of Welch sending "a generous sampling" of work (IR1,

140). Welch still had next to nothing in print, and his reputation could not be relied on to guarantee inclusion. While others such as Kerouac and Ginsberg could comfortably bet on Allen including them the book thanks to the popularity, literary standing, or, in Ginsberg's case, notoriety of their work, for Welch it was his poetry alone that would have to persuade Allen. The two had never met in person, and Allen had little to go on save the poems Welch sent him. Correspondence between the two is largely businesslike in the beginning, but it is clear that Allen soon saw enough in Welch's work to not only include two poems but to also later act as his (often) unpaid agent and literary executor.

———

During this period, Welch had also been tentatively working on a thinly veiled autobiographical novel that provides considerable insight into his own sense of self. The novel, titled *I, Leo*, illustrates the seemingly continual inner struggles Welch had had with his identity and the apparent search for meaning in his life. In a letter to Whalen from early July 1958, Welch describes writing an inscription for the novel while suffering from "the bleakest depression I have ever undergone":

> We all invent ourselves. We invent ourselves by a process we can't control, out of ingredients we cannot choose. If we are miserable enough, and can find the strength for it, we invent ourselves over and over again.— Max Volker. (IR1, 143)

The inscription suggests a philosophical acceptance of fate as the driving force in life, yet there is an underlying sense of hope that strength of mind—should we possess this or be able to generate it—may somehow alter the course that fate has planned for us. There is also the sense that "we" may require multiple reinventions of the self until we become the person we want to be. (Re)Invention is a theme that recurs in much of Welch's work throughout the 1960s, and it featured prominently in the plot outline of *I, Leo*.

Attributing the inscription to "Max Volker"—the first nom de plume in the novel—also suggests the Everyman character, and while *volker* can be translated as the German for "peoples" or "nations," it is likely that Welch consciously choose to reintroduce the notion of "tribes" that he had first broached at Reed after the visit by Williams. Indeed, like (re) invention, it is a term that would feature ever more heavily in his later work. The "author" of the novel is "Frank Stonefield," and the main character is introduced as follows:[10]

> Keeler, became Beuchamp, then Townley, then went back to Keeler at 18. Somewhere in the book, perhaps figuring as the central intelligence, is Lew Welch. All c'est moi. All incidents real ones. Maybe I can kill them all so I can breathe again. (IR1, 143)

Indeed, in the same description, he goes on to specify that Keeler will die "by freezing to death waiting for a ride on the Alcan Highway" having "run out on Mary planning to hike to Alaska" and will be found holding a piece of paper with illegible writing that becomes gradually bigger until it trails off the page into oblivion. Four days later, Welch left Mary for real, and although he did not actually head to Alaska, there is a certain premonitory aspect to his description of not only escapism but also (his own?) death and the idea he often had that writing serves no purpose and little or no meaning can be extracted from it—indeed, he ends by writing "God knows what he was trying to say."

Around the same time, Welch also read at one of Jack Spicer's regular Sunday afternoon poetry readings, which Snyder had apparently described to him as being full of "a real good bunch of people who care about things" (IR1, 142). Indeed, Spicer's gatherings were not confined to his home but were more often than not held at The Place—where he frequently co-organized and hosted the "Blabbermouth Nights"[11]—or at Gino & Carlo's, two North Beach hangouts that were quintessentially "Beat." Spicer surrounded himself with what he called his "magic circle," to whom he expounded on all manner of things, including his contempt for the Beat Generation and all things related, such as Kerouac and Ginsberg, who were regular visitors.[12] Indeed, such was Spicer's hatred of all things "Beat" that he even temporarily refused to have his books sold by Ferlinghetti's City Lights Bookstore.

At one time or another, this group of what has been described as "devoted acolytes" included not only Welch but, among many others, George Stanley, Richard Duerden, Joanne Kyger, David Meltzer, Bob Kaufman, and John Wieners.[13] Although Spicer was not particularly impressed by Welch's poetry (and vice versa), these opportunities for Welch to envelop himself in the very company he had sorely missed in Chicago were critical to his continued development as a poet.

Welch's feelings on the reception of the work he read at these soirees is typically critical, describing the audience as being very disturbed by what (and how) he read and that he had not only infuriated them but also got under their skin (this, however, was partly Spicer's intention in organizing such evenings—the idea being that such readings would "bug the squares"[14] who were flooding into North Beach in search of their slice of "Beat" action). The sincerity of their reaction, however, seems to reflect Welch's own sense of ingenuousness, and he wrote of his pleasure at how, in addition to the infuriation of the majority, John Wieners and Michael McClure had "dug" what he read. Many years later, in 1965, upon being told that Welch was replacing Philip Whalen on her panel during the Berkeley Poetry Conference, Joanne Kyger "implored [Whalen] on the telephone many times to reconsider—'You know Lew always CRIES when he reads and it will ruin the evening.'"[15] Apparently, Welch's ability to "disturb" during his readings was not limited to his audiences.

————

As what must have been a turbulent year drew to a close, Welch was splitting his time between the East-West House and Snyder's shack in Mill Valley, named Marin-an. The latter had become a makeshift Zendo, where Snyder, who had just returned from Japan after spending the previous two years training under Oda-roshi at the Daitokuji temple in Kyoto,[16] led Zen meditation practice and where Welch could make a serious effort with his Buddhism studies. Playfully taking its name from the Japanese word for "horse grove hermitage" (the adjoining pasture was often populated with horses and ponies), Marin-an provided several of Snyder's friends with the opportunity to participate in sittings under his tutelage.

Built many years before by the former owner, Tony Perry,[17] in whose house Snyder and Kerouac had also lived briefly in 1956 (the events of which were later immortalized in The Dharma Bums), the shack very soon became a place of congregation for, among others, Welch, Dalenberg, Saijo, and Robert Greensfelder, who had not only been a fellow Reed student but now also lived in Mill Valley. Despite the relative informality of these short sittings, Welch took the practice of meditation and study very seriously. However, his status as a novice is clear—certainly when compared to many of the others—and he even went as far as to refer to himself as "the Buddha known as the Beginner" (IR1, 153).

Immediately after leaving Mary, he had written of there being "much I do not understand" and that "Senzaki gives me hope" (IR1, 145–146). Reading between these lines, Welch suggests that in order to work on the former, he has placed a certain amount of faith in the teachings of the latter: the Zen monk Nyogen Senzaki, who, having studied under Rinzai master Soyen Shaku (with whom he first came to the United States), was one of the earliest proponents of Zen Buddhism in the United States after his arrival in San Francisco in 1905. As a fledgling student of Zen, Welch, like many of his contemporaries, would have turned to both Senzaki and D. T. Suzuki for a grounding in both practice and thought. Although less well-known than Suzuki—who Gary Snyder once called "probably the most culturally significant Japanese person in international terms in all of history"[18]—Senzaki was key to the establishment of Zen practice on the West Coast, having established what he called "floating zendos,"[19] first in San Francisco and then in Los Angeles in order to facilitate his *zazenkai*.[20] Among his regular students was Welch's roommate, friend, and soon-to-be collaborator, Albert Saijo.

Senzaki's idea of holding his meditation practice and lectures wherever it was possible to do so rather than in some form of institutionalized formality in a fixed location was one that would later appeal to Welch's changing sensibilities regarding external influences on what he felt to be primarily closed practices. The notion of a "floating zendo" was one that, before too long, Welch would propose as an alternative to the fixed nature of zazen (meditation practice) in permanent locations, such as Marin-an. In a statement that presaged much of his later ideology concerning counterculture transience and the need for people to move in order to escape tightening legislation and live their lives according to their own desires, Welch wrote that "the 'Murcan zendo must float. It is harder to hit a moving target" (IR1, 164).

For his part, Albert Saijo had first met Nyogen Senzaki during their internment at Heart Mountain Relocation Center in Wyoming during World War II. When the Japanese bombed Pearl Harbor in December 1941, Senzaki was in his sixties and had lived in the United States for almost thirty-seven years. Nonetheless, when President Franklin D.

Roosevelt issued Executive Order 9066, Senzaki was subject to the same regulations that stipulated that all persons of Japanese ancestry, both citizens and noncitizens, were to be relocated inland and thus out of the Pacific military zone. The stated aim of the order was not only to prevent potential espionage and other threats to national security but was posited also as protecting people of Japanese descent from reprisals at the hands of Americans who had strong anti-Japanese attitudes.

The first trainloads of internees to Heart Mountain arrived in Cody in the first week of August 1942, and when his time came, Senzaki, like 6,500 other Japanese who would be forced to relocate from Los Angeles County, made the long journey northeastward to the concentration camp that would become his home until 1945.

For much of the time he spent at Heart Mountain, Senzaki continued the discipline of regular meditation practice and held *zazenkai* in the small apartment he shared with a young family in block hut 2. However, as the son of a Christian minister, Albert Saijo did not get involved in any of the Buddhist activities on offer in the camp, instead focusing on his schooling and editorial work for *Echoes*, the camp newspaper. Although he saw Senzaki walking around the camp and first learned of "something called Zen" through the parents of a friend who sat with Senzaki, Saijo saw himself as "a dumb teenager [with] no interest in such matters."[21]

Unsurprisingly, Saijo later described his time at Heart Mountain as "awful," in a place encircled with barbed-wire fencing, nine patrol towers, searchlights, and armed guards, in which "10,000 dispossessed people of Japanese ancestry" were confined.[22] Saijo writes a haunting description of his experiences in Wyoming in Carole Tonkinson's *Big Sky Mind*, and although understandably bleak, he did confide in a letter to Marie Hara some years later that, as a teenager in the camp, he experienced a particular sense of freedom from what would normally be expected in such situations—namely, ties to family and community. He further wrote that "there was no longer a need for family"[23] and that Heart Mountain "was not one of them Nazi concentration camps"[24] but instead offered, for him, as a teenager with little to lose, an "odd security and warmth."[25] Indeed, he found opportunities aplenty,[26] and in addition to his schooling and editorial work, the camp afforded him the chance to craft his writing skills, following in the footsteps of his mother Asano, who was herself an accomplished writer of haiku and a regular columnist for two

Los Angeles–based Japanese-language newspapers. Although he would later say how thankful he was that his early poems were lost, Saijo's first serious writing—"stupid stuff" he called vignettes of camp life—was done in the camp.[27]

Saijo graduated from high school at Heart Mountain, and when the US Army lifted its restrictions on drafting Japanese American volunteers in 1943, he signed up and, like thousands of other Japanese Americans, was sent to Europe, where he served in the 442nd Regimental Combat Team. Made up almost entirely of second-generation Nisei soldiers, the 442nd became known as "the Purple Heart battalion," suffering tremendous numbers of casualties but ultimately earning more medals than any other unit in US military history.[28] Like every other soldier who signed up, Saijo—who spent his war years in Italy—played his part in defending many of the democratic norms and values of which he had been stripped during his internment, an irony surely not lost on him or countless others in later years.

———

By the time Welch and Saijo met in San Francisco in the late 1950s, the latter had been a regular attendee at Senzaki's twice-weekly meditation sessions in Los Angeles. Saijo described them as lasting for one hour, with the *inkin* bell rung every fifteen minutes. They started and ended with the recitation of the Four Vows in Japanese[29] before Senzaki gave his "commentaries on different Zen texts such as the Gateless Gate."[30] Welch was certainly familiar with the Renzai monk, and it is tempting to surmise that Saijo's influence on Welch may well have been partly due to Saijo's ability to share Senzaki's teachings in what may be described as an informal *guru-shishya* manner (*guru* is the Sanskrit word for teacher, and *shishya* for disciple). As a seasoned practitioner himself by this time, Saijo would certainly have added an element of experience to the zazen in Marin-an, and it must have been with some sense of comfort (for Welch at least) that, when Snyder returned to Japan in early 1959, Saijo was given the task of running the zendo in his absence—indeed, Snyder wrote to Will Petersen in Kyoto that Saijo was "the heir to my temple."[31] Saijo's discipline must have been the perfect foil for Welch's often contradictory and unreliable behavior. For example, in a letter to Joanne Kyger, Welch wrote of Saijo being "a saint. He runs the Zendo perfectly and builds beautiful things out of old lumber" (IR1, 156) (Indeed, when Welch left for his hermitage in 1962, Saijo presented him with one such "thing" — a simple hand-carved wooden bowl from which he regularly ate and about which Welch wrote a poem/letter included in *The Way Back*.[32]) Welch also praised Saijo's influence in a letter to Snyder, saying that Saijo's job was "to be here, on time, always, like that rock of an Albert, who so often was there when I arrived late" (IR1, 159).

On the occasions that Saijo was unable to attend Marin-an, Welch was given the task of overseeing the zazen, and there came a time when he was given full responsibility because of Saijo's increasing ill health. The smooth transition of responsibilities was important to both Saijo and Snyder; Snyder had written to Saijo from Japan with a set of clear instructions for Welch to follow.[33] Detailing the precise nature of how zazen should proceed, the instructions even go so far as to provide a schematic outline of when, and how often, the *inkin* should be rung before and after each of the four sittings, as well as when the incense should be lit and the tea drunk. The session was also to open with walking meditation and close with a reading of the Four Vows — a copy of which Snyder also provided for Welch. As extra emphasis on the importance of what Welch was undertaking, Saijo's own notes also state, "I quote from [a] letter from Gary: 'Tell Lew to insist upon the dignity & tradition behind the discipline of Zazen'" before adding that Welch should "Be firm" about this.[34]

For as long as his involvement with the zendo lasted, Welch took to this task with the utmost seriousness. Having initially been surprised by the formality of a normal day in Snyder's life in Japan, expecting rather a "studied chaos, an abandonment," he wrote to Snyder, "I agree with you about the importance of the Zendo. Will conduct the sessions with absolute punctuality and strict form and dignity, even if no one shows but me" (IR1, 158).

During this time, Welch seemed to have been willing to drive taxis in the city to earn a much-needed living and to spend his free time running the zazen in Mill Valley too. Yet he was far from happy with this situation, later writing that "Leo 'runs' the formal meditations — often sitting out his vigil alone. Peace descends upon him, only to be wrecked again when he re-enters the world as a cab driver. Each week, over and over again."[35]

Moving out to Mill Valley also afforded Welch the opportunity to acquaint himself with the "sleeping maiden" of Miwok folklore, Mount Tamalpais. Given that Marin-an was nestled in the eastern foothills of this coastal peak, anyone staying there had only to walk out the door to find one of the many trails that led up and through the various valleys and canyons. Snyder had already explored "Mount Tam" extensively, having had his first hiking experience there in 1946. By the mid-1950s, he was slowly developing an intimate familiarity with every trail, ridge, and gully, as well as the ease with which the mountain could be reached: "I was able to go out the door and right up the mountain. . . . I could go out the back door, up into a cow pasture, through the fence, pick up the road to Muir Woods, and then cut across another smaller road (Edgewood) to catch the Pipeline Trail."[36]

Such was the passion and seriousness with which Snyder took these hikes that he grew to understand the complex trail system on the mountain better than most. With Welch's sudden arrival in 1957, Snyder was offered the opportunity to introduce his old friend to the joys of such hikes, and it left a lasting impression on Welch. A few months before he broke up with Mary, Welch wrote to Dorothy that he and Snyder had set up a modest little camp on Tamalpais where they could "sit around and talk and take hikes down to some of the remote beaches."[37] When Snyder had to return to Mill Valley, Welch wrote of his staying up there to put his things in order—a reference to the imminent upheaval to come. Indeed, Mount Tamalpais would later become a significant symbol for him not only in his search for peace of mind but also in his search for the ultimate escape from the world of man.

———

One aspect of many zazen sessions held in Marin-an was the practice of *kinhin,* or walking meditation. Intended as a natural extension of the meditative process, *kinhin* means "to feel a step when taking a step, in other words, to take a complete step" such that "it is not mindfulness of walking, stepping, or even of a step. It is mindfulness of this step, and this step, and this step."[38] And such mindfulness, such heightened awareness, is thus what brings the practitioner one step closer to the wisdom of the Buddha. Bob Greensfelder described *kinhin* at Marin-an as follows:

To relieve the stress of sitting cross-legged—preferably in the lotus or half-lotus position—there's a break every maybe half hour or three-quarters for kinhin [during which] you file out of the zendo and walk around on the porch that surrounds it. That's why zendos have porches around them sheltered from rain. And that's called kinhin. Gary, when he had two or three of us teaching us the rudiments in that little shack, would take us out into the field and have running kinhin in the dark. That was supposed to sharpen our awareness.[39]

Similarly, Welch wrote that "Mister Snyder has us sitting in his shack Japanese style all the while ringing bells and smacking blocks of wood together. Then we run around the woods in pitch darkness falling over fences and otherwise being foolish" (IR1, 149). Despite the implication that these "running kinhin" sessions were not entirely as serious as they were intended to be, it was during one of them that Welch had a vision that would not only greatly influence his writing but also go some way to altering his very sense of himself. Describing the aftermath of an evening sitting, Welch wrote,

I got balanced and rocklike [and] observing spontaneous rambling-talk moving through my head, then realizing that was twice removed from thought . . . then observed the screen it went through before turning into chattering, then saw the first thing itself—a twinkling scum of fox fire darting about over rippled sand. It occurred to me that all one had to do was lift this fire and then let it sparkle where it would and then drop it. When it hit there would be a thought. So I poised it, stopped it, aimed at, and dove. Precisely at the moment of contact Snyder's bell rang and I whooped. . . . The question is: what is going on when one rings a bell. What is a bell? Who rang it? Did I get rung? (IR1, 150)

While the primary image seems to depict the inception of a thought process, it is the questions he poses at the end that prefigure his future use of the bell as a key symbol of his core being. In this case, the "ring" coincides with the moment of thought creation and thus not only supersedes that thought but also becomes its concrete representation. In many ways, however, it is also the moment of erroneous focus on a single aspect of a complete step—thus reducing awareness of that step instead of increasing it.

The bell is a key element of *kinhin*, in as much as it signals the beginning and end of each session but should not necessarily be seen as a means to either, but instead as an integral part of a larger process of mindfulness. Welch's admission that he responded inappropriately ("whooping") shows not only his inexperience but also perhaps his sense (misplaced or otherwise) of exultation at having reached a certain milestone in his practice.

It could also suggest a certain amount of impatience, another affliction from which Welch often suffered. Indeed, while his poetic output increased during this period, so too did his own frustrations about it. In the same letter in which he described his foxfire vision, Welch complained to Whalen about the fact that Donald Allen had returned a number of his poems, writing with obvious annoyance "why do I think I'll ever learn about anything when I don't even know how to sit and my damned legs won't bend" (IR1, 150). Apparently, his impatience also extended to his zazen practice, and his poems are now a metaphor for his lack of success in the attainment of the (half-)lotus position. It was yet another example of his propensity to rapidly change tack from the serious to the critical, and from the inspired to the lost.

———

In March 1959, "Chicago Poem" was published in the second edition of the new Sausalito-based literary magazine *Contact*, constituting Welch's first publication since the *Chicago Review* printed "Aubade" in 1952. It was an important moment for Welch because it provided him with his first printed work on the West Coast and added some sense of legitimacy to his readings—a published poet having a greater standing than an unpublished one, even in the informal milieu of the Bay Area poetry scene. However, his ever-increasing vulnerability to violent mood swings and bouts of melancholy and inactivity saw to it that, for every forward step he took, an inevitable sense arose that it resulted in two steps back. In the almost scripted manner of many previous letters, he wrote optimistically to Whalen about an upcoming reading at the Bread & Wine Mission in San Francisco (by far the biggest he had been offered to date) and a possible reading with John Wieners at San Francisco State Poetry Center. Yet he also described how he had subsequently "sagged into a deep acedia:

lethargic, demoralized, fretful" (IR1, 156) — a state of mind that prompted him to strap on his backpack and walk over to Marin County to look for a cabin.

So as 1958 drew to a close, Welch set about looking for a permanent address in Mill Valley. He spent the first three months of 1959 at Bob Greensfelder's house at 348 Montrose. Although he gave this as his mailing address, Welch was actually roaming all over the area looking for a suitable cabin in which to live, and he had apparently found eight different people willing to allow him to stay with them. He wrote of finding "3 suitable cabins . . . where I can get lumber, cookstoves, etc." — a process that gave him renewed energy and, once again, a positive view on the world, being as he was "delighted with the planet, awake to friends, etc. etc. [with] much writing, some of it good" (IR1, 157).

One such cabin was on a tract of land owned by Erik Krag, a Danish immigrant who, having arrived in America in 1915, had become a successful entrepreneur and owned much of the land in the valley.[40] "The cabin in the gulch," as Welch referred to his intended home, was close to Locke McCorkle's house. Although he was still required to drive cabs in the city, his shifts afforded him the necessary flexibility to be able to maintain Marin-an when either Saijo or McCorkle (who was also called on to do so from time to time) were unable to. And having a place of his own within such close proximity would surely have been something of a godsend. Yet Welch's desire to acquire the cabin is tempered by his usual pessimism — "I want it so bad I know I won't get it" (IR1, 158). Negotiations with Krag were left to Saijo, but the picture Welch paints of the cabin is one that makes it easy to see why it would have been a perfect place to live, and why he was so keen to get it, surrounded as it was with roses, grapes, plums, and loquats.

Within a month, despite any reservations he may have had, Krag agreed to Welch renting the cabin, and Welch immediately set to work clearing the trail and making the necessary renovations, including roofing and cutting back much of the overgrown trees and bushes. However, perhaps unsurprisingly, within two months the Marin County Health Department had received a complaint about Marin-an, and officials started to investigate the tenancy regulations that applied to Marin-an and many of the other cabins. Welch wrote that, to appease the Health Department, they would have to spend up to $500 installing toilet facilities at Marin-an "complex and big enough to serve the Roman Colosseum" and "to provide for the shit . . . which has grown to a pile the size of 10lbs. of potatoes" (IR1, 163). The outdoor latrine that Snyder had previously built ("the lady-favoring roofed pavilion" as Welch called it) was the main bone of contention, and the authorities ordered the latest occupant, Philip Whalen, to demolish it before moving out.

Whalen, near destitute and receiving financial aid from Ginsberg and Leslie Thompson to cover his basic needs, had recently moved into Marin-an on a semipermanent basis, having returned from the relative comfort of middle-class life in Newport, Oregon, with nowhere to live. McCorkle had gone to Mexico, Saijo had been hospitalized with hepatitis in late May, and thus the cabin was free. Whalen's eventual eviction from Marin-an, however, spelled imminent disaster for Welch, who reconciled himself to living as an outlaw and doing no more than was necessary for the upkeep of the cabin until he too was "kicked out for shitting" (IR1, 164).

It would be another three years before Welch would finally fulfill his dream of retreating to a cabin in the woods. However, by that time, the Marin County authorities had seen to it that not only Marin-an but also many of the other cabins in the area had been condemned as fire hazards. By 1961, they had all been summarily demolished.[41]

––––––

Although Welch's reading at the Bread & Wine Mission in March was a substantial step up from the intimacy of Jack Spicer's soirees, by far the biggest of his readings took place on November 8 at the San Francisco State College Poetry Center. Having initially been hopeful of a reading in May with John Wieners and Celeste Wright, it took until November for Ruth Witt-Diamant, then director of the Poetry Center, to arrange for Welch to read there with a local poet and lecturer at San Francisco State College (SFSC) named Frank Dollard. On the night Welch read a selection of poems to a small and (judging by the archive recordings of the event) not particularly enthusiastic audience. Given that *Wobbly Rock* had not yet made it into print (it was published in March 1960 by Auerhahn Press), and his only other published work was the recent "Barbara/Van Gogh" in the *San Francisco Review*, this was an opportunity for Welch to perform in an environment that had been nurturing and showcasing some of North America's brightest and best poets since its inception in 1954.

With the aid of a donation by W. H. Auden, who gave the honoraria from his 1953 West Coast tour to help Witt-Diamant establish it, the Poetry Center opened its doors in February 1954 with a performance by Theodore Roethke, and went on to host such luminaries as William Carlos Williams, Charles Olson, Robert Lowell, and Marianne Moore, not to mention the leading poetic voices on the West Coast such as McClure, Snyder, and Whalen, and East Coast poets like Ginsberg and Corso.

The first of many readings at SFSC over the next decade, Welch's debut is marked by the control and precision of its delivery. If anything, it suggests the antithesis of the weeping, confrontational poet that Welch and others often claimed him to be. Providing careful context where necessary, Welch explains his previous focus on "ear training" and his primary concern for how language "moved" by reading what he called "rinses for the ear."[42] Welch had long felt that his ear was keenly attuned to the economy of language, and he used these rinses to highlight and promote this in others. Indeed, in a letter to Philip Whalen—after reading Whalen's "beautiful novel"[43] *You Didn't Even Try*—Welch goes as far as to suggest to his friend that he has used a redundant "floating that"[44] and missed a preposition in a single sentence on page 208, namely "I can't do anything unless I've got some place that I can sit down comfortably."[45] Welch suggests that Whalen's use of "that" and omission of "in" or "on" after "comfortably" jars with his acute sense of how language works.

His focus on the use of these "rinses" as a means of ear training is an interesting one, and it became a regular technique in the years that followed. Welch wrote long and carefully constructed poems that juxtaposed what he felt were the different phonological aspects of American speech as exercises in differentiating between and identifying authentic patterns and movements. He wrote of language as "abstract pattering" that could be heard (and thus understood) through such clipped constructions as "duffer coat equivalent to parka" or in entire poems such as

> He said there was no music
> in American diction and American scene

at which the lights in the subway tunnel
were not lights passing in the subway tunnel

became lights thrown by
spangles in that lady's hat. (HWP, 55)

Intended "to make you hear the rhythm of your native speech," the
SFSC reading started with "Song in Subway: Another Rinse" before introducing one of his linguistic portraits called "Mrs Angus," his landlady in
New York in 1951. Continuing with what he deemed the "most successful" of his seasonal poems, "Winter" (which he wrote on a bus on a rainy
day in Portland), Welch explains his interest in how Bop music "moves"
and how he is trying to get the poem to "swing" in the same way. He
continues with a reading of another of his Reed College–era works, "A
Memorable Fancy" (with its humorous self-irreverence), and rounds off
the performance with "Chicago Poem," "A Round of English," and the
soon-to-be-published "Wobbly Rock." It is a consummate rendition of his
poetic philosophy, which was outlined in the program notes for the event:

> When I write my only concern is accuracy. I try to write accurately
> from that poise of my mind which lets us see that things are exactly
> as they seem. I never worry about beauty, if it is accurate there is
> always beauty. I never worry about form, if it is accurate there is
> always form.[46]

Having started the year with only one published work and no public
readings to his name, Welch must have felt that 1959 was ending on something of a high note. However, on leaving the Poetry Center that night,
he could never have known that, within a matter of weeks, he would have
found a new source of inspiration and more potential exposure than he
could ever have dreamed.

9
On a Disappearing Road

On November 8, the day of Welch's reading at SFSC, Jack Kerouac left from New York's Grand Central Station on a three-day cross-country train journey to Los Angeles and a scheduled appearance on the *Steve Allen Show*. As he sat in the relative bubble of his comfortable cabin, Kerouac was left to reflect on how his life had unalterably changed since the publication of *On the Road* two years earlier. For one thing, fame had seen to it that he no longer needed to hitchhike or jump freight trains, and for another he was now a significant, if unwilling, cog in the gradual commodification of the Beat Generation. This journey to Hollywood was merely another step on the road to his eventual alienation from almost everything he had previously helped to establish.

Before leaving Northport, New York, Kerouac had ranted to Allen Ginsberg about what was being demanded of him, from readings and receptions to public appearances and publicity blurbs. He was so disturbed by what he saw as the obligations that accompanied his fame that he threatened to cut out altogether and disappear to Mexico, writing, "This is serious. I'm mad. There's no hope."[1] Yet Kerouac also understood his position of relative power as the standard-bearer for this new generation, and in a letter to Gary Snyder from around the same time he admitted that, as everybody was "plugging books on TV now, [he had to] get on the horse and ride away the Sultan" too.[2] Indeed, his appearance on Allen's show reputedly earned him something in the region of $2,000, equivalent today to close to ten times that amount.

The performance itself is now the stuff of literary legend. Kerouac, somewhat embarrassed by the whole affair, answered Allen's slightly irreverent questions before reciting passages from *Visions of Cody*, which he had pasted into the copy of *On the Road* from which he was supposed to be reading. Kerouac, who had refused to rehearse beforehand and had openly expressed his discomfort about the situation, read with splendid authority to Allen's jazz piano accompaniment. As the story goes, when it was all done, Kerouac promptly departed the stage to vomit. His choosing to shun the wishes of the producers (and Allen) by reading unrehearsed excerpts from Visions of Cody was indicative not only of Kerouac's increasing disillusionment with how he felt his work was being treated in the mainstream media but also of his own views that Visions of Cody was a far better representation of his work than On the Road. It was also a clear example of how the importance of the writer outstripped the importance of his words. To the TV executives, simply having a literary celebrity such as Jack Kerouac on the show was worth far more viewers than whatever it was he might have said. While a select few people may have been confused by what Kerouac read that night, it is obvious that neither Allen nor most of his audience were any the wiser to the ruse.

Kerouac's trip to Los Angeles was also notable for a visit to the studios of Metro-Goldwyn-Mayer, where he had been invited to see the unedited footage of director Ranald MacDougall's version of Kerouac's The Subterraneans. However, rather than seeing an uncut work that was true to his novella, he witnessed a blatant Hollywood bastardization in which not only was the mixed-race relationship between Leo Percepied and Mardou Fox changed to an American-French one, complete with George Peppard and Leslie Caron in the starring roles, but also Kerouac's finale was altered to such an extent that the audience was being fed the highly romanticized fodder of stereotypical, happy-ever-after, feel-good movies—further evidence of how his work was being altered to accommodate mainstream ideas and circumvent potential problems with distribution or, more importantly, audience reception and box office numbers. As it was, the film was heavily criticized not only by Kerouac and his peers but also by the cinema-going public, eventually losing more than one million dollars.

After refusing to endorse the movie or provide any form of publicity, Kerouac left Hollywood for San Francisco the next day in search of friendly faces and familiarity (Steve Allen described their parting on Sunset Boulevard in the dark Hollywood night as Kerouac walked off drunk "shouting mildly obscene pleasantries," never to be seen by Allen again.[3]) In San Francisco, Kerouac was scheduled to attend a screening of Robert Frank and Alfred Leslie's avant-garde movie Pull My Daisy. Initially titled The Beat Generation in honor of Kerouac's unfinished play of the same name, the film featured Kerouac as narrator and starred, among others, some of his closest friends: Allen Ginsberg, Gregory Corso, and Peter Orlovsky. Having first been shown to cast, friends, and family at New York's Museum of Modern Art on May 12, *Pull My Daisy* had its public premiere at Cinema 16 in Manhattan on November 11 and was then on the bill at the San Francisco Film Festival.[4] On the night of the screening, November 17, Kerouac substantiated his own growing reputation as a voracious drunk and, having been drinking heavily all day, did not particularly help the movie's reception by falling off the stage during its introduction. Prior to the screening, he had waited outside the theatre with Leslie to give tickets to his friends, among them Welch and his then-girlfriend Annie, who arrived with Joanne Kyger and the painter Jay Blaise in the guise of a chauffeur, stepping out of his car onto the red carpet as if "going to an elegant party . . . dressed as elegant beatniks." In a letter to Stan Persky, Kyger described the Hollywood-esque event as "very grand" with "photographers popping" and Leslie, dressed in a tuxedo, looking like "a dark Harpo Marx." The film itself, she wrote, was "stupendous-ola."[5]

In the end, however, despite winning the festival's award for Best American Experimental Film, *Pull My Daisy* was greeted with "hostility and suspicion or dismissed outright,"[6] prompting the increasingly disillusioned Kerouac to seek what for him was the ultimate refuge—a rapid return to Northport and Mémêre (the name Kerouac affectionately gave to his mother, Gabrielle), where he could metaphorically barricade himself against the intruders, critics, and autograph-hunters who were gradually taking over his public and private life. And this was where Lew Welch and his old army-issue jeep came in.

————

The *Steve Allen Show* was broadcast on NBC on the evening of November 16, and Welch makes no mention of ever having watched it. In the months prior to Kerouac's arrival on the West Coast, Welch had left the East-West House and taken a road trip to Oregon with Dorothy before returning with her to Reno, where she now held a post as chairperson of the Department of Child Development and Family Life at the University of Nevada. However, by the time Kerouac reached San Francisco for the screening of *Pull My Daisy*, Welch had heard he was coming and had arrived in his new Willys MB Jeep (nicknamed "Willy"), which Dorothy had given him as a birthday present prior to their late-summer trip.

Back in San Francisco, Welch immediately moved into the Hyphen House on the corner of Post and Buchanan. Newly established, the house was opened as an annex to the East-West House, which had gradually run out of space. The house was initially inhabited by, in addition to Welch, Saijo, Whalen, Blaise, Les Thompson, and the Black Mountain painter Tom Field, who all helped to clean and paint the apartments. It was furnished with secondhand furniture they either bought on McAllister Street or took from the many abandoned buildings nearby that were being demolished. Saijo describes the social center of the house as being the old wooden kitchen table, around which everyone sat exchanging "a great deal of plain quiet talk or boozing and howling," and it is easy to imagine Jack Kerouac pulling up a rickety old chair, joining the party, and lending his not insignificant drinking prowess to the proceedings.[7]

Welch was now also working at the Bemis Bag factory on Sansome Street as a rubber cutter, a job he apparently enjoyed on account of its artisanal nature—working with his hands in such a skilled manner was something he had never done before. Unfortunately, his lifestyle made it difficult for him to be punctual, and he was soon fired for coming in late too often in the morning. In what remains of a draft for an autobiographical screenplay he wrote called "Bemisbag: A Movie"—in which he features as "redbeard, the hero"—Welch describes the morning ritual of his late arrival at the factory and the manner in which he was subsequently greeted by his boss.

Shot of long concrete building, street with people more or less hurrying to green door. They all go through the door. Street now empty.

Single figure, running. Tall, thin, red beard. Beautiful stride—a professional runner. He is coming from way down the street, past aforementioned truck, sprints.

FANTASTIC WHISTLE (much too loud):

Redbeard runs even faster, jerks open green door, enters.

He runs up the stairs, concrete, green-steel railings, very tight twist takes him only two big up-running strides to make a half-landing.

Door with 3 on it. Room full of tables. Room full of lockers (all swift, we're running). Swinging doors. Card. Punch-clock.

Bam! (Clock) Hand punching it. 8:01 (a tragedy!)

. . .

Portly, stupid-looking man comes out of the door and looks at watch. A tragedy!

One-minute late redbeard gets a funny apron off hook and puts it on. He does not look, even, at little boss watch-watcher . . .

Watch-watcher talks to him. Redbeard nods as if he admits to have heard the words and walks to his desk. He is neither rude nor obsequious.

Watch-watcher worries about this. Everybody is late once in a while, but you ought to be obsequious. So (he ponders) later will fire him.[8]

Despite this stereotypical parody of the worker-manager relationship, the real reason he was eventually fired may well have been infinitely more veiled but no less ridiculous. Seemingly, Welch's decision to have his ear pierced for a costume party was the unspoken reason for his dismissal. "You hear about some nut in the cutting room getting his ear pierced? Who the hell was it?" one of his colleagues had asked (TT, 15). Conservative America reared its ugly head again and reminded him of his position on the fringes of society. After an embarrassed payoff, he drowned his sorrows by getting drunk and then sick on rum and pot—the perfect precursor to what was about to come.

Tired and drinking heavily, Jack Kerouac "appeared to be a binge and determined to party," and with Welch providing transport, they could move effortlessly from place to place and drink until "everyone fell down too tired or drunk to go on" (TT, 5). This apparently continued for some time (Welch writes of being drunk for three days) before Kerouac announced his desire to return immediately to Northport—a desire that Welch was only too happy to oblige. A somewhat melancholy Welch later described the circumstances prior to Thanksgiving that prompted their decision not to wait to celebrate in San Francisco but to depart straightaway:

> There wasn't any reason to stay in San Francisco and Jack wanted to drive across the country again, so I said all right we'll leave after Thanksgiving because the girls want to cook us a big dinner. We cook the Turkeys and they cook the pies. But the night before they all got drunk and Valerie Song threw a half-full gallon bottle of Port at point blank range to crash and break full in the face of Jay Blaise, who'd been needling her. That did it. Violence is sickening and silly. And everything else had gone wrong for a whole week . . . so we decided to leave right away. Missing Thanksgiving, and called Albert over from Mill Valley. (TT, 17)

And so, on November 20, 1959, a somber, rainy night a few days after Kerouac's Hollywood nightmare, Welch, Kerouac, and Saijo packed whatever possessions they needed into their survival kits, had a final meal in Chinatown, and, each in possession of a lucky charm given to them by Blaise (Kerouac was given a St. Christopher medal, Saijo a Mexican penny, and Welch a Jewish coin with a lion and the Ten Commandments), drove into the American darkness.

―――――

Despite Willy having a cracked windshield and an engine already leaking oil, Welch was convinced that his little jeep was more than capable of making the seven-day transcontinental trip. Although Kerouac had initially planned to visit his sister Caroline in Seattle, he was worried about the prospect of having to face potential snowstorms crossing the Rockies and so altered the plans to take a safer and more direct route eastward. Welch was more than happy to drive, irrespective of the chosen route.

Their first stop was in the Mojave Desert, where they slept on the mattress Welch had installed in the back of the jeep for longer trips. Waking in the cold late November morning, Kerouac first suggested a bone-warming jog before going through his morning ritual of standing on his head for ten minutes. They then started off for Las Vegas, where they stayed in a motel modeled after George Washington's mansion at Mount Vernon. In one of the casinos, Welch displayed his "old gambling

taxidriver personality," losing $22 at a table before having to be dragged
back to the motel by his companions to save him from losing even more.[9]

The journey saw them drive out of Nevada via the Grand Hoover
Dam and into Arizona where, with the sun setting, they decided to stop
and steal one of the many little white crosses that dot the edge of the
highway indicating the sites of fatal accidents. Getting out in the falling
light, Welch found himself kneeling in a bed of thorns extracting the
cross from the ground. It was initially intended to adorn the motor grill
but would eventually end up on the wall of Ginsberg's Second Street
apartment in New York, complete with a ball of fluff that had apparently
fallen off a stripper's costume in East St. Louis, where they also stopped
off to spend the night.

Picking up Route 66, they continued through New Mexico ("Green
& silver signs say Refinery Exit One Mile" [TT, 44]), the Texas panhandle
("The last time I was in Amarillo, I smoked my first cigar and Roosevelt
died" [TT, 35]), and on to St. Louis, all the while watching as America
slipped by outside the windows of their "microhabitat chamber" in which
they told stories, meditated, sang and rapped, wrote, or simply sat in silent
reverie observing the passing landscape.

Crossing the Mississippi—that monumental symbol of Kerouac's
America, where the smell of "coal smoke denoted that west ended here
and east began" (TT, 7)—they veered northward through snowy Illinois
to Indiana and into Ohio, each capturing their own personal snapshots
of American life in the form of haiku. Kerouac wrote various little
"poemettes" in a little yellow notebook notable for "sporting the color of
the Sangha"[10] as Welch drove with "finesse, very much as he wrote" and
Saijo "sat in the back of the Jeepster much of the way on a mat meditating
to see if it was possible to meditate away from the tranquility of a zendo."[11]

During the trip, Welch proved himself to be a great talker, spinning
long tales that Kerouac later suggested would make him a "great goody
novelist," and by the time they entered New York City, the three had
amassed a large group of poems that documented the sights and sounds
of their trip.[12]

Thanksgiving had come and gone; arriving in New York, they headed
straight to the Lower East Side to meet Ginsberg and Orlovsky. Welch
wrote to Snyder that the pair were living in "terrible squalor [in] total
dedication to being Beat and Art and Holy," yet despite this, Ginsberg's

apartment building had become something of "a launching or crash pad for [his] extended family of friends and acquaintances," and whenever anyone with even the slightest connection to Ginsberg arrived in New York, they would invariably head to his Second Street apartment.[13]

After dropping Kerouac at the dwelling of another acquaintance in the city, Welch and Saijo spent their first New York night in a cold and deserted parking lot on Staten Island, and for the next few days the pair accompanied Kerouac to various bars and apartments, meeting the great and the good of New York alternative culture, including Robert Frank, Alfred Leslie, and Donald Allen. They accompanied Kerouac to a radio interview and joined him for an evening service at the First Zen Institute on East Thirtieth Street, where the "Buddha on the altar was offered marble cake and an orange. We were served marble cake and tea" (TT, 9).

On December 5, the three of them finally drove out to Northport, where Kerouac showed them the converted attic where he did his writing, his bookshelf filled with copies of his own work in various translations, and "a manuscript of notes, jottings and texts on Buddhism" that would eventually be published posthumously as *Some of the Dharma*.[14] The evening they spent in Northport provided Saijo and Welch with an eye-opening glimpse of the world in which Kerouac lived once the "famous author bullshit" had been set aside and Kerouac could truly be himself, alone with his mother. Much has been written about Mémêre and her influence on Kerouac. Saijo described her as "Everymom. A stout, straightforward, friendly woman . . . who put a wonderfully abundant supper before us," and it is easy to imagine that in such an environment there would have been no place for the airs and graces of New York literary life.[15] This was Kerouac's home, in which he could safely relax and show what a truly beautiful and sweet character he had.

Saijo and Welch spent a single night in Northport before leaving Kerouac and driving back into Manhattan, where they again stayed with Ginsberg. The ensuing days were spent quietly, with visits to Central Park, the Bronx, various museums, the Empire State Building, the Brooklyn Bridge—the stuff of tourism. For Welch, visiting many of these sites was like taking a trip back in time, the major difference being that now he was there temporarily, in the company of friends, and thus less susceptible to the mental turmoil that had characterized his stay there a decade earlier. Gone were the ideas that it was nothing more than a "noisy rock" in which he could not breathe and by which he felt defeated. He was more keenly aware now of his place there, and memories of his first "big freeze" were thus more easily put into perspective. His understanding of what New York represented, not only to him but also to his newfound friends and literary comrades, is clear in his correspondence and his poetry. He shows a sense of acceptance that, with such dreams and on such paths, obstacles of various natures are an unavoidable occupational hazard. While he had struggled to make sense of his position as a fledgling writer in the city the first time around, he was now more confident, more perceptive, and, most important, sufficiently detached from the literary scene there to be unaffected by its claustrophobia and myriad obligations. In a letter to Witt-Diamant, with whom Welch was corresponding about an upcoming teaching vacancy at SFSC, he writes with a mixture of fraternal frustration and personal relief about the fact that "the boys in New York" seemed to be as busy with the administration of being writers as with writing itself:

> The real work of literature which, now, alas, at this particular time seems to be mostly a matter of answering endless tiresome letters from aspiring old maids in Oshkosh, pimply boys in Ontario, girls writing term papers in Florence Italy on Fulbrights, and old faggot writers giving cocktail parties in New York City with the pages of their Paul Claudels not even split with the knife. It is making Jack and Allen very tired.[16]

There were also the movies, magazines, and corporate companies treating the Beat Generation as "a new brand of ice cream."[17] And although far removed from such demands himself, Welch must have been comforted to know that their time in New York was slowly petering out, and that Willy would soon be carrying them westward again—not, however, before one final fling that would cement the relationship between the three in concrete literary terms forever.

Having gone back to collect Kerouac in Northport, they paid a visit to the West Village apartment of Kerouac's friend Fred McDarrah, who lived at 304 W Fourteenth Street. McDarrah had been the staff photographer for the *Village Voice* since the magazine's inception in 1955 and had made a name for himself as not only a cutting-edge photographer but also as the brains behind Rent-A-Beatnik—a business scheme in which he supplied clients with poets for readings in a venture that openly and purposefully satirized the Beat Generation while giving the poets a welcome boost in income on the back of the widespread exploitation that Kerouac had already experienced in Hollywood. One of the most famous participants was Ted Joans who, for $40, would "rent himself out to unsuspecting 'square' New Yorkers and suburbanites who thought they would spice up their parties" by having such beatniks in attendance.[18] Joans then made a point of delivering witty slices of poetry that made fun of the very people he had been hired by—with one of his most famous poems, "The Sermon," providing a perfect example.

McDarrah had first photographed Kerouac the previous New Year, and he was now eager to show him the results. Afterward, Kerouac, Saijo, and Welch began collaborating on a long work featuring several of the haiku they had written on their trip east. While McDarrah captured the events on film, his future wife, Gloria Schoffel, typed the poems as they were being recited, and the group worked their way through numerous bottles of whisky, gin, and beer. When the work—along with the drink—was finished, Kerouac read it aloud, and the entire group went out for a celebratory dinner at the Egyptian Gardens restaurant before going to see Ornette Coleman at the Five Spot nightclub. McDarrah included both the finished work, titled "This Is What It's Called," and several photos from that night in his book *The Beat Scene*. The poems would later be reworked again and published in their entirety by Donald Allen's Grey Fox Press as *Trip Trap, Haiku Along the Road*, a name playfully chosen by Welch in homage to Gary Snyder's debut work, *Rip Rap*.

Welch and Saijo finally left New York on December 12, leaving Kerouac to his dual life of Northport homeliness and New York carousing, which Saijo later said had saddened him to see. The pair then drove at breakneck speed back to Reno, covering the distance in less than three days (despite mechanical problems that prompted hot-wiring) and arriving at Dorothy's house on December 14. A short time later, on December 25, or "Crispness Day," as Welch called it, Welch wrote to Kerouac of their driving back through a beautiful fog in which Willy was "a blasting little circle of speed in a white light crashing into the void ahead," and as the fog froze on the desert bushes it was like driving through "sparkling, coral, purity-white, a white ring of light thrown down from some unbelievable spotlight way up in the Universe" (TT, 57).

In much the same way that Kerouac had experienced the trip east as an enlightening one, so Welch was now describing the return journey as such too. In addition to making clear that Kerouac was now his "greatest real friend since Gary and Whalen" (IR1, 169), Welch's description of the almost transcendental nature of the journey back with Saijo accurately reflects his growing sense of himself and the influences and inspirations with which he had gradually surrounded himself. He was now at the epicenter of a movement that both fed him and bled him. It had provided a renewed sense of what he was capable of and how best to go about it. Collaborating with Kerouac and rubbing shoulders with Ginsberg and company was just the shot in the arm Welch had been needing. San Francisco was undoubtedly an important source of inspiration, but his return to Reno was fueled not only by Kerouac's words of encouragement but also by his methodology. Welch, like countless others enveloped by the myth of writing long spontaneous texts on rolls of teletype paper, decided to begin work on a semiautobiographical work similar to Kerouac's. Both Kerouac and Saijo had been enchanted by Welch's storytelling prowess, and Welch took sufficient encouragement from their views on what he had told them about his life to embark on a project that would focus his attention for the best part of a year. Rather than a continuation of the "Keeler" novel he had previously started, this would be a more focused attempt to write about his life.

Welch decided on "Hard Start" as a working title for his novel after a conversation with Kerouac about what Welch liked in novels:

> —I like novels that start hard, like my
> life did, always seeming to be started
> then starting all over again
> —That's what you ought to call it
> —What?
> —Call it *Hard Start*
> —All right, I will[19]

The intention was thus to focus on the various struggles he had had with the primary influences in his life (both in his younger years and now as a writer)—Dorothy, Lew Sr., himself—and how these had helped him shape his current identity. There is a sense that in making this choice, the work itself would somehow offer an antidote to what the title infers, and that there would be hope and success despite the hardships. However, like many of his other projects, including "Keeler," this one would ultimately suffer a similar fate—uncompromising self-critique and eventual abandonment.

10
It's a Real Rock (Believe This First)

After spending "some very tiresome times in S.F." over the festive period, where "everybody is sulking or flipping" (IR1, 171), Welch returned to Reno for an extended stay at Dorothy's. Dorothy was now a full-time associate professor of home economics at the University of Nevada, and her regular absence from the house gave Welch the necessary space and quiet to make serious inroads into his various projects. In addition to starting his novel in earnest, he had also been asked by Donald Allen if he could send a copy of his work on Gertrude Stein for Allen to read. Having previously committed to including three of Welch's poems in his *The New American Poetry* anthology (although only "Chicago Poem" and "After Anacreon" made the final cut), Allen was increasingly interested in Welch's other work. His inclusion in the publication of this seminal poetry anthology was by far the most important milestone in Welch's career to date, and Allen's commitment to his work was not only a source of satisfaction and pride to Welch but also provided him with a bone fide outlet for his work. Despite later complaints that Allen should have anthologized "Wobbly Rock" rather than the pair that he did, to be alongside such contemporary poets as Olson, Creeley, and O'Hara was a nod of recognition the likes of which Welch could never have received through the publication of chapbooks or the distribution of free broadsides, which were the norm at that time.

Yet by the time *The New American Poetry* was published, in late May 1960, Welch had had his collaborative work with Kerouac and Saijo included by McDarrah in *Beat Scene* (for which he received $5 and about the publication of which he was apparently oblivious) and had also published his first chapbook, *Wobbly Rock*. In late March, Dave Haselwood's Auerhahn Press "handsomely finished" (IR1, 189) his six-part poem "Wobbly Rock," publishing it with a sketch by Robert Lavigne. Welch had asked Lavigne to contribute, and the artist had obliged

Figure 10.1: Preliminary sketch for *Wobbly Rock* by Robert Lavigne, 1960. Courtesy of Estate of Robert Lavigne

with a detailed pencil study of a cliff-face at Muir Beach, complete with aquatic life, swirling gulls, and precariously balancing rocks in an artistic rendering of the poem's fifth stanza.

For Welch, the publication of *Wobbly Rock* finally provided him with a real sense of poetic accomplishment. The appearance of single poems in literary magazines was nothing compared to having your work handset,

published, and available for sale—even if, in Welch's case, the chapbook
was being sold for only fifty cents. The reception it received was also
mostly favorable. Marianne Moore spoke of noticing "the 'hand-crafted'
quality" in the writing (IR2, 110), and Charles Olson wrote to Haselwood
that the poem had "delicious things" going on in it and of it being "very
true and delicate. Very great" (IR1, 198). Yet Welch himself quickly grew
to see the poem as merely a stepping stone to an even closer examination
of the truths he had "learned from Mahayana"—namely, a rendering of
poetry that is truly reliant on entirely local forms and images. Although
the poem takes its primary image from a rock standing in the surf on Muir
Beach, the comparative imagery is often Japanese—thus drawing the poet
out of his immediate surroundings and, in Welch's mind, away from the
types of personal gestures that should define his poetic development.

Despite these later thoughts, however, "Wobbly Rock" remains one of
Welch's most enduring and widely anthologized poems. It is an effort to
crystallize in words Welch's process of meditating on a natural phenomenon such as a balancing boulder and the sense of nonseparation within
the natural world. This "real rock" was a boulder that Welch encountered
on a "foggy and windy day" on Muir Beach, and the different sections of
the poem draw their inspiration from, among other things, his increasing interest in Buddhist practices and traditions. Indeed, in many ways
"Wobbly Rock" can be seen as the ultimate crystallization of the various
poetic techniques and philosophical teachings he had learned in the
period since his arrival on the West Coast. In the same way that "Chicago
Poem" represents everything he disliked about his life in the Midwest, so
"Wobbly Rock" is not only a clarion call for the positive developments he
experienced but also provides a clear example of a new depth of perception coupled with plain and simple language.

———

The first of the poem's six sections introduces the "Wobbly Rock" as
being something concrete. This dictates that the reader will not choose to
become lost in the myth or imagination but will, through the precision of
the language, have no choice but to see the poem as grounded in reality.

It's a real rock
        (believe this first)

> Resting on actual sand at the surf's edge:
> Muir Beach, California (RB, 54)

As the poem continues in the second stanza, the complexity of the imagery increases, and the reader is called upon to understand Welch's meditative ability (or perceived state thereof) as he sees the seashore in terms of a Zen garden. This stanza contains two separate notions—the first is an external discussion of the comparisons to be found between Muir Beach and the Zen garden in the temple of Ryoanji in Kyoto, Japan,[1] in terms of nature and space, and the second is the internal struggle the poet is having to clear his mind and attain a perfect meditative state as he ponders such distinctions.

Welch had never seen such a garden himself—the image he had of it came from the poet Will Petersen, who had published an account of Ryoanji in *Evergreen Review* titled "Stone Garden."[2] However, in the descriptions that occur, Welch tries to make sense of it and see that he is entirely (both in reality and in the mind) in the presence of natural formations. While the conjured image of the Zen garden is based on both his perceived interpretation of secondhand images and the precise work of the monks who created it, essentially the garden is simply rocks and space, and therefore the same in form as that which he sees on Muir Beach. This notion is realized as Welch's mind clears during meditation and he sees things in an instant. Until this final realization, his mind has been a "clutter-image," wherein the process of emptying his thoughts and "waking" to reality is cluttered with conflicting images and confusion.

In what turned out to be the final reading before his disappearance, held at Reed College on March 30, 1971 (the transcript of which was later included in the collection of essays and readings that Donald Allen published in 1973, titled *How I Work as a Poet*), Welch describes the process of only understanding reality in the moment—only after the creation of the garden can the monks fully understand it. It cannot be truly "seen" before—true vision only comes upon completion of the creative process. And, through a similar thought process, Welch sees that Ryoanji is simply an imitation of his present reality, yet at the same time is essentially the same. This notion is supported by the idea that everything is only relevant in the moment. Since Buddhism believes in the nonexistence of past and future, so it stands to reason that the mind will see the completed object only there and then. Thereafter it becomes irrelevant.

The third section places Welch autobiographically and displays the closeness of the bond that he has with the ocean. His current "congenial form" has been reached through him being many other shapes—"all those years on the beach, lifetimes," and the implication of his existence as a continual series of forms points to bodhisattvahood and the multiple rebirths that the bodhisattva must undergo before attaining enlightenment. Furthermore, Welch ends this stanza with a riddle, or koan, in which he ponders the question,[3]

> Waves and the sea. If you
> take away the sea
> Tell me what it is (RB, 56)

While this points to the "unbreakable bond between waves and sea," in his essay *Nature and the Poetry of Lew Welch*, Rod Phillips further adds that this koan coming "as it does at the end of this autobiographical section [of the poem] . . . is also suggestive of another unbreakable bond, between the poet and the sea."[4] However, if Welch intended this to be read as a koan in the strictest sense, then there is no explanation possible, and the stanza must be left unresolved. Given the meditative state of Welch's mind at the time of writing "Wobbly Rock," it is possible that he is challenging himself as a *shishya* to meditate on the "problem" at hand while also asking the individual reader to react to the assigned dictate (by Welch as *guru*) in the same way (thus reinforcing the *guru-shishya* relationship that forms an important part of Buddhist learning). It is arguable, however, that, unlike a koan, this riddle has an answer, and that while Welch is also playfully introducing a linguistic game, the answer is an example of a basic Buddhist precept—the oneness of separate entities, or "nonseparation." Taking the riddle literally, if one follows Welch's advice by taking away "the sea" and meditating on what is left, then "waves and" morphs first into "wavesand" and further into "wave sand" in a clever mirror of the arbitrary movement of one onto the other. The "s" creates the link and thus leaves its imprint on land and creates the unbreakable bond between land and sea.

The fourth section offers a retrospective look at previous events on the beach where Welch is lamenting the ignorance of those visitors who come there only in good weather yet even then fail to understand it. They are fair-weather nature lovers, returning to their lives with the prized catches of a day out, such as shells and driftwood. Welch is aware of the human tendency to gather/collect and then discard, describing the visitors as "feeding swallows." He displays frustration at this: "Did it mean nothing to you Animal that turns this / Planet to a smoky rock?" Yet Welch sees a lone fisherman "far up the beach," describing him as a linear extension of nature and consequently elucidating the nonseparation of humanity from its surroundings.

Welch himself described the fifth part of "Wobbly Rock" as "the satoris"; on one level it is a reworking of the traditional idea that humanity has about the "great chain of being" and their position within it.[5] In describing this section as "the satoris," however, he is also prompting the reader to redefine nature's hierarchy by arriving at a new viewpoint through which reality is more apparent. If we use D. T. Suzuki's definition of satori as "the unfolding of a new world hitherto unperceived in the confusion of a dualistic mind,"[6] then Welch is perhaps using such terms to determine man's random position within the oneness of his environment and the continual misunderstanding of his place there. While Rod Phillips calls this section a "reworking," it can also be seen as an attempt by Welch to display natural interconnectedness rather than a hierarchy, since the latter would still imply separation of some sort. Hierarchies suggest varying levels of existence within the greater group and consequently undermine the notion of nonseparation. If Suzuki's "new world" is what Welch is driving at, then the central stanzas must be seen not as a "great chain of being" but rather as one link, where everything exists on the same level.

It is also interesting to note that Welch fails to give any analysis of this section of "Wobbly Rock" in *How I Work as a Poet* (he otherwise gives limited if important pointers about other parts), a decision that echoes a statement by Suzuki:

> If satori is amenable to analysis in the sense that by doing so it becomes perfectly clear to another who has never had it, that satori will be no satori. For a satori turned into a concept ceases to be itself; and there will no more be a Zen experience.[7]

Furthermore, since in Zen all that can be done in "the way of instruction is to indicate, or to suggest, or to show the way so that one's attention may be directed towards the goal"[8] then this poem is a vehicle with which Welch can push both himself and his readers toward that goal of seeing the reality of existence.

The final part in the poem returns the reader to the beginning and the concrete image of the balancing rock. This section begins with an instance when, during meditation, Welch became one with his surroundings and "lost all separation." This experience is used as a buildup to the final conclusion, in which Welch visualizes and reiterates the nonseparation of what he sees "standing on a high rock looking out over it all":

> Wind water
> Wave rock
> Sea sand
>   (there is no separation) (RB, 59)

In a sense, "Wobbly Rock" is a poem displaying both the interconnectedness of nature from an ecological point of view and an interpretation of Buddhist thinking in its determination to see this interconnectedness as something other than a state recognizable in human terms. Either way, the poem is a vehicle through which greater understanding can be gained, since it shows Welch achieving an enlightened state and reinforcing the notion that in his poetry the two main criteria coexist on individual and inseparable levels.

––––––

The publication of *Wobbly Rock* meant that Welch had to travel back and forth from Reno to San Francisco to meet Haselwood and discuss the various stages of the printing process. However, money had been scarce since Welch's return from New York, and Willy was in an ever-increasing state of disrepair. The journey had taken a serious toll on the old Jeep, and in a letter to Kerouac, Welch said that he needed "$150 to get the charley horses out of him after the long bowl-game over the whole New Ninety States." This was a sum simply impossible for him to amass at that time. Readings provided occasional income but nothing like the amount required to make the repairs. So, Welch resigned himself to arriving in

San Francisco as "the penniless bhikkhu, by bus, not trusting Willy—or rather [not wanting] to seriously break him apart" (TT, 67).

In addition to his transport issues and financial worries, there was a far more serious situation developing—namely his failing health. The intermittent ups and downs he had experienced since his youth were now frequently accompanied by ever heavier bouts of drinking. He frequently writes of suffering from depression, boredom, and acedia—all of which prompted a remedy in a bottle. While his earlier breakdown in Chicago had had serious consequences for his mental well-being, his increasing reliance on alcohol to deal with such breakdowns was now affecting him physically. In a particularly candid letter to Snyder, Welch writes,

> What has happened is, Hope left. I know this to be as foolish a mind-idea as disappointment—related to it, for you cannot be disappointed unless you hope. However, though I finally got to the point where I was almost never disappointed, it was laughably unreal and therefore unfeelable, I continued to hope. Now it is laughably unreal to hope, but I am depressed all the time. There is no way out. Nothing has ever caused me such a total distress. Every other real knowledge I have had gave rise to a new joy, a brightness of present. Not this. I am gloomy and strange. I see things badly. I am not with it. . . . It is real bad, Gary. (IR1, 183–184)

In looking for ways to counteract such downs (or perhaps more accurately to find alternative means of respite from them), Welch dreamed of visiting Snyder and Joanne Kyger in Japan, applied for various grants that would allow him to work and travel in comfort, and planned a spring trip to Oregon with Kerouac and Albert Saijo during which they would head to the Rogue Valley and up to a disused goldmine Welch knew about. By way of preparation, he wrote to Kerouac of assembling his survival kit and how they should aim to "set up the old Benton mine, catch huge fish, live well, and sit right on the very skin of this beautiful planet—having sloughed off all of Mansworld for a little while. I need it terribly" (TT, 67).

Welch had been to the Rogue River previously and knew what the landscape might offer by way of escape. Tentative plans had been made while in New York, and should this project fail to materialize, Welch and the others could always "go into the mountains elsewhere or live a while in the Nat'l Parks. There are thousands in Oregon and all are pretty" (IR1, 185). It is clear from Welch's letters to Kerouac and others that this trip was of vital importance, and his need to extricate himself from "Mansworld" was never more obvious than in a poem he included in separate letters to Kyger and Kerouac in early January. His deepening loneliness and depression in Reno seemed only to diminish in the act of "deep" writing, and as such he wrote of having visions that led him to better understand what his future had in store — both good and bad. Indeed, one such vision, brought to him by his "Muselady," perfectly sums up the trajectory his life was on at that time, and more important, what he needed to do about it.

> First you must love your body, in games,
> in wild places, in bodies of others
>
> Then you must enter the world of men and
> learn all worldly ways. You must sicken.
>
> Then you must return to your mother and
> notice how quiet the house is
>
> Then return to the World that is not Man
>
> That, finally, you may walk in the world
> Of men, speaking (IR1, 171)

"First you must love your body" reads like a manifesto for life. Childhood, adventure, sexual experience, and adulthood, followed by failure/ illness, a return to the safety of the womb, awakening, and acceptance of place. At the time of his vision, Welch was in the womb phase but already understood that abandoning the human world was an essential aspect of his regeneration, and as such the Oregon trip was his first opportunity to "return to the world that is not man" — the natural world, free of human interference and social obligation. This was a key requirement for garnering the understanding and sensibilities needed to survive upon returning. It is also interesting to note that Welch envisions "speaking" upon his return — the conscious choice of an activity that offers an insight into the role that Welch saw for himself as a spokesman for such endeavors.

Indeed, throughout the mid-to-late 1960s, he was a vocal advocate of leaving the city and going into the country. He even gave advice on the best approach to doing this:

> Stake out a retreat. Learn berries and nuts and fruit and small animals and all the plants. Learn water.
>
> For there must be good men and women in the mountains, on the beaches, in all the neglected and beautiful places, so that one day we come back to ghostly cities and try to set them right. (HWP, 21)

In the end though, the trip was canceled. Kerouac later wrote that he had been unable to make the journey on account of his desire to escape the eternal cycle of mail and requests he was receiving at the time and instead go to Mexico. He wrote of trying to explain this to Welch, but that Welch had been so enthusiastic in his correspondence about the trip that Kerouac had been unsuccessful. As it turned out, Welch would have been unable to make the trip anyway, as he had been recently diagnosed with cirrhosis of the liver—a haunting and ironic echo of what had eventually led to the death of his father years earlier and would later be the cause of Kerouac's death in 1969.

Initially describing his condition to Donald Allen as a result of "10 years of too much drink, wrong food, and bad women (either heartless or stupid or hot")" (IR1, 188), Welch later explained to Kerouac that on the trip to San Francisco to handset *Wobbly Rock* with Haselwood, he would "drink about 2 scotches and pass out cold, or suddenly flip out into really wild rages or crybaby scenes. . . . But mainly I was really off my nut—mad, but not in a good way—deranged. Black despair" (IR1, 189).

He wrote of being half-loaded for the best part of the previous decade and that he had supposed the best way for him to approach the "dark, predatory world" was to do so drunk. He was eminently capable of understanding (and describing) the results and effects of such episodes but apparently incapable of doing anything about them. Indeed, all his plans to lay off the drink invariably ended in failure, as his own and other influences inevitably weakened any resolve he had. Even when faced with the debilitating effects of cirrhosis, Welch made light of it by blaming it on his inability to attract sexual partners ("not getting proper love from

America's Frigid Girls" [IR1, 201]). Writing to Snyder in Japan, Welch describes the healing process of his (temporary) abstention:

> No booze at all, lots of sleep, and food 5 times a day. I look better than I have in years. Clear eyes which astonish Albert. But can't stay up more than 4 or 5 hours without a nap. Fatigue something like mononucleosis: general and deep.[9] Wake from drug-like sleep with shaky arms and fast heart. Hungry all the time. (IR1, 192)

Welch's intentions are again clear from this letter. His physical health was now suffering too, and his return to Dorothy was a chance for reha-bilitation away from the distractions and temptations of San Francisco. He was determined to focus entirely on his writing while coming to terms with what he called the "limitations of Leo"—namely, his constantly get-ting fired from work, worrying people, getting hung up on money and women, and doing everything with "forced flair."

Another of these "limitations" was Welch's self-confessed lack of sex-ual prowess. While it is easy to attribute issues of impotence to his growing dependence on alcohol, he had always been conscious of his own sexual inadequacies. In a letter to Dorothy from July 1950, he wrote that "I will no doubt marry someone who has been with me a year or so and gets along well with me in bed. It is the only vulnerable spot I have."[10] And in descriptions originally written for inclusion in "Hard Start" a decade later but eventually omitted from what became I, Leo, Welch not only candidly describes his self-consciousness regarding the size of his sexual organs but also his frustration at not having been able to perform sexually when he was younger. In addition to calling this a "lonely terror," Welch also wrote that "he was almost always impotent. Making love was a very difficult procedure"[11] and how, in searching for a solution to the "sick virility myth," Welch found answers in learning to love himself first.

The abovementioned piece begins by describing Leo Keeler's rela-tionship with his girlfriend Barbara and how he struggles sexually. He writes of his "frightened libido" and how the only thing standing in his way was himself and his apparent inability to love. Only when he faces up to this fact and tries to tackle the issue of impotence may he be able to accept it and move on. In doing so, Welch draws from William Car-los Williams and Catullus as sources of inspiration, citing Williams as a "truth-teacher" and Catullus as dealing with impotence through humor and absurdity:

> One of the truth-teachers was William Carlos Williams. He had a poem which went "Who has not had Venus herself before him, and he cold as stone?"
>
> Another was Catullus whose very witty conversations with his dick put laughter and absurdity to work on Leo's lonely terror. "Look at you this morning, like a proud spear! Where were you last night? For months I have besieged her! And finally, last night, she was mine. And you! Limp as a worm!"[12]

In seeking a solution, Leo turns to Barbara, who also has certain sexual tendencies that may limit Keeler, yet it is through playing these "games" and understanding the "body-hate" relationship that "Leo learned how to love his body." This line clearly infers, "First you must love your body, in games"—where the "games" prompt awareness and acceptance of the self.[13]

In another essay for mid-1960, Welch vents his further frustration at what he calls "primate-constant drives" and even goes so far as to blame God for the "distressing and unmanageable" heights that can be reached through the desire for sex. In "A Farewell Note to God, Should He Exist," Welch contends that sexuality in humans is an impossible and inefficient aspect when combined with the mind and that sexual drive is actually detrimental to the planet and "only allows for a steady derangement" (IR1, 214).

Whether Welch experienced this "derangement" in real life is impossible to say, but in life and in fiction, Welch was unable to father children and, according to his stepson Jeff Cregg, he had long been troubled by this, suggesting that it may have been one of the reasons Welch was such a talented poet: "He [Lew] wasn't real good at anything else, he couldn't get married, he couldn't have kids . . . never bought a house, never put any money away for retirement."[14]

Yet despite everything, he had learned how to love. It may have taken psychoanalysis, understanding partners, and a serious amount of effort, but by the time he met Lenore Kandel later that year, he was better placed as a lover and companion that he had been in years.

———

Dorothy's work schedule was busy and as a result her house was empty much of the time, meaning that Welch was able to work on "Hard Start" relatively unimpeded. By April, he was well into what he called his "huge novel—it has gotten away from me at last: my only job to watch it grow out of this typewriter" (IR1, 192) and had decided to apply to the Eugene F. Saxton Memorial Trust for a literary grant. The Saxton Trust had been set up in memory of Eugene Saxton, who had been literary editor first at Doubleday and then Harper & Brothers, and who had been (partly) responsible for publishing not only such authors as Joseph Conrad and Thornton Wilder but also one of Welch's wartime favorites, Betty Smith. Saxton was similarly credited with providing much-needed encouragement to largely unrecognized writers who lacked the necessary finances to further their careers, which is essentially the situation that Welch found himself in at that time. The subsequent trust was established to "encourage distinguished writing in the fields of fiction, poetry, biography, history, and the essay, as well as outstanding work of reporting, needed popularizations of knowledge, and original interpretations of cultural trends."[15]

Given these criteria, Welch's optimism was high as he set about preparing his application. He planned to use the grant to work on his novel for a few months and then "go to Japan and get my head shaved inside and out" (IR1, 202). The question, however, was would the trust agree that his need was great enough to justify a grant.

———

Welch's application letter to the Saxton Trust is both concise and honest. He makes no bones about his desire to use the money to partly fund a trip to Japan, writing that his "Zen has gone as far as it can, here, now" and that the novel was, in essence, forming itself and in so doing determining the course of Welch's future.[16] He links his desire to deepen his knowledge of Japanese language and culture to the ability of a novel to "write" its author: "This may sound very strange to you, but this novel has been writing me for the past 3 or 4 years. If the novel *must go* a certain way, I know that I must. And the novel must send me to Japan."[17] This desire to leave America was one he had thought of in earnest for a number of years, and Japan was one of the alternatives he envisaged for what was a key pretext of the novel, namely to show "the predicaments and solutions of most of my generation."[18] Acknowledging Kerouac in particular, Welch writes that, although this has been done before, no one has as yet attempted to offer alternatives to the status quo:

> "Alternatives" is the key word to my book. I have had a rare and rich opportunity to observe alternatives. There are many, but America seems to believe there are very few. It distresses me to see the meager little range most of us allow ourselves, and if I write this book truly perhaps a lot of frightened people will give at least a small try. . . . The "Beat" alternative is but one alternative. Madness is an alternative, which, I observed, is actually chosen, sought, by some of us. Being rich and believing in the old American dreams is an alternative. Being able, cynically, to manipulate our great wasteful wealth is another . . . and if you have done enough of them . . . it places you in a position of choice.[19]

Welch openly challenges the committee to see (and even acknowledge) the limitations of their own lives. They, like millions of others in the United States, were part of the very same predicament he himself had been in only a few years earlier, namely the "American Homemaking Bit." The authority with which Welch writes of this is both impressive and slightly patronizing, and one might wonder whether the committee saw his outline as a personal slight on their own life choices rather than as a serious work of literature with similar social value.

It is clear to see the influence of "Beat San Francisco" in the application, with even specific references purposefully placed in order to tap into what he knew was still, for the moment at least, a hot topic. Welch broaches the subject of the East-West House, his cross-country trip with Kerouac and Saijo, and the influence of Zen on his current sensibilities. There is real self-belief in what he wants to say to America and a clarity of purpose in how that should be brought across. But, again, would the Saxton Trust Committee also see and support that?

11
Hard Start

Narrow it down from a whole continent now and focus on First Street in Portland, Oregon. Down the street comes Leo Keeler with a sea-bag on his shoulder, red hair still cut short, a $15 dollar Comoy pipe (his latest kick: great pipes), and dark excited eyes (hunting or hunted you cannot tell) taking everything in.

—Lew Welch, "Hard Start," unpublished draft

The successful publication of *Wobbly Rock*, coupled with the universally warm reception that it received, contrasted with much of what Welch was otherwise doing in Reno during that time. Trying to remain sober, being in a near-constant introspective wrestling match that amounted to continual attempts at reinvention, and writing reams of prose seemed to offer him only partial respite. The level of self-criticism that he felt now provided him with an opportunity to escape into a new alter ego. To Welch, writing on long scrolls resulted in the kind of spontaneous prose that had been so successful for Kerouac and that might in turn lend itself to the better distinction of Lew from Leo, given that Welch was finding it difficult to reconcile his real self with the one emerging on the page. With the self and the other so obviously indistinguishable, as in Welch's case, it must have been challenging for him to separate them. Welch wrote of being sick of "Lew Welch. . . . He is weak, romantic, over-sensitive to others, afraid, wordy, too thin, too proud (really believes he is a Prince), moody, ashamed, unemployable, and vain." In facing himself, Welch invented "Leo" and spoke of a conscious need to use him as a vehicle for his storytelling, but that "Leo" would slowly drop out of the narrative, leaving it to become "little shots of everybody. Lots of places. Lots of people" (IR1, 182). However, in reality, when "Hard Start" finally morphed into what would eventually be published as *I, Leo*, Welch's alter-ego remained a central figure throughout—even though, as Gary Snyder later suggested, the image of "Leo" was one that Welch was never entirely happy with.[1] Apparently it was a manifestation of his personality that he never quite managed to reconcile himself to despite it being ever present in his work.

I, Leo was intended to be a sweeping autobiographical commentary on the life of Leo Keeler in contemporary America. It was to trace Leo's school years, college life, loves and losses, successes and failures, and myriad relationships. It was to be Welch's attempt to make sense of his own life by depicting that of his nearest other. In his Saxton application he had written that the book had the following general pattern:

> Book One, School
> Book Two, Travel
> Book Three, Home[2]

And, drafting his plans for the book's outline in his notebooks, Welch further divided part one—titled "Boll Weevil in Square"—into nine chapters that included "Leo's Lay (Robin)," "Pave the World (Old Men, Danielsen)," "Anna's Dead (Philip & Charlie)," "A Very Odd Night & a Long Goodbye," and "Flight (Seattle, Portland, SF, Banana boats, girls)."[3]

However, even before he had established any sense of identity as a prose writer, he was again undermining his work and seeking affirmation of its worth from others. He wrote to Kerouac, "I don't know what I'm writing about—just go along scene by scene" and "I keep wondering why anybody would be interested in Leo" (TT, 66). He even asked his friend if he ever suffered from a lack of confidence—surely an indication of his own. Coupled with the onset of cirrhosis and his probable feelings of cold turkey, these must have been difficult times for him.

The choice of chapter titles is interesting in itself, especially chapter 1, with its reference to Robin Collins, Welch's Reed classmate and one-time girlfriend of Gary Snyder. Between the book being drafted and its eventual posthumous publication by Donald Allen, the chapter titles were dropped. The reasons for this are likely manifold, but Welch's desire to include details of his/Leo's relationship with "Barbara Small" at all is one that raises questions. With many relationships to write about, why did he choose to use this short-lived affair with his friend's former girlfriend? Was her presence in his life at Reed so important that it warranted inclusion in his autobiography? Collins is first mentioned in unpublished letters from early 1950, and although they had been dating for several months, Welch's contention that "Robin and I are dreadfully mis-matched, but her I need, whoever she is at that time, and I need her as a continuing thing" says much about her importance to him at that moment.[4] Indeed, the frontispiece of *I, Leo* features a line drawing of Collins that Ed Danielsen had done at Reed, thus adding more weight to the suggestion that the relationship was an important one to Welch.

For his part, Snyder wrote long and very emotionally charged passages about his feelings for Collins in his journals from that time, and in a small collection of poems years later (*Four Poems for Robin*) calling her variously "petty, adolescent & foul-tempered," as well as declaring his undying love for her, even after the pair had split up and she had embarked upon a relationship with Welch.[5] In a journal entry that may be as revealing as anything he ever wrote about the relationship between Collins and Welch, Snyder declared that they were "beautiful people . . . tall and slender and laughing, and so fucked up all the way along. Beautiful people, strong and young, and dying fast. The life, the youth of them, is being swept away and they are suffering painless extinction—each in their own way."[6]

Snyder had also previously written that he did not feel "inferior"[7] to either Collins or Welch but that his love for her had awakened a profound sense of himself and that a necessary humility was the result—a revelation that Welch failed to achieve, given not only his reaction to her eventual (and inevitable) departure but also in terms of Snyder's description of how Welch had seemingly bragged to his friend about his lovemaking with Collins. This was a situation that unsurprisingly left Snyder feeling somewhat hurt that the one he so desperately loves "belongs to another now."[8] It is a remarkable reflection of their friendship that in spite of everything Snyder and Welch remained friends after this. However, this may reflect more on Snyder's ability to forgive and/or forget than on any sense of guilt or remorse that Welch may have felt or shown afterward. Indeed, why did Welch not include Snyder in *I, Leo*? Was it down to these feelings of guilt regarding Collins? If so, wouldn't Snyder's obvious omission only serve to accentuate this and remind both men of what had happened? By this time, of course, Snyder was married and living in Japan, so arguably he had moved on and this was more a reflection of how Welch felt about it all. Either way, there seems something underhanded about the descriptions Welch writes of Collins at Reed, without explicit mention of her former boyfriend. Indeed, prior to the publication of *I, Leo*, Donald Allen asked Philip Whalen to deny or confirm the various characters and their true-life counterparts. Regarding Collins, Whalen wrote that "Barbara Small is Robin Collins (At Danielsen's house in Boulder I recently saw that drawing for which Robin modeled & which bears a Chinese inscription in Charlie's calligraphy)." Regarding Snyder, Whalen simply wrote "No, I don't see Gary anyplace, either."[9]

In the end, what finally emerged from the detritus of Welch's time in Reno is a short carefully written prose piece that Albert Saijo described in his review for the University of Nebraska's *Western American Literature* as "one man looking back and saying, let me describe a beginning, let me tell you how it was."[10] Kerouac's doctrine of "spontaneous prose" is clear, if only in some of the most arbitrary methods. For example, Welch adhered strictly, on the one hand, to the idea that Kerouac named "Center of Interest," namely that the writer should "begin not from a preconceived idea of what to say about image but from center jewel of interest in subject of image at time of writing."[11] However, the same criterion advocates, on the other hand, that the writer refrain from returning to the piece afterward in an attempt to improve it—something that Welch did with an almost obsessive compulsion. That his writing was "the most painful personal wrung-out tossed from cradle warm protective mind-tap"[12] from himself is undoubtedly true, but his inability (or unwillingness) to leave them as they were meant that his confessionals ended up crafted rather than entirely spontaneous. Indeed, Saijo likened *I, Leo* more to the prose style of James Joyce—an analogy that is entirely understandable, given Welch's lifelong focus on language and its accuracy.

In the aforementioned letter to Allen, Whalen commended certain sections while deploring what he called the "stuffiness" of others. Yet it is Welch himself who best sums up his own manuscript when he later wrote,

> That novel last spring really has only that value, a personal one, it got a lot of shit out of my head and it gave me a lot of practice at how you can come at things, or out up from things, into words. . . . I don't care about the autobiographical shot enough to keep pounding it into shape. (IR2, 20)

Welch felt that the "I" should be omitted from his stories altogether—a fact that is witnessed in ones he subsequently wrote, including "The Man Who Played Himself," which was published in *Evergreen Review*,[13] and various other prose works such as "Late Urban Love of Peter Held" and "Little Men on Bicycles"—both of which singularly focus on characters that may, in part, resemble Welch but which only employ the *I* persona in its capacity as a minor character. Indeed, this decision to omit himself is one that is developed more fully in his poetry from the Forks of Salmon, where Welch replaces himself with a *he* persona at a point coincidental with an awakening. Although less profound at this stage, the act of distancing himself from the stories may have allowed Welch more freedom to try to achieve his goal of getting his work "to hang just exactly where it wants to hang somewhere on the spectrum between fantasy and reality" (IR2, 20).

Yet, this goal was one that he never seemed able to realize in his own mind. The "Peter Held" story was one he had worked on for almost ten years, rewriting it more than half a dozen times and never quite arriving at a satisfactory conclusion. Nonetheless he intended to submit it to various magazines in the hope of getting paid. Perhaps he felt that readers would be less critical about the spectrum of fantasy and reality than he was himself.

By the summer of 1960, Welch was becoming increasingly restless in Reno, which he had described as a terrible place. He complained of boredom and of being trapped there, and with the recent news that his application for a Saxton grant had been rejected, his need for a trip to the woods was more acute than ever. The bout of cirrhosis seemed to have had a positive effect on his ability to refrain from drinking, and having decided to regain a certain level of fitness, Welch regularly extracted himself from the maternal abode by embarking on as many hiking trips as he could. He wrote to Kirby Doyle, with whom he would soon develop an intimate and important friendship, of "healing. Health fine now. Am *very* hard from hiking 20 miles [a day] . . . with 50 lbs. on me back" (IR1, 217) and to Ginsberg of being "better physically and in spirit than I've been for many years. Lean as a jack rabbit. Have been hiking vast distances, alone, all over this huge, stark, and gentle West" (IR1, 222). His trips took him west into the Sierras and north into the ranges stretching from the Ruby Mountains to the Humboldt National Forest on the Nevada-Idaho border or over the Oregon border to the Rogue River.

Indeed, in June 1960, Welch embarked on a ten-day solo trip that would see him hike into the Idaho wilderness and up the Jarbidge River Gorge. Although he experienced "bitter thots" along the way, his notes on the hike are filled with detailed descriptions of hunting, fishing, and wildlife. He slept in caves and disused mining huts. His notes end with a sketch that underpins his own sense of insignificance: "It's a long long plotless movie. A man in a gorge" (IR2, 4). The image depicts Welch as a solitary figure in an expanse of country so massive as to render him minute and aimless. In a letter to Allen Ginsberg upon his return from Jarbidge, Welch had written of having a particular need for "absolute solitude," the kind that only such expanses could offer him. This "almost pathological" need was coupled with an admission that he had slowly lost the opportunities to talk, communicating almost entirely through his work—the letters, stories, and poems. His stay in Reno had provided him with a definite sense of comfort when it was most needed, but it also removed him from the intellectual milieu that fought for supremacy in his own mind. Living with Dorothy was taking its toll, and he was acutely aware of having to continually juggle the dual existences in his life. Although such moments of solitude had always been part of his life, this was the period in which thoughts of the later, more serious, retreats were first harbored. In the same letter to Ginsberg, he wrote,

> I seem to be entering the years of hermitage, and I welcome them—almost long for it. I, who have always been terrified of loneliness and who once nearly lost the game escaping it. Who could have foreseen this?
>
> My time is all forespent (Williams, "The Wanderer"). What a relief. (IR1, 222)

Not only does he realign himself with his former mentor through the connection with Williams's notion of the self-realization and clarity described in "The Wanderer," but also he introduces the mode of escape on many future occasions throughout the rest of his life, namely by employing the guise of the hermit or the wanderer. Despite these figures being something of trope in poetry, they nonetheless embody the loneliness Welch was embracing and that would, in some way, help him to overcome some of the many demons that had plagued him and were yet to plague him still.

His preparations for these hiking trips were long and meticulous. Often announcing his plans to others months in advance, Welch drew up an exhaustive inventory of equipment and supplies that he would need to take with him. Yet this keen preparation is somehow tempered by the fact that his writing is often conspicuous in its sense of his suffering during the trips and the apparent willingness on his part to endure it. The almost penitential undertone in the writing foreshadows much of what would occur when he finally removed himself from society into Siskiyou County for an extended period in 1962.

Upon his return from Jarbidge, Welch decided to move back to San Francisco. He had spent the better part of nine months in Reno, and his need to get back to the coast and the literary happenings there was clear. While he had written a huge amount at Dorothy's, there was still little to show for it in terms of published pieces. *Wobbly Rock* was selling slowly (and would be out of print within the next year), and attempts to get his work into literary magazines had stalled. Welch needed an injection of inspiration to match the one he had had with Kerouac and Saijo the previous winter. So, packing his belongings, Welch headed back to the city and, once again, into a room at the Hyphen House. The events that followed soon after his return would provide him with inspiration in abundance and an eternal place in the pantheon of literary Beat figures, as the character of Dave Wain in Jack Kerouac's novel *Big Sur*.

12
Big Sur

> Dave Wain that lean rangy red head Welchman with his penchant for
> going off in Willie to fish in the Rogue River up in Oregon where he
> knows an abandoned mining cabin, or for blattin around the desert
> roads, for suddenly reappearing in town to get drunk, and a marvelous
> poet himself.
>
> —Jack Kerouac, *Big Sur*

Welch had not seen Kerouac since their cross-country road trip the previous winter, and in the ensuing period both had undergone profound
changes. For his part, Kerouac was now in the midst of an existential crisis
that had begun long before his newfound fame but that seemed only to
have intensified thereafter. An already hardened drinker, Kerouac was
consuming ever more alcohol as an antidote to the tsunami of attention,
requests, and expectations that were being heaped upon him by what
he saw as a small army of "visitors, reporters, snoopers."[1] He was now a
nationwide celebrity in constant demand—be it for interviews, advice,
or, on account of his sudden (and relative) wealth, financial help. In a
letter to Allen Ginsberg from June 1960, he wrote first of his need to
escape and be on his own for a while ("What I really must do is get off
by myself for the first time since *On the Road* 1957 so I'll take a secret
trip this summer and live alone in a room and walk and light candles . . .
have to have a holiday to rediscover my heart")[2] before enumerating a list
of grievances, implications, and requests he had received from his "innu-
merable friendships," which he subsequently refers to as a "battalion of
money-askers"—including Welch:

> Do you realize what happened just this last week for instance and as
> example: Jack Micheline writes big nutty letter all dabbed with tears
> from Chicago finally asking for ten dollars—Gregory writes "Come

to Venice at once! Money is my friend!" (when I'd told him I might be sent there by Holiday on assignment)—Charley Mills and Graham Cournoyer call me insistently from the Village for money and I have phone number changed—My sister wants to borrow a thousand for her house—(ahead of even the house)—Lew Welch hints he needs a hundred for his jeep. You invite me to fly off to Peru, Gary to Japan, Ansen to Greece, Montgomery to Mill Valley to go live with him mind you, old Horace Mann school chums to reunion, art galleries to their art shows to buy stuff, etc etc.[3]

In the end, Kerouac went back to San Francisco, which was surely anything but the preferred destination to indulge in the "secret trip" he had intended. On the recommendation of Lawrence Ferlinghetti, Kerouac had decamped to the West Coast to retreat to Ferlinghetti's cabin at Bixby Canyon. Ferlinghetti offered the cabin to the troubled author on account of its seclusion, in the hope that the cabin might afford Kerouac some respite from his growing number of nonliterary distractions and allow him to concentrate on writing. Since completing *The Dharma Bums* in late 1957, Kerouac had been unable to finish a single novel; although he had been prolific in his other writings, his increased alcohol and drug use was now becoming something of an issue. The cabin's location under the Bixby Creek Bridge on the Pacific Coast Highway was perfect as a refuge from city life, and Ferlinghetti's many visits there had afforded him just the kind of solitude that he hoped would now benefit Kerouac. However, arriving at Bixby Canyon Bridge in late July 1960 and making his way down the dangerously steep slopes to the creek and the cabin beyond, Kerouac was almost immediately haunted by alarming and sinister associations that would set the tone for the rest of his stay. In her biography of Kerouac, Anne Charters describes the paranoia that overcame him in those first few days at the cabin, and the fears and anxieties that he suffered while there. Yet despite this, Kerouac still forced himself to continue with the experimental writing style he had developed years before in *Mexico City Blues*, only this time, rather than scrawling Bill Garver's morphine-fueled monologues in a Mexico City hotel room, Kerouac sat huddled under the cliff wall and, night after night, despite his fear of the blackness around him, wrote down the sounds of the Pacific in his notebook as its white-rimmed breakers crashed, swirled, and echoed along the shore's edge and into the many caves worn into the seawall.

Yet however committed Kerouac was to writing, and to his need for solitude, two weeks at Bixby was enough, and he once again began to long for the companionship of his friends in San Francisco. Ferlinghetti had agreed to pick him up, but unable to wait the extra few days, Kerouac decided to hitchhike back to the city via Monterey instead. After a long and wearisome journey on which he was ignored by day trippers and families alike (apparently his fame did not extend to hitchhiking), he checked into a skid-row hotel, reconnected with many of his drinking buddies, and embarked on what would become one of the most infamous series of events in Kerouac's Beat Generation folklore.

Aware of his friend's vulnerability, Ferlinghetti advised Kerouac to go back east to the quiet and solitude of Mémère's. However, ignoring this advice, in mid-August 1960, Kerouac descended instead into the bars on North Beach with, among others, Philip Whalen and Welch. The latter pair were rooming together again in the Hyphen House, and although a willing companion up to a point, Whalen, who was not a heavy drinker, would quietly leave Kerouac and Welch to it when it started to go beyond his scope. Nonjudgmental and accepting of both his friends' drinking habits, Whalen "preferred to stay at home and read while Jack and Lew got drunk."[4] As time passed, Welch again became not only Kerouac's drinking partner in San Francisco but also his de facto chauffeur as the pair drove from bar to bar, drinking copious amounts of bourbon and ginger ale along the way.

While it is possible that this companionship was based more on mutual alcoholic prowess than on literary kinship, the two had realized they formed a certain kind of kindred spirit during their time together the previous winter. As Charters describes, Kerouac and Welch were similar in several ways, not least the fact that they were both "voluble, emotionally intense men" who "were fleeing some of the same phantoms"—namely, mothers, wives, and sexual sterility.[5]

During the course of the week that followed Kerouac's arrival in North Beach, Welch and Kerouac decided to pay a visit to Neal Cassady, who was now living in Los Gatos after his recent release from San Quentin after two years on drug possession charges. Kerouac had not seen Cassady for the best part of three years, and had neither visited nor written to him during Cassady's two-year incarceration—a fact not lost on Cassady. Indeed, during his last visit to the West Coast the previous winter, Kerouac had been asked by Cassady to give a talk to inmates during a class on comparative religion given by Gavin Arthur, but for one reason or another Kerouac was so fearful of it that he spent the previous night drinking heavily with Welch instead. Kerouac described the aftermath: "On the big morning I wake up instead dead drunk on the floor, it's already noon and it's too late, Dave Wain is on the floor also, Willie's parked outside to take us to Quentin for the lecture but it's too late."[6]

Worried that Cassady would resent his old friend for his apparent neglect, Kerouac was reluctant to go this time around, but after two bottles of scotch and Welch's nonstop chattering en route, he seemed to have worked up the necessary bravado to face his old friend and inspiration. As it turned out, his hunch was partly true. In her memoir of life with Neal, *Off the Road*, Carolyn Cassady writes that while "Neal showed no particular interest" in seeing Kerouac, he was concerned about the Kerouac's increased drinking.[7] Cassady had seen enough binges through the years to know that the outcome would be anything but good, and he foresaw a bleak future for Kerouac even then. However, when the entourage showed up, Carolyn gave them the warmest of welcomes, despite Kerouac's initially gruff and offhand behavior; the two hangers-on (Paul Smith, who was Welch's seventeen-year-old musician friend from Reno, and an unknown roustabout from Barnum & Bailey's circus) who had piled into Willy were now about to come face-to-face with the fabled "Dean Moriarty" (as Kerouac had characterized Cassady in *On the Road*)—indeed, this was most likely the primary reason for them being there in the first place.

In her memoir, Cassady describes the evening as it played out, painting a lively tableau in which a drunken Kerouac can do nothing other than express "unintelligible roars or grumbles" in his attempts to keep up with the fast-talking Welch.[8] In turn, Welch is described as being "a great delight with intelligent, erudite patter; often the essence of satire. I was impressed by his sharp insights, quick wit and poetic imagery."[9] Neal was still at work on a partial night shift at the Los Gatos Tire Company. When Carolyn called him with the news that Kerouac and the others were in town, Neal, knowing his boss would leave the shop around midnight, suggested the group drive there afterward, giving Kerouac the opportunity to introduce Neal to Welch—something that, given his desire to share people he liked with Neal, Kerouac was happy to do. Returning to Cassady's home after his shift was over, the group spent a long night drinking, woke up the next morning, and headed back to San Francisco—but only after asking Carolyn to accompany them to the nearest liquor store to buy some wine for the journey: a sure sign that Kerouac had no intention of slowing down, despite the realization the next morning that he was on the verge of sliding into an alcoholic depression.

Back in San Francisco, waking in a drunken haze in Whalen's room the next day, Kerouac was apparently well into his own private lost weekend as his drinking rapidly spiraled out of control. As the weekend picked up pace, rather than seeking refuge, and before he knew it (or could do anything about it), Kerouac was being driven by Welch in one of two Jeeps heading south again to Bixby Canyon. In tow were, among others, Ferlinghetti, Bay Area artist Victor Wong, Neal and Carolyn Cassady, and Welch's new girlfriend, Lenore Kandel. Michael and Joanna McClure and their daughter Jane would also join them for the trip. The events of the weekend and the haunting effect they had on Kerouac are recorded in *Big Sur*, which he wrote in Florida the following summer, but for Welch the weekend represented a coming together of some of the country's greatest contemporary poets and literary figures, and he must have felt like he was becoming an integral part of a poetry revolution despite still being largely unpublished. In a letter to Dorothy, he wrote of the weekend that "at one time there were 6 major American poets on the beach together thinking 'one great seawave could almost destroy licherchur [literature].'"[10]

The long and the short of the events that followed in the next few days include Kerouac's descent into an almost complete paranoid breakdown, in which he suffered from serious delirium tremens and saw the vision of the "archangel" appear before him in Ferlinghetti's cabin. In a letter to Carolyn Cassady in 1961, he wrote, "I have gone mad for the first time, while Jacky slept, her kid slept, Lew and Lenore slept, but I had nightmares the like of which I can only barely describe. . . . It was the night of the end of Nirvana."[11]

In addition to these visions, he also began to doubt the nature of his friendships and had delusions in which he imagined that Welch, who he saw carrying a box of needles to administer amphetamines, was out to harm him in some way. The complexities of his relationship with the

Cassadys and his increasing nightmares and hallucinations only served to aggravate things still further, and the weekend ended with him having a holy vision of the cross: "I see the Cross again, this time smaller and far away but just as clear and I say through all the noise of the voices, 'I'm with you, Jesus, for always, thank you.'"[12]

Leaving Bixby immediately thereafter, Kerouac returned hastily to Northport and Mémêre. It would be the last time that Carolyn Cassady would ever see him (or Welch) again.

13
My Love the Fisherman Comes Back

The events of Big Sur must only have served to embolden the already upbeat Welch. Returning to San Francisco, he spent the remainder of 1960 writing, traveling, worrying about his finances, trying to hold down various jobs, and drinking.

When Welch first met Lenore Kandel in August 1960, she was a twenty-eight-year-old fledgling poet who had only recently moved to San Francisco from Los Angeles. On his initial return to the city from Reno in early August, Welch had moved back into the Hyphen House, and it was there that the two met. It seems that their affair provided Welch with a relationship that not only satisfied his desire to engage in something that was beneficial for him on a psychological (and sexual) level but that also was relatively undemanding. He describes Kandel as "a great generous girl who is a sexual genius" (IR2, 13), while stressing to Dorothy that he was too busy for anything particularly serious. Indeed, in the same letter to his mother in which Welch had written about his visit to Bixby Canyon, he also wrote, "I'm slowly getting over the embitterdness I was gradually feeling about women—a great relief" (IR2, 7).

When the pair met, Kandel had already had a number of chapbooks published by Three Penny Press in Los Angeles, and it would not take her long to establish herself among the poets and artists in her new home. Although her greatest achievement, *The Love Book*, was another five years in the making, the early work she produced in North Beach already reflected the newfound influences around her. Indeed, Ronna Johnson suggests that Kandel's poetry bore the marks of both Ginsberg's "contempt of conformity and anti-authoritarian contentiousness" and the "nihilism" sometimes found in Welch's work.[1]

It appears that, in the early stages of their relationship, Welch moved back and forth between San Francisco and Reno, splitting his time and energy between working for necessity and working for pleasure. In this same period, Kandel's poems began to appear more frequently in little magazines and journals, and she steadily began to find a voice as her work employed denser and less redundant use of language alongside more unconventional imagery than the freer coffeehouse vignettes she had written in LA. Her poems during this transitory period often contain what Donna Nance called "a sense of fragile hope combined with anticipated loss"—aspects that can also be found in Welch's work at that time.[2] Yet while Welch's influence may not be overtly evident in her work, as Kandel moved toward a mix of the more unashamedly direct yet often mystical language of her later poems, his sense of concision and authenticity can be traced. Indeed, in an interview in 1979, Kandel said that Welch had helped her to get rid of "the lead-in that I tended to use, a sort of shuffle. He said to get right in. And I got a little more ruthless with the blue pencil."[3] And although Welch had initially balked at the idea of Kandel associating herself with the predominantly male poets in North Beach, he finally conceded that she was as good a poet as he was.

It was also during this period that Welch took a trip back to Reed College, with the poet Jerry Heiserman, to give a reading there. This would be the first time he had been back to Portland since leaving for New York a decade earlier, and he returned with a sense of gratitude to the city that had occupied such an important place in his life. Indeed, in many ways it had made him the poet he had become. The trip not only afforded Welch the opportunity to reacquaint himself with many of his former friends, including Lloyd Reynolds, Ed Durham, and Carol and Bill Baker, but also gave him a chance to make new ones.

Shortly after their arrival, Heiserman introduced Welch to a young artist who was then in his junior year at Reed, Robert Ross. A friend of Heiserman, Ross had studied at Reed during his freshman and sophomore years before spending a year at Cooper Union School of Art in New York. Already familiar with Reed, Ross was a regular "in Lloyd Reynold's calligraphy and graphics shop in the topmost garret of the Administration Building" and "had access to [Reynolds's] old hand-fed Chandler & Price platen press, two full cabinets of type plus typesetting paraphernalia."[4] After meeting Welch, the two became firm friends and, in preparation for the reading—which was held on November 2—Welch asked Ross if he could design an advertisement featuring Manjusri (the Bodhisattva of Wisdom) riding his lion, saber in hand, atop a lotus leaf.[5] The resulting

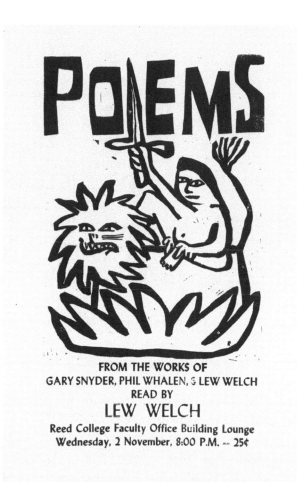

FROM THE WORKS OF
GARY SNYDER, PHIL WHALEN, & LEW WELCH
READ BY
LEW WELCH
Reed College Faculty Office Building Lounge
Wednesday, 2 November, 8:00 P.M. — 25¢

Figure 13.1: A linocut
poster advertising
Welch's poetry reading
at Reed in late 1960.
Courtesy of Robert Ross.

poster, printed using a linoleum block and hand-set type on the C&P press, announces that, in addition to his own work, Welch would also read poems by Snyder and Whalen. Recalling the event some years later, Welch remembered it as a fairly intimate affair at which he managed to sell only a small number of copies of *Wobbly Rock* and that this was yet another illustration of the fact that making a living as a poet was nearly impossible.

Welch and Heiserman stayed in Portland for a number of weeks, having rented rooms in an old Victorian building where Welch's attic room was sparse and in which he wrote "strange and philosophical" poems that seemed to reflect the current state of his creative mind. Indeed, three weeks after the poetry reading, Welch, Heiserman, and Ross organized

Figure 13.2: A flyer for the Circus Poeticus Magnetiquo. Courtesy of Robert Ross.

the Circus Poeticus Magnetiquo—arguably the highlight of their stay and what Ross later called a "one-night-only extravaganza."[6] Welch later described it to Kerouac as a "huge circus . . . DaDa outrage" with "naked girls and clowns and lion make-up" (IR2, 23).

The event was held at the Thirteenth Avenue Gallery, an exhibition space opened in 1959 by Paul Hebb where "art students could exhibit" and activists "could make a social statement."[7] The gallery was a place where students, beatniks, young families, musicians, and other assorted people could congregate to tap into what was something of a novelty in Portland at that time. The gallery hosted exhibitions and marches, raised money for local organizations, and provided a place for the exchange and bartering of secondhand furniture and other items. As such, it was the perfect location for the bizarre collaboration that saw several of Portland's underground figures take on various roles in the circus. Welch and the others employed the services of Reed College students as well as local musicians and artists to create the event, and they worked for hours in Reynolds's shop to create the poster that was eventually plastered around the city to advertise the event.

Such was the nature of the circus that everyone involved took on the moniker of a character that made it impossible to know who was appearing unless you were in attendance. Welch was (predictably, in retrospect) "Leo the Magick Lion," Heiserman was "Clown Luna," and Ross played the "Mysterious Wild Ether Oracle." Other participants included Mel Lyman, the banjoist and later self-proclaimed prophet, as "a faggot superman"; Paul Siegel, a Reed student and friend of Ross as the circus ringleader "Dr. Roony"; and a female acquaintance as the "giant fern goddess." Music was provided by notable Portlanders such as Mike Russo Jr. (whose father, the renowned artist Mike Russo Sr., was also in attendance) and Ron Brentano (dressed as a mummy), plus an assortment of "Jass" bands.

Welch remembers the event as a "huge parade with gongs and cymbals and a dead loon hanging upside over the stage (just below a huge Nazi flag burned at the edges)" (IR2, 23) that made its way to the gallery in a procession of cars behind a flatbed truck on which a band played. As Hebb later wrote "ladies . . . wore Christmas tree decorations on their nipples—and nothing else."[8] If the initial reason for Welch going back to Portland had been the reading and the opportunity to leave his poetic mark there, it was surely the circus, and Welch's role therein, that ultimately left a bigger impression on the city.

———

Money, as always, was a key issue for Welch, and selling only a few copies of *Wobbly Rock* at Reed had hardly helped to alleviate his concern. Having recently sold a short story to *Evergreen Review*, he then complained that their payment rates were too low. Rather than see the publication as an opportunity, Welch characteristically found an excuse to whine about the financial shortcomings of such a success—for surely *Evergreen Review* constituted a success for a poet who was still largely unpublished. While Kerouac and company were apparently being paid huge sums by the likes of *Playboy* ("I understand they pay a grand"), Welch had to make do with "$5 a page" from *Evergreen Review* founder and editor Barney Rosset. Yet, despite the repetitiveness of his whining, it is not without substance. Welch was often merely surviving from week to week or month to month. He spoke of arranging "dinner engagements" and "hanging tough" as his financial situation continually hovered just above the breadline, and as such, his desire to increase his earnings is entirely understandable.

Yet, as had always been the case (and would continue to be so into the future), Dorothy's influence was also just out of sight; with almost every letter he wrote to her referring to a donation or financial gift that she had sent him. These payments ranged in value but were conspicuous by their regularity. While his feelings for his mother fluctuated (eventually descending into an apparently real and deep-seated hatred by the time he died), his dependence on her financial generosity certainly did not, and during the early 1960s he needed her more than ever. While Philip Whalen wrote that, contrary to his own perilous financial plight, "it appears that Lewis's mother has rescued him from the dangers of starvation," once again,[9] Snyder summed it up best years later when he suggested that it was precisely this financial dependence and comfortable middle-class background (the silver spoon) that made it difficult for Welch to be as productive as his friends and peers. While Snyder and others had to work to make ends meet and achieve their goals, Dorothy was always just one letter away, and it is easy to imagine that when Welch worked, he did so with this knowledge firmly in his mind. Indeed, this is manifest in the regular admissions in his own correspondence of an overall lack of drive—the acedia, the fluctuating output, the dreams, and the retreats.

However, in the same way that his correspondence with Dorothy waned during his marriage to Mary, so his relationship with Kandel seemed to inspire a similar distance between mother and son. It is striking that, when Welch was involved with anyone on an intimate level, so his reliance on Dorothy seemed to wane. In many ways, any serious female presence in his life seemed to constitute a kind of surrogacy—a replacement for Kali. Yet, as history tells us, in all cases such surrogacy was purely temporary, and his relationships inevitably broke down. Indeed, after his disappearance, Whalen, in trying to make sense of the situation, wrote in his journal that Welch had "a hard time accepting the love of any person. He kept falling in love, & enjoying that feeling, but I do not believe he often felt the love which other people expressed for him."[10] Indeed, of his girlfriends and the love he had for or accepted from them, Welch once wrote,

> I cannot understand, control, or relate to women. It just brings me down. I cannot enjoy them physically unless I'm flowers-up-the-ass in love, and then I don't enjoy them because it all gets unreal and I hate them for killing something in me they never wanted to kill, and they hate me for moping around. (IR1, 182)

———

As 1960 drew to a close, Welch was trying to hold down any one of the various jobs he found himself doing while simultaneously being as poetically productive as he could. With few imminent publications to speak of, he set about working on a follow-up to *Wobbly Rock*. His autobiography had stalled, short stories had been sent to various magazines with little success, and his application for the Saxton Grant had been denied almost as quickly as it had been submitted. The rejection by the Saxton Committee must have been a blow to his confidence at that time, but in a letter to Ruth Witt-Diamant he writes that perhaps it was something of a blessing:

> In a way (and this really isn't sour grapes) I'm just as glad—for I'm sure their presence in the affair would have, to some extent, influenced the writing. . . . I think it's better this way: a job nobody asked me to do—a job I am freer to complete on its original terms—where accuracy alone can be my guide.[11]

In reconciling himself to this rejection, he again turned his focus to language and how it should be captured. It was an ideology he had never wavered from, and that must have given him some sense of solace and justification that his methodology was true regardless of what others thought.

And so "Hard Start" was shelved, and Welch set about working on a new collection of poems tentatively titled "Earth Book." Although the collection never materialized in this form, the essence of what Welch wanted to include in the book is best summed up in one of the only two poems he published during that time. Written during his trip to Portland, the poem captures his increasing sensitivity to aspects of environmental decay and humanity's need to look to nature rather than technology for solutions.

Published in *Nomad* magazine, "Portland Oregon 1960" is a short poem in which Welch contrasts the dereliction of the city with the bounty of nature. It bemoans the fact that a decade after he first witnessed and wrote about the demolition of the Kamm Building to make way for expanding urbanization, the city and state authorities had not yet let up in their desire to remove old buildings and replace them with freeways. He uses a quote from Isaiah 40:4, "and the rough places shall be plain," to draw an ironic parallel between the eradication of obstacles standing in the way of the Jews on their exalted return from Babylon and the Oregon State authorities who were demolishing their own obstacles (such as the Victorian mansion in which he was staying) on their journey to modernization.

Yet far from being as incendiary as what he wrote about Chicago, this poem offers not only a sense of perspective but also a reminder that there is beauty in dereliction, and companionship. Echoing in some small way Snyder's "A Berry Feast," Welch writes, "Berries made fountains of fruit and thorn on / rubble, iron fence and gate, on / basement steps to basement now just / berry pit" (RB, 60). It further provides the first inkling of what Welch would go on to use as one of the key images in his writing about the battle for supremacy between man and nature, namely that of shoots and plants, and roots and tentacles cracking through the sidewalks and slowly deurbanizing the urban. In addition, Portland and other such cities are rendered insignificant in the face of such imagery:

> You looked so beautiful picking blackberries,
>   long black hair in the summer sun, black eyes
> glancing up to where I stood
>
>   In a landscape of ruined buildings,
>   on a small green rock
>
>   wheeling about the heavens. (RB, 60)

These final lines are clearly Welch's way of belittling the impact of humanity on nature and that, as the world turns, it does so in a universe far greater than anything that might be occurring on it. It is a philosophical ending in sharp contrast to the other major project Welch had worked on during the latter half of 1960, and which seemed to have caught and retained his attention, namely "Din Movie."

First mentioned in letters to Snyder in May 1960, this project would mutate over the next five years into a long poem accompanied by music called "Din Poem." It would encapsulate much of his views on the situation in the United States at that time and crystallize many of his ideas regarding not only urbanization but also religion, advertising, language, and society.

Welch's view of contemporary America had long been typified by the notion that everything that occurred there was in some way related to an overwhelming noise—like the violent soundtrack to modern life. Welch used various aspects of this noise to create a work that more resembles a performance piece than a conventional poem, including as it does songs, advertising slogans, dreams, religious mock presidential addresses, and quotes and lines lifted from other poems. It is a poem firmly rooted in the performative aspect of literature, and as such, should also be seen as a medium that helped Welch veer outside of an exclusively "beat" orbit.

In typical linguistic style, Welch opens the poem with a veiled nod in the direction not only of Walt Whitman—"Tizuvthee, I sing" clearly evoking Whitman's "I hear America singing"—but more directly to Samuel Francis Smith's 1831 hymn "My Country, 'Tis of Thee," which is widely recognized as one of the earliest and most patriotic songs written in praise of America. As Smith's celebratory stanzas—sung to the tune of "God Save the Queen"—list the myriad reasons that America is praiseworthy, so "Din Poem" immediately launches into a scathing attack on the very things Smith so lovingly lauds—America's people and the cities they call home.

In a fashion similar to "Chicago Poem," Welch initially employs a confrontational tone, writing about how America is "despised" and likening it to strontium, which was widely seen at the time as undesirable and dangerous because of its widespread use in nuclear testing. Welch then emphasizes America's isolationism, and its major cities are referred to as "intolerable up-tight dirty noise New York, rusty muscle Chicago, hopeless Cleveland Akron Visalia alcoholic San Francisco suicide" (RB, 101), before the tone alters dramatically to be more whimsical and jauntily melodic, with Welch satirizing America's greatness by employing advertising slogans in the form of a "supermarket song." Well-versed in the creation of such slogans, Welch combines his advertising and musical backgrounds to poke fun at the disproportionate rise of consumerism and the role played by corporations in promoting and selling their products.

Continuing through a series of stock phrases, religious messages, racist slang in the form of another song, and snippets of random conversations overheard on the street, Welch's primary image then centers on a vision he had while standing atop the Empire State Building in New York City. Following on from the violent imagery he used in "Chicago Poem," this section of "Din Poem" offers a grim prophecy of a city being consumed by itself and its inhabitants. From the viewing platform, Welch thinks he sees a small section of nature living and breathing under the skyscrapers; "a large green rectangle, Central Park, lies flat, clean-edged, indented" (RB, 107). He describes a piece of tape or a bandage being removed from the city to reveal this greenery, this purity. But in reality, those clean edges give away the park's imperfections—that it is a man-made imitation of how people think the planet might look having never been in nature to see and experience it for themselves. It is the parable of Emil Garber and the Lake Wisconsin deer all over again.

Welch observes the robotic patterns of everyday life, hears the "hum" of the city caused by "all our voices, and all our millions of lives, foolish and passionate and tiresome as the yapping of our million dogs (the last sound balloonists hear at 80,000 feet)," and stands amid the crowds joined by unspoken rules and rituals.[12] He is horrified by the intensity of the subway—a subterranean world that, despite the dirty air, nonetheless takes him away from the epicenter, away from the homogeneity of the people. Distant yet inevitably connected. He finds his escape in the simple screw-top brandy bottle in a sterile, white-walled apartment with the doors locked. It is almost a vision of self-incarceration in a room in a psychiatric institution—the only plausible antidote to urban life.

This "cure" was the one he most often turned to at moments of crisis. Any previous remedies, including the psychological ones he had been given by Joseph Kepecs in Chicago, were never enough to help him deal with the complexities of American life and his role within it. "Din Poem" reveals one such remedy—a "potion"—that was intended to "fix [him] up to be, actually *be*, a Native of a World!" and which would help Welch to understand and assimilate, to "*be* one of them. Think the way they do, see as they see" (RB, 108). It was clearly an attempt to bring Welch into line with conventional thinking and behavior, but at moments of clarity such as he had experienced on the Empire State Building, the effects of this potion flickered out and a sense of wakefulness enveloped him that created a clear vision of "this vast, insane assignment" of conformity, in which everyone is more or less drugged into participating. The "brassy" taste may imply mood-stabilizing drugs such as Thorazine, which was widely prescribed in the 1950s and 1960s to medicate manic-depression and other psychiatric disorders. Is this then a veiled reference to the bipolarity that Philip Whalen mentioned? An admission disguised in a poem? Or simply a comment on the human condition in Welch's America?

The development of "Din Poem" was spread over a number of years and underwent the same rigorous critique to which Welch subjected everything he wrote. From its initial inception in 1961 to the first public reading in May 1963, elements of the poem were added, discarded, and amended until Welch was finally happy with it. Written on the same tele-type scroll that he had used for *I, Leo*, "Din Poem" is as close as anything in his oeuvre that highlights both the care and attention he took to create a poem, as well as the range of subject matter that reflects every aspect of how he saw himself at that time as a citizen and a poet.

His ear for the use of language on the street is as keen as it ever was as he offers us glimpses into the lives of his fellow citizens: glimpses into their problems, their love lives, their prejudices, their mundanity, their frustrations. For Welch, authenticity came not in floral constructions or in the forced use of literary devices but, like his literary mentor Gertrude Stein, in reproducing the voices of America as they were. He saw it as his duty to "let America speak for itself," and years later in 1967, when he read "Din Poem" at the Magic Lantern Theater as part of a series of readings in support of Jack Shoemaker's Unicorn Bookshop in Isla Vista, Welch explained his philosophy thus:

> I sometimes see myself as being stuck with the problem of letting America speak for itself. It's a very, very difficult task sometimes. . . . You see, the job of the poet is to tell the tribe certain things. See in any herd animal there has to be a few members that watch and warn while the others feed and breed. And so there are certain times when you have this duty. So I've had to let America speak for itself in my work for some reason.[13]

Welch was acutely aware of not only the duty he had but also the burden of responsibility that the poet carries in embracing this duty. At that same reading, he used Allen Ginsberg as an example of a poet who he felt was being side-tracked by a duty to teach Americans "the joy of really chanting." Welch claimed that in spending most of his time "doing that Krishna thing," Ginsberg was essentially working and that "he's got all kinds of things that he would rather be doing than that now but that seems to be his job."[14] It is arguable whether Ginsberg would have seen it as such, but it is illustrative of how Welch felt about poets being required to invest in their trade for egalitarian purposes. In much the same way that Stein had written "let me tell about the character of the people of the United States of America and what they say,"[15] so Welch listened to what he heard around him and used it not only to substantiate the points he was making in the poem, but also, like Stein, to capture "the rhythms of America, of American speech."[16]

An earlier poem, "You Can't Burlesque It Anymore, —1956," originally titled "Let America Speak for Itself," is very much a forerunner for this section of "Din Poem" in its use of newspaper headlines, advertising slogans, and everyday speech to illustrate Welch's duty as spokesperson-poet. Yet unlike its successor, "Burlesque" is unassuming in tone. It neutralizes the voice of the poet through its matter-of-factness. These are not Welch's words, but words borrowed from, among other things, posters, buses, newspaper editors, and postage stamps. Welch's job is to simply arrange them in such a fashion that he maximizes their effect.

Conversely, the soundbites he chose for "Dim Poem" lean more toward negative stereotypes that seem to underpin what Welch increasingly saw as America's dissatisfaction with itself. They are as much his voice as they are America's, being accusatory, suspicious, and incendiary:

> "You never say you love me anymore."
> "I told you once, I told you a thousand times"
> "You've been late four times this month."
> "What's that girl doing down the hall, anyway?"
> "Where you workin' fella? Get in the car." (RB, 104–106)

One section of the poem that offers insight into Welch's love of language as music is "Supermarket Song." From his days at Reed when he had first tried to put Shakespeare to music, Welch had long understood music's association with poetry, and those early dabblings were later developed into full-fledged scores that Welch performed with some aplomb. However, during the first reading of "Din Poem," when he was accompanied by Kirby Doyle at San Francisco's Batman Gallery in May 1961, the poem was markedly different from the final version—largely because "Supermarket Song" had not yet been developed into the song that it later became. Indeed, not until 1965, when Welch collaborated with the Sausalito-based artist Chris Roberts, was the poem given its musical accompaniment.

The reading at the Batman Gallery was held on the opening night of "Constructions," the first significant solo exhibition by the artist George Herms. The posters for both the reading and the exhibition were designed by Wallace Berman, who, according to Berman's wife Shirley, had met Welch by chance when he had been called to collect the Bermans and three other friends in his Yellow Cab. Welch had initially refused to take them because he was permitted to carry only four passengers, but after Shirley invited him to join the party they were going to, Welch quickly relented, and another friendship was duly born.

Although this was Berman's first venture into poster-making of this sort, he was already a well-established assemblage and collage artist, photographer and filmmaker, and founder of *Semina* magazine; Welch became a regular in his circle of friends. Alongside Welch and Doyle, Herms's opening also featured a short film by the painter Paul Beattie (titled *Film Poem* on Berman's poster but very possibly *Scenes from the Tap City Circus*, the experimental portrait of Herms at work that Beattie had made in 1960), as well as a performance by the musician Bill Spencer, who had been a session musician on the jazz circuit in Los Angeles and was a close friend of Herms.

In terms of "Din Poem," the addition of the annotated musical score lent it a music-hall feel that helped underscore the clearly powerful message that Welch was trying to convey. The story goes that both "Supermarket Song" and another of Welch's scores, titled "Graffiti," came about after a drinking session on one of the houseboats in Sausalito that ended with "about eighteen songs . . . one for every sign of the zodiac" (HWP, 68).

———

Although the opportunities to participate in poetry readings gradually increased in the early 1960s, Welch continually struggled to balance his creativity with his financial obligations. Since leaving Montgomery Ward, he had worked various jobs for various periods of time, often being laid off on account of his drinking, time-keeping, and apparent lack of loyalty; sometimes he simply quit. His most consistent means of work had been driving cabs in San Francisco for the Yellow Cab Company. Not only did this seem to suit his lifestyle, but also it provided him with an almost continual audience and source of inspiration. He would recite poetry to anyone interested and used experiences and passengers in his cab as the basis for some of what he felt to be his best poetry at the time. Yet the worker-poet balance was often out of kilter, and Welch frequently bemoaned his treatment at the hands of the company while seeming to understand, at the same time, that he must continue to do it in order to finance a grand plan he had hatched to visit Japan or fund some other extravagant trip.

In the end, although no such trip ever materialized, his cab driving yielded a small but significant collection of vignettes called "Taxi Suite," which he frequently submitted to magazines in an effort to increase his published work. The poems, which Robert Duncan described as "poems of wakened consciousness" by "one sentient being in many miles of darkness in the middle of the night,"[17] range from candid conversations with passengers to a collection of observations in the manner of the ancient Greek lyric poet Anacreon. Indeed, in one of Anacroen's poems, "Youth and Age," the author whimsically recounts the positive influence of youth and wine as he ages, beginning the poem with "When I see the young men play." It is intended to be humorous, a challenge to the more mature not to be outdone in drinking and cavorting by their youngers.

Similarly, Welch's poem begins with "When I" and is also intended as a frivolous observation on his experiences as a cab driver, yet such was his desire for accuracy that he doubted whether certain lines should be deleted to save the poem from becoming bogged down or heavy and thus lacking the necessary humor and missing the point. In a letter to Whalen, Welch asked, "Would it be better if 'all may command me' and 'guided by voices' were deleted? At least one? Seems to bog down in there" (IR1, 144). Yet there is a definite accuracy in such lines as "beguiling me with gestures" or "a revelation of movement comes to me. They wake now" (RB, 21). His relationship with the various passengers ranges from hunter-prey and servant-master to facilitator-benefactor and savior-saved. Yet despite its obvious humor, Welch still conveys a sense of seriousness, a sense of the role of the cabbie as essential to the fabric of normalcy in society. This seriousness was not lost on him, and in an effort to counterbalance this, he suggested to Donald Allen that only when "After Anacreon" was placed alongside the other poems in "Taxi Suite" would it achieve the desired effect. The poem ends with a line that is not only poignant in its meaning but also concise, evocative, and masterfully written:

> When I drive cab
> I end the only lit and waitful thing in miles of
> darkened houses. (RB, 21)

The next three parts of "Taxi Suite" are excerpts from short conversations Welch apparently had with passengers. Originally titled "Rider Poems," they are strategically placed not only to emphasize the humor and provide a lighter tone after the seriousness of "After Anacreon" and the philosophy of the closing poem, "Top of the Mark," but also to embolden. Welch felt that, without the "Rider Poems," the other two lost a great deal of their impact—the intervention providing a necessary "connection with fact."

The passengers in the "Rider Poems" are the Nurse, Mrs. Angus, and the Mailman. In and of themselves they are simple anecdotes, but as linguistic sketches they are yet more carefully constructed examples of how Welch used poetry to showcase the speech of everyday people in everyday situations—to transfer, so to speak, the language of poetry from the halls of academia into the back seats of taxis. The first, the Nurse, is a rapid, largely unpunctuated poem that can easily be read as real-time speech. Then comes Mrs. Angus, who, one imagines, speaks in a soft Irish brogue and employs local dialect and grammar to define her observations. Finally, the Mailman who, unlike the Nurse, converses in a slow and deliberate manner in responding to Welch's questions. He too uses the language of the common man, a mix of clipped sentence endings, pride and patriotism, and comfortable languidness. In using such language Welch wanted to make his work accessible to the working man—the taxi driver, the barfly, the longshoreman. Reading "After Anacreon" to cab drivers in a local poolhall during a break from work, he was grateful to hear them respond not, as he had perhaps expected, with insults about him being "a goddam queer" but by exclaiming how he had captured what it *is* to be a cabbie.[18] And how his words had created familiar images.

By June of 1961, Welch and Kandel were apparently preparing to get married. They had planned a "Zen wedding with gongs and flawr-petals & all that."[19] The problem, however, was that Welch was unsure about his own marital status. In a letter to his mother, Welch wrote that their plans were "bogged down because I can't discover whether or not I'm divorced" and that "I don't tickerly see why John Law has to get into this relationship, but Lenore has never been married and it will simplify things like cabins in boats if we plan to go to Japan etc. so why not?"[20] He knew that no such proceedings had been instigated in San Francisco County by Mary, but beyond that he was in the dark. However, rather than attempt to contact his former wife or parents-in-law himself, he wrote to his mother, knowing instead that she would most likely do it for him: "Am anxious to [get married] without contacting Garbers if possible, & think that it is possible. We'll see."[21]

Welch was only too aware of Dorothy's sense of decency and must surely have calculated that, should he wish to marry Kandel, Dorothy would have given her blessing only if she was sure of his being divorced. And, as predicted, in August Dorothy took it upon herself to indeed investigate whether Mary had ever filed for a divorce, writing to Emil in Chicago to inquire about it.[22] Her letter is both deferential and guarded, outlining her own anxieties about the unfinished business of her son's marriage without mentioning Welch's plans to remarry. Indeed, she goes as far as to suggest that an eventual divorce would be beneficial for Mary's future happiness while chastising her son for the "sloppy" way in which he was living his life and not seeming to care much about his marital status. Furthermore, she apologizes on behalf of Welch and expresses regret that the marriage had failed. However, Garber's answer, although undocumented, would have been irrelevant before too long anyway, because by this time Welch and Kandel's relationship was already under some strain because of Welch's self-proclaimed inability to "love anybody except in general" and his constant bitterness about lack of money. It seems that even Lenore was unable to deal with this, and by the spring of 1962 she had given up on him altogether. Like many other such instances in his life, the specter of Kali had once again risen up and cast another long shadow over him.

His relationship with Kandel seems to be indicative of a never-ending search for the type of love that Welch felt would somehow supersede that of his mother. However, it could also be concluded that it was a fruitless search, because such was his dependence on Dorothy that it was almost irreplaceable. Welch had a particular need to be loved unconditionally—in some ways the type of love that only a mother can provide—and consequently, it surely became almost impossible for any of his girlfriends to provide it. Although he wrote at the time that Kandel had left him on account of him being "bitter and ugly" and a struggling and largely penniless poet, in the interview with Meltzer many years later he was more candid about the breakup, stating that it was he who had left her because she was "corny and our life was not getting on together because of that."[23] In retrospect, however, it is very hard to believe that Kandel could have been in any way "corny," given what was to come with the publication of *The Love Book* in 1966 and the huge impact it would have in establishing Kandel at the forefront of what Jennie Skerl called "the proto-feminist evolution of women" transitioning between Beat and hippy cultures.[24] It is perhaps more down to Welch's need for something she could not provide him than any corniness on her part. Indeed, in that same interview Welch concedes the following about himself: "I have an exquisite kind of fineness in my life that she could not meet. She could not meet the other stuff I needed with perfection. She was a good help mate. A goddamn good wife. But I didn't need it. I needed some other goddamn thing."[25]

It appears that Welch himself was unable to pinpoint exactly what that "goddamn thing" was, let alone explain it to Kandel. He simply said that he "had to go . . . had to split" and that, as they both wept on the staircase, Kandel knew that he had to leave but did not know why. His own memories about the reasons, even years later, seemed similarly confused when he asked, "How can you give up the most beautiful girl in San Francisco, who you need?"[26] Indeed, Kandel may have asked the same question herself. In any case, in a letter to Don Crowe shortly after the breakup, she writes of both her sadness and hope at what has passed and what may come from it: I'm "going to try tha lsd [the LSD] but not yet — there is still a small lingering sadness from breaking off with Lew — don't want to go that far out till I'm more clear. Lew is in Big Sur and steadily drunk but he says he is going to write again and I hope so."[27]

14

I Saw Myself a Ring of Bone

Oh that is just the tall red man howling again.
Howling in his hills of Sur!
— Lew Welch, typescript of "A Poem for the Fathers of My Blood"

When Welch made the short journey south to Lawrence Ferlinghetti's cabin in July 1962, it had been almost exactly two years since he had last been there. This time, however, he was not part of some thrill-seeking entourage hanging onto the proverbial coattails of Sal Paradise and Dean Moriarty. Suffering from various nightmares of his own and needing an escape from the aftermath of his breakup with Kandel, Welch found that the Bixby Canyon cabin was to offer the same sense of respite that Ferlinghetti had hoped it would provide Kerouac. However, by the time Welch returned to San Francisco less than a fortnight later, he would not only have undergone a similar existential crisis as Kerouac but would also be accompanied by Rimbaud-like nightmares of suicide and pseudo-erotic visions. He experienced a spiritual revelation of such magnitude that it precipitated another fundamental reevaluation of himself.

In much the same way that Kerouac had captured the events of summer 1960 in *Big Sur,* so Welch found his own outlet in the cathartic power of letter-writing. Despite having spent so much time over the previous three years completely enveloped by the San Francisco poetry scene, Welch seemed to be at his lowest ebb yet. His correspondence had been voluminous, and as had always been the case, he was unafraid of showing his emotions or inner turmoil on the page. What is more surprising, however, was that when he chose to describe what had occurred at Bixby Canyon, he chose not to write to his closest friends or to his mother but to someone he admired greatly but had had relatively little contact with, compared with the rest—the poet Robert Duncan.

The pair had first met in New York after Welch's cross-country trip with Kerouac and Saijo, and they had entered into a correspondence in May 1960 after Welch had sent Duncan a copy of *Wobbly Rock* with the inscription "Why is there no oyster-closer? Where is the very long, slender needle?" Both of these lines were taken from Gertrude Stein's *Tender Buttons* — a fact that would not have been lost on Duncan and was somehow reflective of Welch's sense that to use Stein was to legitimize his work in the eyes of poets such as Duncan. Welch's book was accompanied by two letters, the first of which he wrote candidly and somewhat aggressively, only to recant in the second. Dated a day apart, when read together the letters provide a clear example of how Welch's mind was working at the time. The first letter is a mild attack on Duncan and how he had been part of a system of poetics that, in Welch's eyes at least, was partly concerned with punctuation and poetry as a cause of anxiety because of what poets perceive to be "rules." Welch admonishes Duncan on account of him "misleading the young (and pure)" and demanding instead that everyone should look to Stein for a clearer sense of what should be done.[1] Welch also mentioned that he was writing "staggering beautiful prose that is impossible to punctuate correctly" and therefore he had invented a system of only using ": and — and . (all but the last being cementrical)" — apparently a self-constructed remedy for poetic anxieties. His second letter is far more measured, beginning with an apology and going on to explain how he had "been writing for several days, deeply, and had arrived at that madness where all that was contained in my mind was freely available."[2] This "madness" was then defined as Welch not only feeling that he knows all but also that he is "all-known." As a result, he uses his own frailties to ironically illustrate what he called "my point of contention with you and your influence on young writers."[3]

However, in his reply Duncan seemed to have overlooked the first letter and responded in such a positive manner (with little or no reference to what had been said in either letter) that Welch felt it was among the most generous of any of the replies he had received regarding *Wobbly Rock* ("I am amazed how well the poem was received, but your reception had a particular force to it"). To have earned the respect of such a prominent poet was critical at a time when Welch felt that he was resurfacing in poetic terms after a period of "cheating." The poem was a "statement of place after the bitterness, the hurt, and the shame," and Duncan's reception had thus been extremely valuable.[4] And all the more so given Welch's criticism of him.

By the time Welch went to Bixby Canyon he had also featured alongside Duncan in Allen's anthology *The New American Poetry*, and their correspondence had developed such that the draft letter Welch wrote while staying there is revealing not only of his utmost respect for Duncan but also of his desire to bare his soul to someone who had been through a similar experience. The letter, which was neither completed nor sent, starts by simply stating his reasoning for writing:

> Dear Robert, It is odd that I should want to write this letter to you, since I know so many of us better, who I respect as much. But at this moment my spirit, what's left of it, needing to speak, can only speak to you.
>
> I must speak to one who has gone through something as black as this, and made it somehow—or at least is still there, operating his Human Being (your phrase) decently, as a poet. (IR2, 51)

Although much of what Welch was experiencing at that time would be explained in greater detail in the draft letter he wrote to Duncan at Bixby, in the weeks and months leading up to his departure to Big Sur, Welch had written to others of sliding into prolonged periods of acedia; he later told Dorothy that he had been having "a violent time of it spiritually" (IR2, 51). His need for solitude was apparent, but he may have underestimated the eventual effect it would have on him. His reference to acedia in an earlier letter to Snyder was frivolous yet intentional. Referring to himself as the "Buddha known as the sleeper," Welch was now experiencing something more akin to spiritual torpor. Drawing comparisons to Charles Baudelaire, in a later poem titled "Acedia," his reference therein to T. S. Eliot's study on the French symbolist might just as easily have referred to himself.[5] As Baudelaire's ennui was "the result of an unsuccessful spiritual life," so Welch's acedia may have amounted to essentially the same. Attempts at zazen had failed, and his hiking trips into Marin County with Allen were invariably accompanied by Jim Beam, in bottle form. Aware of the prescribed Benedictine cure for such a state of mind—"hard work for the body frees the soul"—the millstone of Baudelaire's ennui was apparently claiming another victim nonetheless. And before opening up to Duncan, Welch turned first to two of the most powerful anchor points in his life: Dorothy and psychiatry.

By late spring, his relationship with Kandel was over, and his torpor had lifted just enough for him to see the need to reconnect with his "Self." He had been undergoing more psychiatric therapy—a regular part of his life since his sessions with Kepecs—and this once again raised the issue of his "easy acceptance of love," a deeply rooted aspect of his personality that was forever having an effect on his relationships with women and one that can be traced back to his childhood and that unshakeable bond he developed with Dorothy. His continual reliance on her support—both financially and emotionally—was surely something that only served to exacerbate his continued sloth. She had once again provided him with not only a new car and new clothing but also with a new typewriter, all of which would feature heavily in the coming weeks and months and all of which served to eradicate any serious need on his part to find alternative means of sustenance.

In July 1962, Welch wrote to Dorothy insisting that he was recovering from this latest breakdown and that it was absolutely essential that he extricate himself from society and spend some time alone. While the time he spent at Bixby may have been a precursor for his prolonged stay at Forks of Salmon later that year, it was a momentous stay in its own right and provided Welch with what was his most vivid and sustained period of self-analysis and revelation since Chicago.

———

Welch drove to Ferlinghetti's cabin on July 11. His initial letters from Bixby are conspicuous for their matter-of-fact tone. He is both defiant and accepting of his situation. Writing to Dorothy of future plans, of "growing calm," and of taking baths at the hot springs, his tone gradually changes after a week at the cabin, as he begins to place himself outside of his physical body and into the body of the "Poet." He philosophically writes that "breakdowns for others are breakthroughs for the Poet" (IR2, 51), a theme he would go on to develop in great detail in his letter to Duncan.

Welch's choice of Robert Duncan as his recipient was apt. Although the "black" place that Welch refers to in his opening paragraph would suggest that Duncan had undergone similar psychological traumas to Welch, the reality was that in Welch's eyes Duncan's expertise was almost entirely related to the dichotomy presented to the poet as a Human Being. Duncan's desire to investigate the innermost workings of the psyche and relate it to how Poet and Human Being interact was not only a struggle that Welch found himself embroiled in at Bixby but also one that had occupied him for some time. As he had previously found direction in the accuracy and truth of Gertrude Stein, Welch now found in Duncan that, from the turmoil of his mind, he could better understand the purpose of meaning and transpose it into his writing. This was not the traditional psychiatry of the mind but the search for balance. His broken spirit needed to be repaired, and Welch explained his inner turmoil to Duncan in almost confessional terms:

> I have reached what I hope is nearly the end of 5 months of a total withdrawal of all love, all spirit, all hope, all vision—my whole being drawn into a fist pounding bitterly against Everything, for it seemed that Everything is dedicated to only mocking MAWKING all that I know is good, and here. I cannot shut out the din anymore. I am afraid. . . .
>
> For though I am mad, I am mad in the way Poet is always, must be, mad. It is the difference (as Jung put it to Joyce, trying to convince him that his daughter was in grave danger, and Jimmie said "Nonsense, what you describe is what I do all the time. It is even a game we play together.") And Jung answered, "Yes, Jimmie, but you dive to the bottom of the river. She is sinking."
>
> The poet can never sink, and while sunk, be Poet. His diving is always a dive, even if to do it he must sniff the vapors of his oracular cave—or otherwise drastically wrench himself open that the whole river flow through. . . .
>
> And so I have not sunk. And the sickness I've been through had nothing to do with the bottom of any river. It was a worse thing: the deliberate closing of myself.
>
> And I found that whatever it is that chooses to flow through me is so powerful it will destroy me if I resist it in any way. That I must open to it or die. And the death will be a suicide. (IR2, 51)

After almost a week of total solitude, Welch's "radiant vision of openness" prompted an epiphany in which he opened by degrees and allowed everything around him to flow through the "mess of gates" that he suddenly realized make up the Human Being. Having dived into the proverbial stream of Jung's analogy, Welch was enveloped by his Self and the understanding that all that flowed through him was "transformed. Different on the way out." While the process was painful—he experienced violent and vivid nightmares that echoed those Kerouac had at the cabin two years earlier—almost prophetically, they featured Dorothy and a mutation of friends who Welch referred to collectively as "my brother" (in reference to his long dead sibling—a deep-seated trauma that he may never have fully dealt with and that, again, echoed Kerouac's lifelong feelings regarding his deceased brother Gerard). However, the process also conspired to induce gratification of a pseudosexual nature with Welch writing of "fucking the world" in some kind of indefinable "black satori."

In the immediate aftermath of his revelation, Welch was struck by bouts of almost continual and uncontrollable weeping that he felt as desperate wonderment rather than pity. He wrote to Duncan that he was "struck by what is meant [by] 'crying in the wilderness'" in reference not only to Kerouac's description of Japhy Ryder in *The Dharma Bums* searching for "ecstasies in the stars" but also, again, to Isaiah's biblical contention that the wilderness was a metaphor for spiritual emptiness. This was in Welch's own mind a manifestation of the acedia from which he had previously admitted suffering. His vision was all-encompassing. Indeed, the meaning of his weeping in relation to some sense of spiritual wilderness evoked Rimbaud's statement that poetry (and thus the Poet) should be understood "literally and in all senses"—Welch's vision was an epiphany to be understood on multiple levels too, with his job as the poet to "write it down and down and down" in plain and colorless language such that it reveals itself and substantiates nonmeaning in the process. He was afloat in the stream of his own consciousness, tears swelling the current and enveloping him in this radiant vision. The next day he wrote, "I saw myself a ring of bone in a clear stream, and vowed, never, ever, to close myself again." It was the seed of what would later become one of his most important poems, encapsulating a new philosophy that would draw a line between acceptance and suicide—a line along which Welch precariously walked for much of the rest of his life.

I saw myself
a ring of bone
in the clear stream
of all of it

and vowed,
always to be open to it
that it might flow through

and then heard
"ring of bone" where
ring is what a

bell does (RB, 77)

As a philosophical notion, the "ring of bone" that Welch's vision portrayed is both a physical property of the human being and the internal
enlightenment that affects the complete body. In some senses, he likens
the "ring of bone" to a metaphorical ringing of consciousness, which in
turn refers to the very spirituality that he felt had deserted him in recent
months. In the literal sense, the bell implies the strict sounding of the
Zen *inkin* bell that is an essential part of zazen and is further suggestive
of a short verse by the sixteenth-century sage and poet Kojisei, from R. H.
Blyth's *Haiku*, which Welch had read at Reed: "At the sound of the bell in
the silent night, I wake from my dream in this dream-like world of ours.
Gazing at the reflection of the moon in a clear pool, I see, beyond my
form, my real form"[6] — and which is clearly aligned with what Welch had
written in his letter to Duncan.

Welch's "real" form was often ambiguous, but at Bixby he saw himself
in that vision of openness as walking the line between suicide in its literal
sense and the continual psychological challenge of killing the demons
that were forever pushing him toward it—the closed mind, the gates
that block the path to all the receptors the Poet needs to survive, and the
pain of such openness. He offers a vision of complete acceptance of pain
as the only viable alternative to being closed to the flow of life. Yet that
pain is in itself seemingly so unbearable that hosts of other poets have
succumbed to it and killed themselves. He names Vladimir Mayakovsky,

Hart Crane, Vachel Lindsay, and Dylan Thomas as some of the many poets who removed themselves from the pain by means of "drownings and gunshots and the irreclaimable madnesses," while stating at the same time that their suicides were tragic because they had "missed the truth of it by a quarter of an inch." His realization of openness, while painful, was this truth that his contemporaries had failed to see but that, ironically, he would also lose sight of before too long.

PART II

The Mountain

Why do we dream of lost cities when
Lost groves grow beyond the next rock knoll
the next and next

—Lew Welch (RB, 49)

15
The Journal of a Strategic Withdrawal, Part 1

The road to redemption finds its source in many places. For Welch it was Highway 99 leading north from Sacramento toward gold-mining country in and around the Shasta-Trinity National Forest and Siskiyou County. As the road leaves the urbanization of north-central California behind, the landscape grows in stature. The countryside is dominated by imposing and majestic mountains such as Mount Shasta and Lassen Peak, scattered with a plethora of crystalline alpine lakes. Forests of cedar, pine, and oak thicken and expand as human life is slowly marginalized by Mother Nature. Leaving the interstate in Yreka after more than three hundred miles ("Dig, it's *Yreka*, not Eureka" [IR2, 82]), the road turns southwest through the sleepy former gold-rush hamlets of Fort Jones and Etna and deeper into what is now known as the Marble Mountain Wilderness. Driving on until the road forks north to Somes Bar and south to Cecilville, one finds a magical depth to the quiet that is almost palpable. Left and right, the forest floor is swathed in a thick and resplendent undergrowth, and "trees around are doug fir & sugar pine & yew. Many small oaks & maples. Dogwood. Dogwood now lavender" (IR2, 79).

For Welch, departing the city and seeing the majesty of Shasta rising from the terrain around it seems to have offered him a moment of timely realization. A moment to revere the mountain and its ancient presence, yet perhaps also to question its dormancy. Although Mount Tamalpais had long been one of his principal sources of inspiration, Shasta provided a reminder of nature's power to endure—a power that, given all that had gone before, was something Welch dearly needed. In a poem titled simply "Shasta!" Welch lauds the volcano's longevity and endurance yet questions the fact that it now stands "so cold," having once been so powerful and destructive that it could

spit fire forth molten
rock & gold
splash hills with it, char trees,
incinerate (RB, 179)

He seems to cry in both reverence and frustration: "all I can say is
Shasta! I cry Shasta I cry Shasta!" Might this be Welch challenging himself not to lose his own power? Not to have strength subjugated by the
metaphorical clouds and mist in his mind? The consciousness he gained
at Bixby needed now to be maintained, and the positivity of Shasta may
well have offered another source with which to achieve that.

Welch finally arrived in Forks of Salmon behind the wheel of the new
Chevy Dorothy had bought for him when it became apparent that Willy
was beyond repair. Why Welch chose such a place to extricate himself
from society is easy to see: comprising little more than a post office, a
one-teacher elementary school, a general store, and some houses, cabins,
and shacks, the modern-day town has a population of fewer than two
hundred; when Welch arrived in late August 1962, it would likely have
been considerably less. After the toils and tribulations of his time at Bixby
Canyon, his primary intention was to "go up to the Salmon River and live
in a mining claim cabin & catch big steelhead and never see people &
drink good spring water and not worry if the bomb goes off MEANWHILE
writing all truth into imperishable pomes" (IR2, 59).

With his car packed with most of his worldly possessions ("food and
supplies, a bed roll and necessaries, typewriter, fishing gear, and a .30-.30"
rifle), Welch began by finding a suitable place to camp. Close enough
to the community but far enough away to feel that the solitude was real,
he initially settled on a spot two miles upriver. From there his aim was
to search for a suitable property to live on more permanently. The mild
autumn weather would soon give way to harsher winters (indeed, the
coming winter would prove to be one of the harshest in years), and the
need for a solid and reliable cabin was already on his mind. Within weeks
of his arrival, he wrote to Snyder and Kyger in Japan with his usual (albeit
short-lived) optimism about having his eye on a ten-acre claim with a
"spring, huge cedars, enough gold to pay my way." The land was owned
by a Polish American couple, Pete and Sally Novak, and Welch wrote
of them being "one of the sweetest couples" he had ever met and that,

given their financial troubles caused by mounting medical bills, lack of work, and being "systematically bilked by this 'nation of finks (IR2, 63),'" he could take the claim for a mere $500. Novak was even going to teach him how to mine for gold. Yet, as with many of Welch's plans, the basis was sound but the execution was anything but. Having written to Snyder about the impossibility of raising such funds, he subsequently wrote to Kirby Doyle that he had come to an arrangement with Dorothy, who was providing him with $100 a month (on the proviso that he went "to Japan, the nuthouse, or the woods") and that, if he lived as frugally as he now was, he could raise the money that way and pay the Novaks in monthly installments of $50. However, he began to worry about this straightaway, and the fluctuation in his optimism is never more apparent than in that letter to the Doyles, writing first, "I'm going to buy!" and then a few dozen lines later countering that enthusiasm with "I will scratch it out . . . and I might fail" (IR2, 65–66).

All the while, having not yet found a cabin, he camped in relative solitude under the California stars in a "beautiful forest where I have salmon and eggs every morning for breakfast" (IR2, 63) and where he could sit cross-legged with his typewriter first on his footlocker and then on a board he nailed to a tree stump. For all intents and purposes, he was happy. He had found work building cabins and firefighting. He was writing a book titled "Earthbook." And, most important, he was now free from the "vulgar and dangerous din" of urban America.

———

Naturally, the events of Bixby Canyon and his part in the breakup with Kandel were still weighing heavily on his mind ("I got so bitter & ugly even Lenore gave me up & she's most patient" [IR2, 62]), and the first letters he wrote from Forks of Salmon are again conspicuous for their sense of catharsis. He attributes his failings, on the one hand, to the milieu in which he continually finds himself and cannot reconcile himself to and, on the other hand, to his being born into a situation akin to F. Scott Fitzgerald's "silly world" of snobbery and pseudo-wealth. Although he writes of his escape from this world through a *fortunate* accident," it is a world that had forever left its mark on him. His misery and bitterness were constant aspects of his life, whether the result of failed fishing ventures, loneliness, publishing issues, relationships, or jealousies.

In an entry in his notebook from August 1962, Philip Whalen writes of Welch's envy at the successes and lifestyles of others and suggests that his decision to leave the city was partly an attempt to finally seek such success on his own terms rather than wait for it to happen or to simply complain about it:

> Lew to go to live in the Trinity Alps, goodbye to the city at last, etc. In his note to me, Lewis is very jealous of Snyder & others because they do what they please & live where they will, in spite of the tyranny of the system, the force of circumstance or whatever adverse natural forces which block us from enjoying the world as we ought to, as we DESERVE to enjoy it—although what are his {or my} qualifications which would earn for us such deserts [sic] he does not say.[1]

Yet this self-imposed hermitage seemed to offer a serious sense of respite from this pattern. Welch knew that with "a small amount of strength, discipline and clarity of purpose etc" (IR2, 59), he could potentially undo the shackles of his current state—becoming "deserving." In an unpublished poem from the period, he revealingly writes,

> Come, then, to this cabin. To heal, to try to
> Figure it out.
> All winter long
> Staring at it!
> Hurt, composite face. Man, victimised.
> My face.   Yours.
>   "Though it break my brain . . ."
> Gautama.[2]

Welch understood his need to heal and make sense of his wider predicament when he left for the Forks. And perhaps this poem also shows his readiness to look closely both at himself (as a victim of himself?) and as an outsider to whom objectification provided distance and more clarity of purpose. His use of Gautama Siddhartha's analogy that enlightenment can be attained only by breaking through the surface of the mind to free what lies beyond provides a sense of the role that Buddhism still occupied in his life and may also point to a continual need for such inspiration to achieve the lofty (and ultimately unattainable) goal he set himself. Indeed, *Hermit Poems*—the collection of works he wrote at the Forks, published in early 1965 by Donald Allen's Four Seasons Foundation—is laced with references to various Eastern philosophies and can be read as a collection that establishes clear links to ancient Chinese hermit poets such as Tao Ch'ien and Su Tung-Po. These Chinese poets were in some respects on a journey similar to Welch's—that of the intentional recluse shunning "the empty pursuit of professional life" and returning to their mountain homes.[3] One additional aspect of Welch's journey, however, was his initial intention to reduce his alcohol intake and somehow break the cycle of fear, sickness, and anxiety that he had been on. This was more than just a "struggle to free himself from the constraints of official life" such as that faced by Ch'ien;[4] rather, it was a profound endeavor to alter the course of his life. And key to this was finding a place he could call home. A place to "put the typewriter and plant trees." A place that would act as the bedrock of his existence.

Yet this need for a home is surely twofold. On the one hand, Welch had an image of himself that is clearly rooted in his past: "The figure of a Prince without land or power, borrowing a corner of someone else's estate" (IR2, 62). This is obviously an image born, first, of his parent's divorce, and of Dorothy's continual uprooting of the family and moving them up and down the California coast to serve what were largely, Welch at least felt, her own needs. And, second, a product of his dysfunctional childhood, in which a normal home was something his sister and he never truly knew; his longing for this is entirely understandable, even if there is little evidence of his ever having looked for it previously. Indeed, is it not true that his marriage to Mary afforded him the opportunity to gain exactly what he said he was now missing? Hadn't this been the case in their marital home in Chicago? Or in San Francisco? Hadn't Lenore Kandel also "made a beautiful home for [him] out of nothing"? (IR2, 73).

There is an idea in these letters that Welch is again looking for any possible way to make sense of his self-destructive nature or to understand his propensity to lose himself in alcohol. He seems to have acknowledged (partially at least) his need to face the issue head on and to work at it honestly, yet he also admits to it being so deeply rooted in his very self that he does not really understand it. Only when he can "know what other men know when they say 'this is where I live'" (IR2, 62) can he know something of worth.

However, this need to have a home is also inextricably linked to the nature of his poetry and his use of American speech to establish an identity that must also be substantiated by the inhabitation of a particular place. Only when he had found such a place could Welch truly begin to work on finding an answer to the many questions relating to the matter of "this is where I live" (IR2, 62).

––––––

The manner in which he eventually found a place to live is conspicuous for being both fortuitous and comical. Having wandered through the forests and meadows to no apparent avail, following what he described to Snyder as "spooky hints," he got a temporary job helping an old local build a cabin. On returning from Yreka with the old man's belongings, they were greeted by a huge forest fire. Recruited almost on the spot to help fight the blaze, Welch suddenly found himself in a truck full of forestry engineers who had been surveying the area for a road that was to be built there. They pointed him in the direction of a disused and unclaimed cabin they had used during their preliminary surveys the year before, so Welch now had the opportunity to find somewhere to escape the ever-harshening weather. After three days of firefighting, Welch finally had the chance to search for the cabin and wept when he saw it, such was the "beauty, dignity and fine history" (IR2, 69) of the place.

Getting to the cabin proved difficult however. Leaving his car parked in the valley below, Welch had to trek along a dirt road, cross a creek, and hike up an "impossible switchback trail" to the mountain meadow where his "fine old shake cabin" was located. When the time came to move his belongings there a few days later, Welch decided instead to forge an alternative trail, but lost his way, fell, and inexplicably allowed his pack to roll down a steep embankment and disappear into the ravine below. On finding it again the following day, he ran into a hunter who told him that the cabin had once been built by a "Wobbly . . . who wouldn't do one damn thing to aid the finks" (IR2, 70)—a fact that only enhanced Welch's sense of serene verisimilitude despite his obvious embarrassment at having made such a potentially catastrophic blunder with his pack.

In a letter to Kirby Doyle dated October 12, 1962, Welch describes his joy at having found a home so beautiful that it reduced him to tears every time he looked at it. Surrounded as it was by virgin forest, the cabin was

> made by nailing big cedar shingles over a frame of poles made by 3-to-6-inch diameter pine trees. This one was made by a craftsman such as the world will likely never see again. The doors and windows are made of oak planks apparently split, not sawn, from native oak . . . the inside is all golden with the natural woods, crisscrossed with delicate bracing. It is like living in a Vermeer. (IR2, 73–74)

The craftsman in question was a former Wobbly named Lawrence Meyer, who had built the cabin years before, naming it "100 Mile Cabin" thanks to its distance from what was then the most heavily populated settlement in the region, Eureka. Meyer had been a member of the International Workers of the World (IWW, commonly known as "Wobblies"), the global labor organization founded in Chicago in 1905 as an alternative union movement that emphasized the socialist (anarchist) aspects of unionism and sought to undermine the rise of capitalist ideologies in industry whereby human labor was replaced by machinery and workers were "sunk in the uniform mass of wage slaves."[5] As industrial developments grew, so too did discord between unions and employers, and it is reasonable to imagine that, like Welch, Meyer chose instead to vacate the urban milieu and move into the country, where he could avoid paying taxes and live in relative safety and quiet. It was, after all, a time in which Wobblies were regularly attacked and persecuted for their antigovernment stance. And Meyer naming the cabin such may also be seen as a symbolic statement of withdrawal, much like Welch's.

Indeed, Welch was a proud, Red, card-carrying member of the IWW himself, and he very much adhered to the ideologies of the Wobblies in that, as the poet-turned-Sufi journalist and author Stephen Suleyman Schwartz stated in *Hey Lew*, he understood the "tradition, especially in the West Coast transportation industry, of such 'two card men,' who backed up their regular unionism with I.W.W. membership."[6] As a longshoreman, Welch was also a member of the International Longshore and Warehouse Union (ILWU) at a time when the union was undergoing some "bruising and highly-visible internal quarrels."[7] Yet true to the nature of such unionism, Welch was a stalwart advocate of the kind of physical labor required on the waterfront. As a fisherman, he had admired the work ethic of his Swedish counterparts, and the notion of "right" work was one that remained close to his heart throughout his life. Indeed, it could be argued that his sense of what work meant was born in the cabs and trawlers and warehouses of the West Coast.

Taking over Meyer's cabin was not without certain requirements, however, and laying a claim to the Wobbly's former abode also meant that Welch had to fulfill his share of mining activity in a nearby quartz mine. The problem was that Welch did not know where it was, and the alternative—paying an annual $40 sum for a ninety-nine-year lease—was out of the question, given his ongoing financial issues (despite working at various times during his stay at the cabin, Welch was forever complaining about his lack of money or thanking Dorothy for providing him with whatever she could afford to send). In the end, however, he did neither.

For as long as he lived in his cabin on the meadow, Welch called it Rat Flat, in an ironic homage to the "horrible infestation of rats scampering over face, clawing at old shakes, nervously scuttling like mad things" (IR2, 76). The weather was now turning cold, and incessant late October rains had arrived, making his find of the cabin all the more fortuitous. Winter in the Trinity Alps was sure to be a challenge, and his need to set up the cabin so that he would be as comfortable as possible was acute. And the first task was to get rid of the rats.

Setting traps, spreading poison pellets, and promising "unspeakable torture," Welch set about killing the entire population, raving at their "scratch" and "nervous skitter twitch" and going so far as to attack them with only a hammer—the result of which was nothing more than a broken handle. The entire episode is recorded in a somewhat whimsical poem titled "Buddhist Bard Turns Rat Slayer," in which his initial mad desire to exterminate them (*"Kill, Kill, Kill Shrieks Wordsmith"*) is replaced by a mild sense of loss:

> found one in my trap: immaculate
> gray fur, white breast, white
> little paws, short tail, mountain-sweet as
> everything else is here
>
> . . .

The Cabin
                almost too quiet
        ever since (RB, 180)

———

With a new typewriter given to him by Dorothy, Welch's output at the Forks was as prodigious as it had been in some months. Aside from his musings on life without the rats, the poems from Rat Flat are all carefully constructed portraits of his newfound calm in surroundings that were wholly conducive to healing. The works that finally made it into *Hermit Poems* are conspicuous in one sense for their serenity and emotional stability. They contain a clarity of vision that is both celebratory and emboldening and are surely among the best of Welch's oeuvre. However, they are not without a characteristically somber edge.

Not only are the poems highly personal but also they contain a spirituality and affinity with the nature in which he was immersed. In many ways, the works included in *Hermit Poems* are observations on daily life stripped of all superfluousness and obligation. Over the course of the collection, Welch's continued devotion to Buddhism is clear, as is his knowledge of the poetry of reclusion that was prevalent in medieval Chinese literature. Indeed, his self-imposed hermitage echoes those of the greats of East Asian poetry, such as Tao Yungming, Su Tung-Po, Han Chan, and Milarepa. Often the retreat was a return to a previous dwelling induced by untenable situations or conditions they no longer felt were satisfactory. This return was often to a place or situation that existed prior to the poet securing a government position or a role within the ruling class that fulfilled some sense of obligation or ambition—a situation that was certainly applicable to Welch before he left Chicago, and to some extent also during his time in San Francisco.

Tao Yuan-Ming (who became known as Tao Ch'ien—Ch'ien meaning "in hiding," thus "The Recluse Tao") is widely considered to be the first poet to turn his back on professional ambition, as early as the late fourth century. The reasons for this retreat are to be found in Ch'ien's aforementioned struggle to free himself from the limitations of life as a government official, which was the expected path taken by the educated classes from which he came. Unlike legendary Tibetan Buddhist yogi

Milarepa, who withdrew from society for primarily spiritual purposes, Tao Ch'ien's withdrawal has been described as a "protest against the eminently unworthy ruling class" of that time.[8] As a result of his retreat and subsequent reclusion on a small farmstead, Tao Ch'ien has been credited with establishing the gardens and fields poetry, a subgenre of Shanshui—the mountains and rivers poetry established some decades earlier and the inspiration for much of Gary Snyder's later work.

Ch'ien's poems have a rich and profound spiritual depth rooted in Taoism that describe a poet who has not only returned to a situation in the countryside akin to that of his childhood but also to a more acute sense of self. This was an endeavor that Welch strove to emulate during various periods of his life—not least his strategic withdrawal to the Trinity Alps. The fruits of this endeavor can be seen in both the development of his ideas and the wealth of imagery and iconography portrayed in the poems that he produced while there.

The journey that Welch takes us on in *Hermit Poems* is a cyclical reflection of his retreat and return. The collection is an obvious reference to the Chinese poetic tradition that Welch was emulating, and the works document his journey from initial self-imposed reclusion to a sense of egolessness or, as it is known in Buddhism, *anatta*. This transformation of sorts is reflected in the everyday descriptions of his life at the Forks, as he strives to find greater clarity and vision in himself and his surroundings. In the penultimate—and most revealing—poem of the ten poems included in the collection, titled "The Image, as in a Hexagram," Welch even goes so far as to abandon the previous first-person narrator in favor of a third-person "He" who "locks his door against the blizzard" (RB, 76) only to reappear after winter stripped bare and having shed all superfluous possessions and ideas. He has thus undergone a cleansing of the mind (and body) necessary to achieve such egolessness.

"The Image, as in a Hexagram" clearly evokes a verse from Tao Ch'ien titled "Home Again Among the Gardens and Fields," in which Ch'ien talks of having fallen "into [a] net of dust"—a situation he describes as a "blunder lasting thirteen years" (read: Welch as junior advertising executive, Chicago)—before returning to a place where there is "no confusion within the gate, no dust, my empty home harbours idleness to spare."[9] If we can equate Tao Ch'ien's "dust" to Welch's "din," then they both symbolize a deep dissatisfaction regarding elements of their respective lives, namely government and society. Where Tao Ch'ien's "net of dust" may symbolize a suffocating presence, so Welch's American "din" was likened to hopelessness, finality, and even death. Yet, while Tao Ch'ien appears to have been more at peace with the "net of dust" in which he had found himself, Welch's feelings about the "din" that surrounded him eventually manifested themselves more profoundly in the final line of "Din Poem," in which he uses a line from Guillaume Apollinaire's poem "Zone" — "Adieu, Adieu, Soleil cou coupe" (or "Farewell, decapitated Sun" [RB, 110]) — to make his dissatisfaction abundantly clear.

It is also important to see that the notion of "idleness" ("my empty home harbours idleness to spare") in such poetry means "profound serenity and quietness," or being in a meditative state such that daily life becomes the "essence of spiritual practice." In Welch's case, his daily life for much of his retreat was centered on his attempted spiritual purification and search for self through writing, subsistence living, and zazen.

Another element of the aforementioned poems worth comparing is the gate motif, often used in Shanshui poetry to symbolize both the physical surroundings the poet finds themself in and a metaphoric sense of awareness and contentment in reclusion. As Ch'ien writes of "no confusion within the gate," thus reflecting an already serene mind, so Welch writes of locking "his door against the blizzard," in what we might take to be a symbolic shutting out of his problems as a step toward such serenity — another important element of Ch'ien's sense of "idleness."

The gate motif is also reflected, albeit differently, in the calligraphic *enso* symbol Welch uses to illustrate his poem "Step Out onto the Planet." When closed, the *enso* may symbolize mindfulness and perfection, which in turn reflects the contentment of mind and place and thus serenity. However possible it is to find such moments of serenity and clarity in Welch's poems, he often contradicts himself — showing humor, self-deprecation, and the internal mental struggle he was still having, to find himself. The following poem is a perfect example of this. In initially comparing himself to the Chinese poet Su Tung-po, Welch subsequently undermines that comparison by using key Chinese poetic symbols as motifs:

> The Empress herself served tea to Su Tung-po,
> And ordered him escorted home by
> Ladies of the Palace, with torches.

I forgot my flashlight.
Drunk, I'll never get across this
rickety bridge.

Even the Lady in the Sky abandons me. (RB, 74)

First, Su Tung-po was a government official who, on account of his non-conformist attitude toward the policies of his superiors, was exiled to the provinces and subsequently became a subsistence farmer living in relative solitude. Where Su Tung-po drinks tea in the poem, Welch is drunk on wine.

The use of alcohol was seen by the ancient Chinese poets as a useful tool in their practices, and the drinking of wine is very often mentioned in poems. Indeed, Tao Chien's poem "Drinking Wine" is preceded by an explanation about the relationship between wine and poetic output—a relationship echoed in another of Welch's lines, "I can finish the rest of the wine, write poems till I'm drunk again" (RB, 69). Wine was seen as a way of "clarifying awareness" in the same way that tea was drunk to heighten awareness, practices that it could be argued Welch was aware of but, given his growing dependence on alcohol, was incapable of controlling. Indeed, the above poem serves to illustrate that drunkenness results in Welch's inability to cross the bridge that leads back to his cabin. In various letters to his friends, Welch often emphasized that this bridge was *"on the other side of the river."* As such, the bridge could also symbolize the metaphorical obstacle that Welch needs to negotiate in order to achieve "serenity of mind"—where the bridge is the way to "the Other Shore," in the most widely used and influential of the Buddhist sutras that he was aware of, the Prajñāpāramitā Sutra. However, it could conversely constitute the vehicle by which this nirvana is gained:

The Bodhisattva who wishes to set free the gods and men,
Bound for so long, and the beings in the three places of woe,
And to manifest to the world of beings the broad path to the other
    shore,
Should be devoted to the perfection of wisdom by day and by night.[10]

While Welch was far from capable of fulfilling the role of bodhisattva —indeed he called on the Bodhisattva of Compassion, Avalokiteshvara, for guidance in his own search, calling himself Vimalakirti, who was a Mahayana *upāsaka* or lay practitioner—this necessary devotion to wisdom was certainly a key objective. The potential failure to achieve this objective was ever present, and the contradictory nature of his mind is further substantiated by Welch's own reference to his abandonment by the moon (the "Lady of the Sky"), which in Shanshui is an important symbol of enlightenment and in Buddhism represents *bodhichitta,* the "mind of enlightenment." In this case, not only does the moon deny Welch the visibility that he needs to reach the other shore but also it accentuates his own carelessness in forgetting his flashlight, a fact that further separates him from Su Tung-po, who was given palace escorts to accompany him home.

The image of a drunken Welch stumbling through the dark forest is a far cry from the composed nature set forth in the Prajñāpāramitā, thus illustrating the length he still had to travel. Indeed, in an unpublished piece he likely wrote at the Forks in 1963 called "Commentary on Prajñāpāramitā," it is clear that Welch is aware of the task facing him, being "trapped in the human mind" and seeking the miracle of escape through the "proofs of *discontinuity,*" where *discontinuity* implies the breaking free from the "universe of continuity" in which humans live.[11] This in itself further suggests the cyclical and unending nature of *samsāra*—the Buddhist ideology that life is a continuous cycle of birth, existence, and death, broken only by the attainment of nirvana. Welch concludes in his commentary that this search alone will be what "frees us from THE USUAL EXERCISE OF THE human mind"—thus continuity and entrapment. The question as always, however, was, to what extent would Welch succeed in achieving this?

————

Although small in terms of the number of poems that Welch felt were good enough to make the final cut, *Hermit Poems* has a peculiar density in terms of subject matter. In addition to his musings on Eastern philosophy, the book contains among his earliest and best poems dealing with nature and with what Welch increasingly felt was the diminishing relationship between nature and society. While many of the poems are celebratory and establish a clear sense of his immersion in his surroundings, a certain ambiguity remains in some, none more so than in "Step Out onto the Planet," in which Welch challenges the reader to look beyond what they choose to see and to be more investigative about the wider world around them. It is a call to look for the details. For the unknown. For clarity. However, there is no answer. As in "The Rider Riddle" (discussed in chapter 16), only the individual can arrive at a satisfactory answer.

> Step out onto the Planet.
> Draw a circle a hundred feet round.
>   Inside the circle are
> 300 things nobody understands, and, maybe
> nobody's ever really seen.
> How many can you find? (RB, 73)

While clearly related to the metaphorical search for more substance in our individual lives, the poem also contains a veiled reference to the Tao, in that Welch specifies a specific number of "things" that can be found. Similarly, the Tao suggests that ten thousand things or, more simply, "myriad things," are ultimately begotten by reason. Yet of these things, how many do we, or can we, actually see? In reducing this number and localizing it, is Welch then asking us to use our reason more accurately? Have we seen some of these things but not understood them? Or are we incapable of seeing or understanding?

Forks of Salmon provided Welch with the opportunity to practice what he preached: he drew the metaphorical circle around himself, and the details of what he found can be easily seen in his poems. "Preface to Hermit Poems, The Bath" is a perfect example of Welch's ability to see much of what he had surrounded himself with. Whether the vision of "frost-flowers, tiny bright and dry like inch high crystal trees" or the sound "spring-rain tin-roof clatter" (RB, 67), the accuracy in the writing accompanies the challenge of finding those three hundred things.

Yet the poems also contain reverence and a sense of sincere praise for these things. Welch observes and is inspired. He establishes a link between the cyclical nature of animal life and the writing of poetry. There is no escaping the beauty of nature, and therefore poetry comes easily as a result. Similarly, this reverence extends to his use and reuse of the resources at his disposal: "I replaced my rotten stoop with a clean Fir block" (RB, 71). There is also no room for waste, for neither financial nor ethical reasons: "I seldom let a carrot go to seed, and I grind up every kind of grain" (RB, 72). These principles would stay with him during his time at the Forks and also years later, when he strongly advocated that we vacate the cities and connect more closely with the vast expanses of nature that surround us.

However, his life at the Forks had another side that was equally important to Welch — that of his role within the community, as a neighbor and a friend. A member of the tribe, so to speak. Having moved there, Welch immediately saw the benefits of getting acquainted with the locals. After all, without the local firefighters, he never would have found Meyer's cabin. Also, when the winter began in earnest, he was advised by the old Wobbly's best friend, Dan Wann, to avoid the impending snowdrifts and move into the Meyer's final abode along the river, "a tiny cabin of CCC sections with ceiling 6'2" high" that was also a tranquil place to write.

And write he did. To Whalen: "Am WRITING again. Many new poems. Good ones" (IR2, 61). To Snyder: "I'm writing like crazy" (IR2, 80). To Doyle: "Huge productions! New works!" (IR2, 91). To Allen: "I am writing steadily and well. . . . So far I have about 30 pages that hold together fairly well" (IR2, 86). His plan was to compile not only a collection of poems but also "a long rambling prose work" about his life on the river called "A Place to Put the Typewriter." In the end, however, this intended prose work appears never to have materialized, and the "Hermit Poems" were eventually whittled down to a mere ten poems. This was the first real example of, to quote Robert Creeley, Welch being an "intensive perfectionist, hard on himself in this respect since he felt that many of his own poems were lacking and so refused their publication."[12] Indeed, many of the poems Welch wrote at the Forks have not survived, while others were relegated to the pages of travel journals or remained as rough sketches unworthy of further development.

By the time Allen published *Hermit Poems* in early 1965, as number eight in his Four Seasons Foundation's Writing Series, the collection had been transformed from "a group of about 20 short lyrics" (IR2, 124), titled "The Hermit Songs," to its finished form: ten tightly gathered poems reflecting a period of hermitage but telling only a fraction of the story.

16
Can One Spend His Life Drunk Like Li Po?

"Life in the world is but a big dream;
I will not spoil it by any labour or care."
So saying, I was drunk all the day,
lying helpless at the porch in front of my door.
— Li Po, in Arthur Waley, *The Life of Li Po*

Alcohol has long been the medicine of choice for artists and writers alike. And much has been written of its influence—as both a stimulus and a debilitation. Its use can be traced back to the Greeks and their tradition of reciting poetry during banquets and symposia, and to the Chinese poets of various medieval dynasties. Indeed, the benefits to poetry and its composition under the influence of alcohol was stated as early as the third century by the Jin Dynasty poet Ruan Ji, who wrote in his poem "Xiuxi Yin" that "Fleeting worldly affairs have a ridiculous urgency; / quiet and sad feelings are wasted heartache. / How does one resolve sadness?"[1] Simple: "The answer is wine" and its ability to extract the poet from the world and its problems. By going "into the hills, we can immediately forget about our worldly affairs." Wine allows the poet to take "pleasure in the mountains or streams" unhindered by the world of man, and once drunk, "a cup of wine can bring 100 stanzas of poetry." Ji was associated with the Seven Sages of the Bamboo Grove, whose ideology, like that of many of their contemporaries and successors, was to escape the rigors and conventions of court life in favor of a life in nature where they could enjoy a certain peace while writing, drinking, and enjoying their freedom and a sense of unbound spontaneity.

Clearly there are similarities to Welch, not only in his escaping to the hills, but, as mentioned, in his almost ritualistic daily tendency to finish any wine at his disposal and write poems until drunk again. Indeed, throughout his entire collected writings—whether poetry, short stories, or correspondence—the looming specter of alcohol is never far away. In some of his earliest surviving letters, written during his military service, the presence of alcohol is already apparent. He wrote to Dorothy of his off-duty drinking escapades on leave in Amarillo with Gilbert and other cadets. While his increasing dependence on alcohol was bad enough, it was a recipe for disaster when coupled with regular bouts of depression. Welch's personality had always been such that he could swing from melancholy to euphoria and back with a relatively limited amount of alcohol in his system (he once wrote to Snyder that he could "survive best on half bottle of hardstuff (gin Scotch etc.) per day" (IR2, 31).

Certainly not uncommon among alcoholics, meanness and a violent streak can often be the natural consequence of alcohol abuse, and in some of Welch's unpublished letters (especially to Dorothy), he often finds himself apologizing for remarks he has made to her or excusing himself on account of some act of apparent mean-spiritedness. His deepening dependence on alcohol as a means of counterbalancing the demons and angels in his life and work at times even superseded his spirituality. Welch's focus on and participation in serious zazen practice was often also lost, as was an element of discipline. While he had been happy to partake in intense meditation at Marin-an, he openly rebelled against the idea as time passed and his alcohol intake grew. While he would have jumped at the opportunity to sit under the guidance of Suzuki Roshi in previous years, by the time he had returned from the Forks, the master's sessions at Sokoji Temple—one of the premier Soto Missions on the West Coast at the time—elicited only disdain. In discussing his own zazen participation, Donald Allen also described how Welch was often too hung over to accompany him, and that Allen would encourage Welch to go along nonetheless. However, Welch apparently now took exception to the Zen practice of hitting people on the shoulder with a *keisaku* stick if they fell asleep, stating "I'm not going to let that little Jap hit me."[2]

The use of the *keisaku* to ward off sleepiness or lapses of concentration during meditation is a traditional aspect of zazen that Welch was fully aware of, and Suzuki's wielding of it was no different than that of any other *jikijitsu* or zazen master (including one of his prior inspirations, Nyogen Senzaki). Indeed, at Marin-an, he had not only written of the beauty of the sticks that Snyder had sent him from Kyoto but may also have been responsible for wielding it as an essential form of discipline himself. Welch's previous proclamation that he was "a Zen Buddhist partial to the formalism of the Rinzai Sect" (IR2, 161) now seemed like hollow words. The formalism of zazen and the *keisaku* were apparently less important than the regular need for liquor.

One crucial aspect of Rinzai Buddhism—indeed the aspect on which it bases its practice—that continued to attract Welch, however, was the emphasis on using Zen koans as a means of raising awareness of the self. During "koan-study," monks are given riddles or cryptic and often paradoxical questions to consider during zazen. The objective is not to solve the koan on a literary or contextual level but to encourage and engender more insightful and intuitive thoughts on the nature of their own existence. Successful presentation of the "answer" will thus result in further koans until Buddha-nature is achieved.

This search for Buddha-nature and a clearer understanding of the self was a quest that Welch never gave up on, even as his dependence on alcohol increased. And writing his own koans was his way of undertaking this. Not only did he introduce a riddle into *Wobbly Rock* but also he regularly peppered his work with others, assuming the role of what he called the "Red Monk" to offer comments on them. While not representing a specific Buddhist figure, Welch's advisory alter ego may well have been based on Tibetan monks who traditionally wear red robes. This incarnation was Welch acting as a kind of *roshi* or spiritual guide in order to pose questions that his readers could ponder while meditating on his poetry. In the notes he wrote to accompany the riddles, Welch stated that they were "the first American koans. They are koans for beginners, making no claims for Perfect Enlightenment, but those who solve them will discover deep spiritual insight" (IR2, 127). One such riddle was "The Rider Riddle":

> If you spend as much time on the mountain as you
> should, She will always give you a Sentient being to ride:
> animal, plant, insect, reptile, or any of the Numberless Forms.
> What do you ride?
>   (There is one right answer for every person and only that
> person can really know what it is)

By way of assistance, the "Commentary by the Red Monk" offers us a focus for our meditation:

Manjusri rode a tiger. One, just as fierce as he, rode a
mouse. There is no one who can tell you what the answer
is. The Mountain will show you. (RB, 128)

The note immediately provides the reader with a certain amount of
context. Manjusri—one of the oldest and most important bodhisattvas in
Buddhism—is seen as the embodiment of the *prajña*, or transcendental
wisdom, found in the Prajñāpāramitā Sutra. The image of Manjusri riding a tiger is a striking one, and as such engenders a certain sense of
who he is—and, given his vehicle, what we might want to be or aspire to
become.

However, the deity riding the mouse is perhaps the image Welch
intended to use to focus our thoughts more acutely. Legend has it that
Ganesha—the elephant-headed Hindu deity that is represented in
various forms in Buddhism—rode a mouse while fulfilling his role as the
Destroyer of Obstacles. Aside from functioning as a vehicle with which
to clear a path to an unburdened mind, Ganesha is also seen by many
gurus as a symbol of exalted intellect and wisdom—facts surely not lost on
Welch in his use of this image as part of a koan. "The Rider Riddle" is thus
not only a test for the mind but also a means of eradicating the obstacles
that block *our* path to enlightenment. Welch himself would later write
that while he desperately wanted his vehicle, or *vāhana* as it is known in
Sanskrit, to be a "cougar, (I, Leo, etc.)," in the end it was on the wings of
a turkey buzzard that he made his final journey.

———

Escaping the *keisaku* in favor of the fishing rod was an apt alternative
for Welch. Since he was a boy, Welch had seen (and enjoyed) fishing as
one of the few things that provided him with a connection to his father.
Welch Sr. had first shown his son the basics of fishing, as he did with
hunting, and it is tempting to think that, like his first rifle, Welch also got
his first rod and tackle from his father. In the years that followed, fishing
became one of the mainstays in his life, particularly after relocating to the
West Coast. Whether on lone trips to the Rogue River or working on the
salmon fleet, fishing elicited a certain meditative quality that attracted
Welch. Indeed, it could be argued that fishing offered him a viable and

more easily attained alternative state to the more disciplined practices related to zazen. In a letter to fellow poet Larry Eigner, Welch states his intention to use lines from one of Eigner's poems ("I have to use your lines . . . there are all types / of an animate gaiety" [IR2, 38]) as a dedication to a section of a long poem he was working on about commercial salmon trawling. He notes that fishing has finally freed him and that he would try to refrain from including his customary bitterness and allow the content to reflect his love of the craft rather than his frustrations surrounding any of the peripheral issues plaguing him at that time—primarily concerning notions of work and finances but surely supplemented by his increasing alcohol (ab)use. Indeed, in long prose fragments he wrote in the early 1960s about salmon fishing, he laments the "many years working at foolish, exhausting, tiresome, humiliating jobs" and how he sees the need to use this backdrop for his treatment of the "real work that I wish to speak of: how it is to do a proud thing which earns you your house and your food and your woman; how it is to actually go fishing, that is to catch the fish for people to eat" (IR2, 43–33).

What remains of these fragments (titled "Fishpome")—both published and unpublished—was written between June and September 1961, at a time when Welch's relationship with Kandel was at its height. Indeed, in a poem from the same period that is suggestive of her relationship with Welch, Kandel wrote an ode to her lover, describing not only the nature of his physical return but also offering a moral perspective about the lengths fishermen are required to go to for their quota:

> my love the fisherman comes back
> smelling of salt drying
> his ocean arms embrace me and
> I taste the death of seals on his
> thin mouth
> they eat the fish—and so
> he shoots them with a .22
> sighting among the green waves
> so do we, I think,
> wondering if the seals
> will ever invent gunpowder[3]

"Fishpome" itself offers no such dilemma but rather praises the sea and the bountiful life it holds and also engenders a thankfulness in Welch as he stands on the bow of the boat at sunset. In his ledger from the summer of 1961, Welch wrote a section that would serve as a possible ending for the poem. In it, we find a clear sense of reflection and melancholy as he takes stock of those aspects of life that can be put into perspective through fishing:

> Stood on a bow at sunset in a
> clear calm, the sun
> a red ball squashed on
> the bottom
> had a
> Three pronged sea-weed with
> Tufted ends and
> Prayed:
> With this flower I
> Thank you sea, I
> Thank you my beautiful
> Plant, I
> Thank you my girl Lenore
> My
> Many beautiful friends
> Into the never ending complaint,
> The
> Whining of my kind,
> Let it be known there was
> One who was glad and grateful . . . [4]

However, this praise is tempered in other sections, becoming somewhat more arbitrary in context as Welch describes the details of baiting or the visitations of birds or porpoises to the boat. The positivity is nonetheless reflected in his use of Eigner's "animate gaiety" to cover the multitude of beings that populate not only the sea but also his life as a fisherman on it.

"Fishpome" also includes yet another example of the paradoxical nature of Welch's relationship with Buddhism. In describing the process of shooting, gaffing, and gutting the salmon, Welch posits the statement,

"You can't do that and be a Buddhist," only to answer, "So much the worse for Buddhism" (IR2, 40). Was zazen now definitively replaced by fishing? And Buddhist ideology replaced by the grateful necessities of commercial trawling? For the foreseeable future at least, it seems that fishing and the sea would be Welch's next chance to escape. The ocean his next hermitage.

———

Looking for a reliable trawler on which to fish, however, was not as easy as Welch had thought it would be, even in a city with a huge commercial fishing port. Having again refused to take a job at the post office, and with alternative employment hard to come by, Welch had been working as a de facto janitor at the Batman Gallery, earning whatever he could and, as always, looking for the next big break to establish some sort of financial security. Indeed, in March 1962, to coincide with an exhibition of works by Bruce Conner at the gallery, Welch wrote to Dorothy that he owned twelve of Conner's drawings and had consigned some to the gallery.[5] He hoped that they would sell during the exhibition, as Conner was "real *in* right now with the New York crowd" (IR2, 48), and if they were to fetch the going rate, he would make close to $300—a significant amount at that time.

However, fishing was not a purely financial venture. In the summer of 1961, he had written to Dorothy that while fishing was arduous, frustrating, and painful, he also found it "beautiful & wild," likening it to "a state of near nirvana . . . gladness as remote from ecstasy as it is from fear" and thus once again using Buddhism as a means of illustrating the meaningfulness of fishing at that moment in his life. Indeed, he saw fishing as the potential cure for his financial problems and realized it gave him the feeling of "never having my life be my own." In the same letter he wrote that to own a small boat would afford him the opportunity to "make a good, free living—doing something of a well-made man. Rare in America . . . How I'd love to have my own boat, my own life" (IR2, 33).

At that time, though, Welch was at the beck and call of the various captains and owners, and as such had to take what he could get. Don Crowe, who Welch had almost begged for work in a letter at the beginning of the year, initiated Welch into the industry, and in response to Crowe's warnings of hard work and bad smells, Welch wrote, "I am so tired of being without work, money, social position! So what if I'll stink?

Crab will have to become Lenore's favorite perfume! Tell that man I'll work till I drop!"[6]

And so, after learning the ropes on a boat that Crowe had leased and helped to overhaul, Welch got his first real job as a boat puller on the *Annie G.*, a "double-ender, formally a schooner, keel laid in 1925," which despite his initial enthusiasm turned out to be "a filthy unhappy mismanaged boat run by a psychotic." Yet even though he tried to temper his joy at having found a new focus for his energies, fishing was obviously a profession that appealed to Welch for all manner of reasons, not least the fact that the "work is connected with things I know are real: weather, animals, tide, fatigue, cranky tools" and that it "demands all that men can do, is a challenge to every resource, requires being *with it* totally" (IR2, 34).

As a boat puller, Welch was required to haul the catch on board and was often subject to the inadvertent violence of fishhooks, gutting knives, or other assorted equipment on his "poor, swollen hands." It is easy to see him in the role of indispensable crewman in the mold of Johnson, that "fine chap . . . [and] best sailorman in the fo'c's'le. He's my boat-puller" in Jack London's novel *Sea Wolf.*[7] But even though salmon trawling was subject to the many other vagaries of life at sea, it was a demanding and unforgiving experience that Welch seemed to relish. He optimistically wrote to Snyder of "a changed life, another crack at it from another direction, a transformation, a shed skin & exposed vitals" (IR2, 34), and he is clearly prepared to expose himself again not only to the elements and the inconsistencies of such a life but also to their physical and psychological consequences at a time when he felt "very strong & well" and was in a stable relationship with Kandel, who he said "makes everything work for me" (IR2, 33).

In August 1961, with his experiences of the *Annie G.* behind him, Welch found himself working as a baiter on a salmon trawler out of San Francisco called the *Chico.* The boat was owned by Bill Yardas, a fisherman from Oregon Welch had known for a number of years; he wrote of how this was far more in line with what he had envisioned fishing to actually be like. Writing to Snyder, he lauded the adventure and excitement of trawling for "huge fat old salmon" without the accompanying "20 hour days, daily flips, disasters, frights, tizzies & the like" (IR2, 36). By comparison, working with Yardas was like a dream, and Welch later wrote a detailed description of it in a fragmentary essay titled "On Salmon Fishing," in which he describes the daily (and nightly) routines that accompany trawling not as nightmares or inconveniences but as real work in which all manner of life races by "going at every speed at once" (IR2, 45).

Although Yardas also remembered working with Welch, their relationship developed into a great deal more than a working one in the years that followed. Indeed, Welch spent some of his final weeks with Yardas and his family in Oregon during Easter 1971, only a few weeks before he disappeared. Yet looking back to the *Chico*, Yardas wrote of Welch introducing him to the concept of ecology over a meal of cioppino and red wine while at anchor behind the Duxbury Reef off Point Reyes. Yardas asked Welch what he would have done differently had he had his time over again, to which Welch replied that he would have liked to have been an ecologist. It was a simple and genuine answer at a time when Welch was at his happiest. The work was real, as was the friendship. It had inspired him not only to write about the joys of fishing but also to dream about becoming a bona fide fisherman himself. About owning a modest trawler—to be called the *Capricorn* after Lenore's star sign—before moving on to something more robust and better equipped. In a telling moment of self-reflection and forthright honesty, Welch admits to being happy in his own skin. Happy at the prospect of living for his own ideals and impatient with his own misery. He writes that it is "amazing how everything is falling together for me at long last" and that fishing is "absolutely necessary for all kinds of reasons" (IR2, 37)—not least, one suspects, as a powerful source of inspiration that also provided him with a sense of accomplishment and a oneness with nature around him. Welch was part of an indefatigable cycle of life and death. And perhaps it was this notion that he retained when, only a few months later, his life fell apart and he went to the Forks of Salmon in search of not only himself but also a greater understanding of his own place in that cycle.

17
The Journal of a Strategic Withdrawal, Part 2

We are conscious of those within range.
— Lew Welch and Kirby Doyle, "Second Joint Statement"

During the 1962 holiday season, Welch returned briefly to Palo Alto to spend Thanksgiving with Dorothy. He had recently heard that fellow poet and editor of *Floating Bear* magazine, Diane di Prima, was intending to publish his "Rat Flat" letter to Kirby Doyle in the magazine's first edition. On returning to the Forks, he stopped off at Doyle's home in Larkspur; their friendship as strong as ever, they hatched a plan to collaborate on a magazine project called "Joint Statement."

The idea seems to have been a spin-off from their apparent practice of what Michael McClure once described as the pair creating "an infamous proclamation containing a group of demands and threats, laughing their heads off as they composed it."[1] The plan was to issue a one-off magazine in which both poets made statements on various subjects under different pseudonyms (including two that Welch had used previously, "Jimmy Vahey" and "Leo Keeler"). It also appears to have been intended as a classic Welchesque money-spinning venture, doomed to fail as always but not without the usual optimistic bravado. Welch suggested that by using "various fake names . . . we not only ensure that the magazine will contain really first-rate stuff, but also we can assure ourselves most of the loot" (IR2, 95).

Welch and Doyle had known each other since the late 1950s, when Welch was living in the East-West House. Doyle and his then wife, Didi, had an apartment on Sacramento Street, close to where Didi had opened a bookstore called the Golden Bough on Fillmore in 1956. Doyle had been attending Kenneth Rexroth's classes as part of the curriculum at San Francisco State College's Poetry Center since that same year and had published poems in the college's literary magazine, *Transfer*, as well as in the special edition of *Evergreen Review* that was devoted to the San Francisco Renaissance. Similar in nature and background, the pair had an immediate chemistry between them that lasted for the rest of Welch's life.

Although six years younger than Welch, Doyle was in many ways the perfect match for him. Both were of Irish descent. Both believed firmly in the use of accurate colloquial speech in their poetry. Both were hopeless romantics at heart. And both were very fond of drinking. They collaborated on the "Din Poem" reading at the Batman Gallery, and Doyle had also featured in Donald Allen's *The New American Poetry* anthology. Like Welch, Doyle had been working on an autobiography on teletype paper, spending much of 1959–1960 on what would eventually be published almost a decade later as *Happiness Bastard*. And, just as Welch extricated himself from society at regular intervals, so too did Doyle, spending more than a decade off the grid, reexamining his life and language through what Charles Olson described as "discovering this discarded thing nature."[2]

During Welch's time at the Forks, Doyle offered like-mindedness at a time when what Welch needed most was a mirror in which to see himself—and what better mirror than a kindred spirit. Welch described them working together as a "huge outrageous twin-genius love shout" (IR2, 88) and complimented Doyle on his ability to apparently capture in writing what Welch himself felt that he could not—an accurate and beautiful "vision we all have of this generous, easy planet" (IR2, 89).

It also seems that Welch found in Doyle an outlet for his own issues of loneliness and instability, writing apologetic letters that bemoaned his own insecurities about, for example, whether Doyle had enjoyed a visit to the Forks:

> I worry that maybe when you were here you were disappointed, not by the planet but by me—the way I move on it, even here . . . [and] that you didn't enjoy your stay here for lack of the real mountain guide I am when mortally ill. (IR2, 91)

Welch also speaks of finding himself in a metaphorical "rest home" where the "phantom nurses and doctors are very lenient," thus excusing his own laziness and allowing him to arrange his various therapies in such a way that he "may or may not accept" them. It is a loosely organized day-to-day existence made up of "looking at things, helping others, writing, working prodigiously at real things like getting in the wood" (IR2, 89–90).

And despite his being "mortally ill" in one paragraph, Welch was once again able to change tack completely and, through the sharing of a positive anecdote, paint an altogether different picture of not only life at the Forks but also his own state of mind. In the same letter to Doyle, from December 21, 1962, he lauds the simplicity of codependence and the generosity of his friends, stating simply, "What more could a man possibly want?"

The first proposed "Joint Statement" was mooted in February 1963. Welch suggested that, after placing an ad in *Evergreen Review* to generate production costs, and appointing Wallace Berman as art director, he would start working on a piece titled "Free Trips, Available to All," which would then be used in the magazine. These ideas were clearly meant to be taken lightly and were developed "in the middle of an exquisite natural high . . . brought about by boredom & loneliness." These "great free trips" included

The next time there is a warm, light rain go to the park and lie down on your back in some nice grassy spot. Look directly up into the rain. Try not to blink when a raindrop is coming directly at your eye.

And

Submerge your head in the toilet bowl.
Open your eyes. (IR2, 95)

The frivolity in these statements is clear, and it is easy to see Doyle and Welch "laughing their heads off" when writing them. However, despite the plan to release a magazine, the second "Joint Statement" they issued was a single signed mimeograph that simply stated, "We are conscious of those within range." It was written at a time when Welch was especially attuned to his surroundings and may well relate to this heightened sense of (self)-awareness that he felt he was developing. It was also around this time that he finally completed the definitive version of "Ring of Bone," the poem that was born in Big Sur but that needed this hermitage to reach its conclusion. By including it in the same letter to Doyle alongside the "Free Trips," Welch shows both his humor and his seriousness, and as such, the second "Joint Statement" can be read as a recognition of Welch's own consciousness regarding the myriad characters that he found himself to be at that time—whether "Leo" or "our friend Lewie" or any other of his incarnations.

———

In April, Welch and Doyle read at San Francisco's International Music Hall as part of the "Voices: 1962–63" poetry festival, which saw readings over the course of nine months by the likes of Philip Whalen, David Meltzer, Andrew Hoyem, and Michael McClure. Welch's reading with Doyle had been in the planning for some months and would not only provide him with another opportunity to read some new work but also to do so with Dorothy in attendance for the very first time.

His relationship with his mother had been largely conducted at a distance since he had left Reno and returned to San Francisco in 1960. However, like all good children (especially those in continual need of financial sustenance), Welch visited his mother at Thanksgiving and Christmas, sent birthday cards and updates on his progress in the wilderness, and on occasion expressed his gratitude and flattery with a poem ("Thanx for the roast and Sunday"). Surely, therefore, finally seeing her son in action must have constituted something of a triumph for Dorothy—affirmation that all her support was somehow justified and had not (yet) been in vain. In a letter from April 22, Welch wrote to her, "I didn't properly tell you how pleased I was that you could hear my reading & that it moved you so. It was great triumph. I am now called 'Lewis of America' by all my friends."[3]

Upon his return to the Forks, Welch set about working on a longer piece of poetry titled "Spring Rain Revolution at the Forks." He felt that his work lacked structure and, of the many obstacles he faced when writing poems, one of the most acute was his sense that poetic form is created in the mind but often proves difficult to transfer onto the page. In addition to the works that would later become *Hermit Poems*, Welch had also written a short essay on the manner in which he felt poetry should be written, or rather "discovered," by the poet.[4] The essay offers not only a clear insight into his ideas about the poetic process but also a vision of

Welch as a poet at the Forks. He had often been accused of possessing a level of self-criticism regarding his poetry, with an unwillingness to let it go until it adhered to extremely stringent internal guidelines. And such essays provide greater clarity into why this might have been.

Beginning with a simple statement on the invention of a poem as a product of "the poet's fact of mind which was set in motion by a nearly unobstructed vision of All That Is Out There," in this essay Welch is reiterating his earlier notions that poems require clear vision and insightfulness to take shape. In some ways Welch is also providing a blueprint for reading his work, giving his readers an inside look at how the poems are developed and on what basis they should be read and experienced by the reader. He describes how the poem is an already established entity that only "discovers what happens when this nearly undifferentiated vision starts settling itself in the mind" of the poet. The vision itself is "wordless, like surf," and provides nothing more than the basis for the poet to watch "the process of interpretation and integration." He further suggests that the poem is akin to a reporter who, while watching the "behavior of steel particles advancing, retreating and joining together in the presence of a new, magnetized, particle" provides coverage of the process much like a queen bee that is involved in the "furious activity" around it while being detached from the process itself. In this way Welch contests that, rather than constitute a "performance," the poem "performs" instead—striving for completion through the process of its own creation. And in so doing, it presents itself to the reader devoid of any meaning but rather symbolizing a birth that awakens an acute sense of "excitement and joy and a deep sense of accomplishment in the mind of the reader," due to it having caused the reader to perform in such a way as if he or she had written the poem themself. The essay suggests that the poem is an independent entity that forms in the mind of the poet only to re-form in the minds of its readers again and again, thus re-creating itself with every reading. It is a process that reflects "the exact picture of 'thinking,' not of joined, already-invented thoughts" but, rather, of discovery.

Struggling to find a suitable structure for "Spring Rain," however, Welch turned to Rainer Maria Rilke's *Duino Elegies* for inspiration. For him, Rilke was an example of a poet who had been able to construct a form that allowed his longer poems to be connected into a series such that the "'deeper meanings . . . darker thoughts' can gather as you go

along" (IR2, 106). In this way, Rilke used his poetic structures to create a momentum that retained a feeling of serenity before exploding at
particular moments throughout the poem. In some ways it is possible to
see that Welch's intention at times was similar to Rilke's, inasmuch as
he also wanted to be able to do what Steve Silberman once observed
about Rilke: an "ability to sustain a tremendous amount of abstraction in
a poem, without disrupting the reader's ability to envision the landscape."[5]
If anything, "Spring Rain" was intended to provide a landscape in which
language, however abstract, could create a "Big Sung landscape scroll" in
the minds of readers, and in which certain images would also explode off
the page,[6] thus allowing the poem to be "heard" in the mind.

Welch may also have had a natural affinity with Rilke anyway, given
the fact that the latter also battled with severe bouts of depression and
regularly suffered from creative impasses as a result. Indeed, in some ways
Welch's successes were similar to Rilke's, especially the *Elegies*, which
represented something of a triumph, given that Rilke only managed to
complete it after a lengthy hiatus brought on by depression.

Spring was slow to begin at the Forks, and the late snow and rain only
dampened Welch's mood still further. The Forest Service would not be
recruiting until June when the fire season started, and Welch was again
neither writing nor working. Indeed, in a letter to Whalen he wrote that
he was "tired of thinking and writing and reading. . . . I don't have a single
line of poetry in me and I'm glad" (IR2, 112). In addition, he admitted
to once again being tired of himself and of his desire not to be given a
lookout job but to be on a trail crew instead—a sure sign that, in his current state, company beat loneliness.

Until the opportunity for work arose, however, Welch contented
himself with panning for gold and spending time with the artist Jack
Boyce, who had also left San Francisco and was living close to Welch in
Cecilville, where the local bar regularly provided a refuge where both
poet and artist could drown their mutual sorrows and discuss their individual positions in a counterculture neither seemed entirely comfortable
in. Writing to Snyder in February 1963, Welch described Boyce, who
would later marry Joanne Kyger, as "a fine fellow . . . intelligent, hip &
disgusted by everything and a great lush" (IR2, 97). The pair would drive
to visit one another, get drunk, solve the problems of the world and, in
Welch's case, sleep for most of the next day. Welch was clearly impressed

by Boyce and spoke of how he felt that exposure to the right circles back in the city could help Boyce realize his potential as a painter. Having gone to Siskiyou County in an apparent attempt to escape the city for a while, Boyce was apparently "spending all his time trying not to be a painter"—something that Welch felt should (and could) be remedied if Boyce were given the opportunity to see another side to the artists, writers, and musicians that contradicted the "dirty, shiftless, hate-filled, doped up unhiply, lack-genius side of this whole generation" (IR2, 97) he had thus far apparently been exposed to.

There seems to be a certain irony in Welch championing and encouraging Boyce to shake off the shackles of his creative listlessness, given Welch's own situation. Yet to some extent it may have worked. Within a couple of months, Boyce was painting again and stretching new canvases in anticipation of new works—a fact that Welch was only too happy to share in his letters. While it might also have been possible that Boyce's productivity rubbed off on Welch, he seemed not only as tired of himself as ever but was also sometimes gripped by an overpowering vision of "the huge agonized face of all victimized humanity, tyrant & slave of every place and time" all with their own "atrocity stories, each as unnecessary & pointless & appalling as the next" (IR2, 118). This image was then juxtaposed against that of a "nice little green planet," which, upon merging into that of the aforementioned face, seems to question the legitimacy of its inhabitants in the style of a policeman. It also implies that these atrocities—like "a movie stuck on a single frame"—are irrevocably linked to our relationship with the earth and cannot be so easily erased.

18
The Mountain Man Returns

The soldier returns. The shy inhabiter of rooms, returns. The husband returns. The frightened girl. The boy who cannot tell, just yet, how right he is.

—Lew Welch (RB, 92)

Welch finally left the Forks of Salmon for good in late November 1963. He had been working on and off as a ranger clearing trails for the Forest Service since before the summer and was looking forward to winter in San Francisco. Since arriving at the Forks eighteen months earlier, he had endured four seasons of relative solitude and wanted to get back to society before he became "as crazy as Han Shan" (RB, 84). Of course, this solitude was tempered by frequent visits to and from friends and acquaintances both in San Francisco and the local community, and the summer ended with a long visit from Dorothy, who was still living in Reno. Despite this, however, his need to return for an extended period was clear.

Despite various ups and downs, Welch's poetic output had been steady. He continued to work on "Spring Rain Revolution at the Forks," which now contained a carefully compiled collection of works that represented how his hermitage ended and his return began. In this collection Welch packs his belongings, makes the necessary preparations, leaves the Forks, and spends his first night back in San Francisco.

Welch's idea for this new poem was simple. In terms of form, he envisaged that it would "be nearly 50 pages long. Structurally, it will move like [William Carlos Williams's poem] *Paterson* (that is, looser than *The Cantos*) but tighter than *Paterson*. . . . Each unit is a nearly separate poem" (IR2, 124). The poem would have a musicality akin to Thelonious Monk, who Welch once credited with providing him with poetic meter. Welch further said that "Monk and I are listening to the same source. It is the source of American speech. Americans talk like that and you can hear it if you listen to Thelonious Monk."[1] But Welch was again beset by self-doubt and the kind of biting self-criticism that had plagued his creative process for years. In a letter to Henry Rago, the editor of *Poetry* magazine, Welch wrote that "Spring Rain" describes

> a year of mine in the woods, last year, during which I tried to figure out in terms of everything—starting with the fact at hand: the oatmeal, the wood stove, the piss can, Avalokiteshvara, the essential shuck, the basic con, tyrant and slave of every place and time, my own loneliness. (IR2, 125)

However, in the same letter to Rago he also bemoaned that fact that the poem "rages into rhetoric" and "regularly appears to me as a work of the wrong kind of derangement" (IR2, 125). While Welch strove for musicality in a more free-flowing sense, his need for structure was a constant obstacle. And he understood this better than anyone—to write a series of sonnets and simply add them to the sequence was a blessing, but this was not possible, given his idea that contemporary poetic forms determined that structure only presented itself in its finished form after the fact. This philosophy must have been difficult to reconcile for a poet so hung up on trying to control such processes. In the same way that William Carlos Williams continued to offer inspiration, Welch was also more than familiar with the work of other poets from whom he could take structural pointers. Indeed, in trying to work "Hermit Songs" into the longer piece, he suggested to Rago that they could "sit as Rilke's *Sonnets* to his *Elegies* (certainly the process of their composition is the same)."

His return to society is all the more ironic given that, unlike the poets and yogis of which he had earlier written, Welch is leaving his reclusion and returning to the very object of his initial frustrations. Despite his prior insistence that he would "never, ever go back there again" and that everyone should "get out," at a certain point he concedes that his fatigue and loneliness necessitated the return. Yet in this collection Welch strives to cement the sense of *anatta* he had introduced previously. Again he employs the "He-persona" as narrator, but the tone and imagery of this collection undergoes a profound change as we see the poet preparing to return to his previous life and praising the things that he feels have sustained him during his stay.

His trepidation about returning to society is clear. For all the clarity of mind he gained at the Forks, it is more indicative of the reality of his experiences there that the opening poem, titled "He Prepares to Take Leave of His Hut," ends with a somewhat prophetic admission that he has neither found any Buddha-mind nor seems capable (or willing) of even trying to anymore:

> . . . that is just a
> pretty imitation of a
> state of Mind I don't possess
>
> or even seek, right now, or
> wait for anymore. (RB, 81–82)

Thus, necessity trumps perseverance. And it is left to the bodhisattva to continue to lead the way. Welch likens himself to Dante on a journey through some personal hell and humbly asks Avalokiteshvara to be "as Virgil for Dante" and also guide him.

While this admission of dependence on the guidance of others is perhaps unsurprising, there is an underlying sense of defeat that accompanies it. To be a member of the "Ancient Order of the Fire Gigglers," like Han Shan and Wittgenstein, was to have successfully shed the ego and

> . . . walked away from it, finally,
> kicked the habit, finally, of Self, of
> man-hooked Man
>
> (which is not, at last, estrangement) (RB, 84)

Yet, Welch appears to have been unable to do this.

The serenity that Welch captured in the earlier poems that became *Hermit Poems* slowly gives way to the reality of city life and the new perspective Welch now has of it—yet not before we are given clear vignettes of what he is leaving behind. In one final rendering of a Shan shui image, Welch writes,

Cloud-shrouded gorges!
Foggy trees!
I can't see the ridges anymore! (RB, 87)

It is tempting to see these lines as descriptive, lyrical images that represent Welch's own mind as he faces the consequences of heading back to the city without a clear sense of what awaits him there. Upon departure, Welch simply writes, "Shut the shack door." Yet unlike the previous evocation of the gate motif, this action has far more negative finality and may thus be seen to symbolize the very opposite of the Shanshui motif, in that it precedes a journey in which he leaves the door firmly bolted behind him and travels "400 miles through valleys of larks" back to San Francisco. As Shakespeare's "larks at Heaven's gate sing," so Welch leaves paradise; in so doing, he also leaves behind much of the serenity and contentedness he had gained while there.

―――――

Arriving back in San Francisco must have held mixed emotions for Welch. Although his initial intention was to spend only the impending winter in the city before returning to the Forks, somewhere deep down Welch had already realized that this was a permanent vacation. Writing to Donald Allen prior to his departure, Welch admits to wanting to save enough money to enjoy at least a few months in the city because otherwise another winter there would have driven him crazy. However, this desire to save money was, as always, tied to his ability to hold down a job. Only two months after his return, Welch was already complaining about his unemployability and that his solution was to apply for a position teaching Native American kids in Alaska. He had apparently already given up on returning to the Forks and once again set his sights on getting enough money to make the now fabled journey to Japan: a dream he had long held but would never achieve.

In the face of financial issues, Welch looked to find some sense of achievement in his writing. He had returned from his retreat with a significant amount of poetry, and he now had the job of organizing and arranging it into what he felt would be a publishable collection. Although Welch had missed the seventeenth annual San Francisco Arts Festival in October, one of his poems had been chosen to be part of a folio collection of eight large broadsides featuring collaborations between poets such as Ginsberg, Duncan, Whalen, and Ron Loewinsohn with various celebrated artists. Welch's contribution was "Early Summer, Hermit Song," which he had instructed Donald Allen to use and which was beautifully illustrated by William Weber and printed by Lawton Kennedy. The other poems were printed in various print shops throughout the city, including the Grabhorn Press and Auerhahn, and with each collection selling for $12.50 (individual issues were also available for $2.50), the print run of three hundred was a potential money-spinner for all the poets involved—Welch, given his circumstances, as much as any other. Allen later wrote to Welch that his broadside was "popular" and had sold well. "Early Summer, Hermit Song" was the opening poem in *Hermit Poems*, thus serving as a test for the material he had written at the Forks. Contrary to his often-self-deprecatory attitude toward new work, this turn of events must have been something of a triumph for him. Not only was it to be his first broadside, but for it to be part of such a collection was surely a significant boost.

Two months after the release of the folio collection, and perfectly timed to coincide with his return, Welch was asked by Andrew Hoyem and Dave Haselwood if he would like to take part in what would turn out to be his biggest poetry reading to date: the Auerhahn Press's benefit reading to showcase new work by six of its published poets: Welch, Ginsberg, Hoyem, McClure, Whalen, and Philip Lamantia.

Scheduled for November 26, 1963, the event was held at International Music Hall, which was located next door to the Batman Gallery and just around the corner from the East-West House. The venue was very much at the heart of the counterculture scene in San Francisco at that time, and it was the perfect place for such an event. Charging a $1 entry fee, Auerhahn cofounders Haselwood and Hoyem were looking to raise money for the continuation of the press; having poets with the pulling power of Ginsberg on board was more or less a guarantee of attracting a large audience. For his part, Welch read a mixture of old and new material, using the event as an opportunity to read some of the poems he had written at the Forks—parts of which were mooted to be included in a new magazine by Allen titled *New Review* (which would later become the Writing Series).

Allen had already been working as Welch's agent and editor for several years, and had tried to get Welch's work as widely published as he could.

Despite having accepted (and published) his short story "The Man Who Played Himself," *Evergreen Review* had continually rejected the poems submitted by Welch (or Allen)—much to the pair's mutual confusion and disappointment. In addition, other magazines were either unable or similarly unwilling to use Welch's work. To that end, Allen suggested that Welch approach Rago at *Poetry* magazine.

At that time, *Poetry* was one of the foremost literary magazines in the United States and represented, in some respects, the antithesis of Allen's more avant-garde tendencies with regard to poetry. Established in 1912, *Poetry* has published works by a wealth of major American poets and as such would surely have provided Welch with a boost should Rago decide to accept any of his poems. Rago himself had been editor-in-chief of the magazine since taking over from Karl Shapiro in 1955 and had overseen the publication of not only more established poets such as William Carlos Williams, Carl Sandburg, and Marianne Moore but also the underground publishing works by the Black Mountain poets, the New York School, and, of course, the Beat Generation. Indeed, in the letter Welch sent asking whether Rago would consider his work for publication, he praised Rago for the "more exciting magazine *Poetry* has become recently" and hoped that Rago would "find these poems usable" not only because of the prestige of featuring in such an eminent magazine but also because, quite simply, Welch was "broke" (IR2, 123).

The poems that Welch sent for consideration included several of what he was still calling "Hermit Songs," and Rago, much to Welch's delight, was willing to publish them.[2] In replying, Welch candidly wrote that the "acceptance of my poems brought about the most productive days I've had in several months. Thank you. I had not realized how badly I needed something to get me off my ass" (IR2, 124).

19
"Bread vs. Mozart's Watch"

By the spring of 1964, Welch was once again on a high. He had decided to stay in Mill Valley rather than return to the Forks, and he was looking forward to Gary Snyder's return from Japan.

Snyder, who had been in Japan since 1959, had always been something of an inspiration to Welch, even though Welch was often jealous of Snyder's successes, both poetically and personally. To mark his return, Snyder had been asked by Don Carpenter to participate in a reading with Philip Whalen that was planned as a means of not only celebrating Snyder's return but also promoting their work and providing what Carpenter hoped would be a more professional event than many poets in the city were used to. Snyder, Whalen, and Carpenter had originally met in Portland in 1953 when Carpenter was on leave from his post in the US Air Force, and Snyder in particular had first encouraged him to get involved in literary pursuits. Carpenter later arranged a small reading in his Portland basement at which both Snyder and Whalen read, and although Carpenter described that reading as a disaster at which "it was too crowded, too hot, and when we passed the hat to pay for the wine we only got $1.75. A window got broken and somebody trampled my nasturtiums," the saving grace was the poetry.[1] If such poetry could be performed with better planning and aforethought, it would surely be worth it. And so the reading that would become known as the Freeway reading was born.

Carpenter's idea coincided with a similar notion that Donald Allen had around the same time to pair Snyder and Welch, and so the two would-be impresarios joined forces and decided that Carpenter would put on a reading featuring all three. As Carpenter said many years later, "It was a wonderful combination. . . . They had totally different personalities and wrote totally different poetry."[2]

Planned for mid-June, the reading was to act as an accompaniment to a radio broadcast in which Snyder, Whalen, and Welch would discuss poetry and the role and position of the poet in contemporary society. Welch envisaged a massive turnout for the reading and saw it as an opportunity to not only further bolster his own reputation but also make a few bucks along the way—after all, Carpenter had promised to do all the work, with the poets getting all the money.

The event was held on June 12, 1964, at the Old Longshoreman's Hall on Golden Gate Avenue. Welch described it as a funky old place that had served as the base for leading members of the International Longshore and Warehouse Union (ILWU) during the infamous Great Maritime Strike in 1934. Welch wrote to Rago of it having "a monstrous mural (very well done) showing policemen beating women and children to death—the fuzz mounted on horses with hooves like Percherons, which hooves are crushing the heads of brave men" (IR2, 127).

Carpenter had rented the hall for $75; given that it could hold more than six hundred people, the potential for making a profit—a rarity for poetry readings—was clear. The trick would be to attract enough people to fill the place, and since most readings rarely pulled in more than a couple of hundred, it must have been with some trepidation that Carpenter set about organizing this event. The key to success in his mind was to keep the entry fee low and maximize the advertising. Handbills were made and printed by a friend of Welch, with five hundred being plastered up all over North Beach.

At the same time, Donald Allen had suggested that the three poets produce a signed broadside of their own work that could be sold at the reading for extra profit. As all three had studied under Lloyd Reynolds, the idea was to create calligraphic works in their own styles. Welch chose "Step Out onto the Planet," Snyder "Nanao Knows," and Whalen "Three Mornings." Three hundred of each were printed, and if preparation was any indication, the reading had all the makings of being a huge success.

To bolster publicity, Carpenter had also asked his friend, the journalist Ralph J. Gleason, if he could spare some inches in his weekly "On the Town" column in the *San Francisco Chronicle*. Carpenter had hoped for nothing more than a brief announcement at the bottom of the column, given that Gleason had never before shown any particular interest in poetry events. However, after an afternoon chatting and listening to jazz with Welch, Gleason offered Welch an entire column not only to publicize the reading but also to offer an introduction to the plight of the poet and how readings such as these were the first steps toward changing it.

As the poet most vociferous in bemoaning his financial status—despite being the least of the three affected by such issues, Welch was the perfect choice for the column. Published the day before the reading, "Manifesto: Bread vs. Mozart's Watch" is a personal outpouring that only Welch could write. He starts by lauding the poet: the job, the nobility, and the necessity. But he soon writes that he cracked up on account of the weirdness of working so hard and still "starving to death." There is no money for poets, he wrote. Society likes to pay its creative souls not with money but with gifts ("don't pay the guy, that would be too vulgar a return for work so priceless" [OBP, xi]). So the poet receives trinkets, like Mozart received watches. Welch's reference to the composer was carefully chosen—and extremely apt—given that Mozart had once told his father that the "rewards for his musical efforts were watches—gold watches to be sure—rather than money."[3] He went on to sardonically say that "I now have five watches. I am therefore seriously thinking of having an additional watch pocket on each leg of my trousers," so that it might discourage others from giving him yet another one.[4] This column was Welch's manifesto to the same end: come to the reading but give money rather than watches.

He ends the piece by once again exulting the poet and outlining his vision for possible solutions. And to encourage the masses to attend, he promises "utopian Visions built before your very eyes! Poems! Delights! . . . Cheap. $1" (OBP, xii).

———

Welch's article in the *Chronicle* may have been perfectly timed to coincide with the Freeway reading at the Old Longshoreman's, but it was far from a new idea in his mind. During the previous spring, Welch's letter to Robert Duncan from the Forks had elaborately outlined his thoughts on the financial plight of the poet and his plans to draw attention to it on a wider stage. Having spent a couple of weeks in San Francisco in February 1963, Welch wrote to Duncan of an evening he had spent with Philip Lamantia, Kirby Doyle, and David Meltzer in which the four poets decided that "this is precisely the time to do a book which discusses, directly, the absurd financial difficulties of American poets."[5] Their idea was to create a book with statements and manifestos that would make a case for "patronage" and the creation of a situation in which poets such as they could survive without the accompanying agonies of enforced impoverishment. While the letter reads like a manifesto, it also demonstrates a mix of hope and desperation. It is hard to see anyone other than Welch introducing the plan to the others, and his role as self-appointed editor is clear—to use the book as a means of approaching those sections of the art world that would be able to support the likes of Welch and colleagues to the tune of as much as was needed to cover their annual expenses—approximately $1,500 per year, in Welch's eyes. He makes it clear that this venture will serve no real purpose if it is only aimed at fellow poets and their equally impoverished audiences. Instead, he was looking to target the upper echelons of society, those he felt were already generous in their patronage of the arts and would therefore surely be interested in aiding the struggling poets of the United States. He mentions, among others, Nelson Rockefeller, the Kennedys, Frank Sinatra, Willem de Kooning, and Leonard Bernstein.

The book was to be titled *Bread* and was likely the natural continuation of the idea he had previously had with Doyle, to release a magazine highlighting this same plight in order to generate some sort of financial gain. Yet by now he was also aware that if he was to go ahead with it, the plan was just as likely to fail as it was to succeed, and he is uncharacteristically accepting of this, writing, "I will either succeed in getting myself enough bread for at least a year or two . . . or I will fail, quit thinking about this dreary subject, and run guns."[6]

Years later he spoke of the venture as also being something of a reaction to the system of providing or refusing grants to writers and artists—a situation he knew all too well thanks to the rejection of his Saxton application in 1960. Of the projected $10 million that he felt would be required to support the poets that were living (or would live) in the United States, there would be an annual budget of $500,000 plus a hospital fund so "poets could have babies and fix their wives' teeth and other things that we need. The rest would be doled out."[7] He went on to suggest that the grant system was flawed because

> most grants are for a book that is going to be written or not written, or it's for going to Italy and doing something. Nobody just gives you bread. I don't need any grants. I don't know what I'm going

to be doing next year. What I need is $4,000. Dollars! Like maybe I'll just sit here and spend it all on bourbon, but that's my goddam right. If I am a poet I need bread to go just like a car. You have to put gas in it. So that's the idea behind Bread, Inc.[8]

Written from the Forks, the letter to Duncan was, ironically, written during the period of Welch's life as a poet that he was most financially secure. Dorothy had agreed to fund his hermitage, providing him with not only a hundred dollars every month but also buying him a car, a typewriter, and other assorted necessities. Thus, in addition to promoting the needs of others, *Bread* (or Bread, Inc.) may also have acted as some sort of contingency plan for his own continued solvency.

Welch's proposed manifesto was initially titled "Munny and Mozart's Watch" and is clearly an early version of what he went on to write for Gleason's column the following summer. It was one of two articles that were discussed that evening in San Francisco, the other being Lamantia's proposed article, called "Pouch," in which he would "talk about (among other things) the South American custom of making Poets diplomats, as Neruda was, apolitical, with orders to meet the Poets of the world."[9] Welch's article implies a greater role for the poet and thus also indicates in a broader sense what *Bread* may also have been intended to do: namely, provide the poet with a means of making money and promoting a situation in which they are more than poets but also representatives of an entire culture who should be rewarded rather than ignored. In a similar letter to Snyder, Welch wrote of the perceived need to "challenge & change a totally unexamined cultural attitude" of antipathy toward creativity, and that true poetry cannot be written if poets are "forced by poverty to sing through rotten teeth" (IR2, 104).

What the others would contribute to the book was unclear, but Welch's letter, like the one to Snyder, was an obvious attempt to persuade Duncan of its worthiness in order that he might also get involved.[10] Indeed, Welch is anything but backward to this end, writing,

Of all the possible people on my list of contributors, you have stood longest and, I believe, most deeply and honestly, as Poet. To me you have always proved that freedom and authority are things which are taken, which will never be "given" especially by some enemy, imagined or real.[11]

With this "enemy" in mind, Welch appeared to be offering himself as a commander-in-chief, or more poetically a "Lawrence of America on Racing Camel Corvette (or Scarab?) with saddlebags of gold for poetry, for Bread!" (IR2, 105). And although in the end nothing would come of the book as he and the others had envisioned it, the idea of bread and poetry was still in the air, and as his article, the KPFA interview, and the performance at the Old Longshoreman's Hall attest, Welch was still very much the "Lawrence" figure championing the plight of the poet and taking the fight to the people on as many fronts as he could.

———

June 12 was mild and dry. At the venue, Allen had set up a stall at the door with the broadsides, and Carpenter had bought tickets to hand out on payment of a dollar. Expectations were high, and nerves were frayed. Carpenter remembers that "the three poets were reacting in their own ways to stage fright, Lew jacked up and visibly nervous, Gary tense and short-tempered and Phil Whalen calm and quiet."[12] Carpenter had arrived hours in advance only to witness, in amazement, a group of people already waiting. Gleason had come to cover the event with two record company representatives who were there to record it, and celebrated North Beach literati peppered the hall, including Jack Spicer, who was "swollen with drink." Indeed, the evening was later somewhat jealously and disparagingly described in Stan Persky's *Open Space* magazine (under the heading "Whoregon") as "the worst poetry reading I've ever been to," with Welch being described as nothing more than a "taxi-cab hack with a midget brain [who] proved that every sentient being is a lake."[13] However, when Welch took to the stage, the hall was packed, anticipation was at a fever pitch, Allen was out of broadsides, and Carpenter, despite any misgivings he may have had, was about to witness arguably one of the best and most important poetry readings to have ever been staged in San Francisco.

Welch was funny, intimate, and tuned in. In his reminiscences of the event, Stephen Vincent described him as having "a charming and hilarious wit" and reading "exquisitely."[14] Whalen was relentless, focused, and witty; a "troubadour of local truth." And Snyder was thought-provoking, awe-inspiring, and mesmerizing (although by his own admission he felt that the reading had not been "an unqualified success").[15] Three totally different poets at the top of their game. In his essay "Poems in Street, Coffeehouse, and Print," Vincent perfectly sums up the importance of not only the reading but also of the participating poets:

> Freeway was enormously important to what was to occur in the next few years. It was a declaration of space and position. The space was both the City and the country, with a definite West Coast fix. And the poet's position became that of public person. The reading put the poet back in the position of responding to the City in an actual way, letting the poetry move as the City does, responsive to the edges, to the corners, to the voices that flood our City lives. Built out of a democracy of eye and ear, the poetry would help create a culture where language would have a genuinely liberating function. There was definitely a politic and ethic to the new stance: It was the poet's community responsibility to make accurate perceptions, not false metaphor.[16]

In the aftermath of the Freeway reading there was a renewed acceptance of the poet as a voice. Carpenter's event had created a shift in the response to and acceptance of such readings, and what Welch, Whalen, and Snyder had achieved—whether they liked it or not—was a newfound position as public responders and de facto orators of truth and conscience. During the years that ensued, readings and publications became commonplace and, in many ways, huge cultural events such as the Berkeley Poetry Conference, held the following year, were made possible by the legacy of Freeway.

When all was said and done, the event was not the financial payday the poets and organizers had hoped for. Each poet earned the better part of $100, with Carpenter more or less breaking even. Yet Welch for one was happy enough with his takings: "Hell, a hundred bucks for half an hour's work? Not bad!" And Whalen's modest goal of hopefully earning enough from the evening to "buy a new pack & sleeping bag & grub for my own use in the mountains" must surely have also been met.[17]

In addition, the reading set the stage for even more publicity surrounding the upcoming broadcast of the interview that Jack Nessel had conducted for KPFA the previous month. The Berkeley-based free radio station aired the "On Bread & Poetry" discussion, billed as a panel discussion among the three poets on subjects similar to those Welch had dealt with in his column in the *Chronicle*, on July 11.[18] Moderated by Nessel, the conversation ranges from the role of the poet in society and the inspirations each drew from in their own work to the assassination of JFK and developments in jazz. Some moments are humorous, others serious and philosophical. Welch talks unreservedly, Whalen calmly interjects, and Snyder corrects and makes pointed assertions like that of the statesman or teacher. The clear understanding between them, and their mutual respect and friendship, allows each to have the freedom to speak their mind. There is a sense of collective individuality, with each having a unique perspective on the communal issue of poetry and its composition and importance. Indeed, after the recording, Nessel described the trio as

> coming on like scholars. No shit, for a long time I thought you were putting us on, like maybe you were a road cast from Evergreen Review or something, the real Lew Welch in the woods, Snyder still in Japan, Whalen safe and hiding. (OBP, vii)

Nessel was surprised by the insightful intellect on show, and he had underestimated their visionary sensibilities. Indeed, in writing about the broadcast, Welch saw that Nessel's somewhat stereotypical view was indicative of a greater problem faced by poets in general—the public perception that they were frightening hippies who were uncommunicative, offhand, and detached from reality. The hope lay in the fact that the reading, article, and radio broadcast could help change that. And Welch for one relished the challenge, writing to Rago, "How great to have our job at a time like this—eh, Henry?" (IR2, 128).

"Our job"—the job of poet. While it may seem contradictory, Welch actually now saw his position as a job and, as such, worthy of payment. Although poetry may have been slowly permeating the collective consciousness of a generation, poets like Welch were not exactly reaping the rewards. Payment rates for even the most well-established magazines were low, so money was scarce. Luckily for Welch, though, Dorothy was still regularly willing to provide him with money, clothing, and various other necessities—although one cannot help thinking that perhaps he saw a new typewriter or overcoat in the same light as Mozart saw another gold watch.

20
Lost

It's enough to be home,
At last, Magda, for
When you found me I
Wasn't even wandering
Anymore, just lost.
— Lew Welch (RB, 218)

In early 1964, Welch, now thirty-seven years old, was living in Mill Valley again. He had temporarily taken a room in a house on Shoreline Highway that he described as "a circus," with people coming and going and parties being held almost every night. Despite all the free liquor and good conversation, Welch also began to wonder why he had returned to the city at all. His same need for a "home," which he had previously felt when arriving at the Forks, was again disturbing him. He could give no valid reason either way for staying in the city or returning to the wilderness, relying on his instinct and being true to his feelings about subconsciously wanting one more than the other. He characteristically joked about this by saying that it was no doubt down to "the thousands of more or less satisfactory lays" in the house but the lack of what he really craves — "one good love-shot" (IR2, 126). Was his return from the Forks therefore one born partly of a need for love? The kind of love that he had had before but could not hold on to? It may certainly have been a mitigating factor, yet little did he know that his "one good love-shot" was just around the corner.

At the time, Welch was working as a busboy at the Trident Restaurant in Sausalito. Already a long-established fixture on the waterfront, the Trident had been a regular haunt for artists and musicians from the Bay Area. Frequented by the likes of Janis Joplin, the Grateful Dead, and Bob Dylan, the Trident was as much a celebrity hangout as anything else throughout the 1960s and 1970s, and it must have been strange for an established (and relatively well-known) poet such as Welch to wait tables and tidy away other people's leftovers. However, given the possibility that the restaurant was likely to be the focal point of some wild parties, his motives may have been more hedonistic than financial.

In addition to the Trident, another of Sausalito's celebrity hotspots was the No Name Bar. Equally popular with the glitterati of the film and music scenes on any given evening, it also had a less salubrious side, attracting "a group of fairly down on their luck [types] drowning their sorrows in booze," of which Welch was a regular.[1] It is a short walk from the Trident to the No Name Bar, and it is easy to see Welch finishing work and heading south along Bridgeway to spend what money he might have earned there. He had always been a hard drinker, and his life in Sausalito proved to be no exception. With nothing in print, and few prospects of that changing in the foreseeable future, the opportunities to read in San Francisco were crucial, as was the time he spent with his friends there. Yet even more important was the chance meeting he had in the No Name Bar on a July evening shortly after "On Bread & Poetry" was broadcast.[2] Welch's meeting with Magda Cregg that night went some way toward altering the course of his life, giving him a sense of home, purpose, and love that he had rarely felt before.

———

Magda Cregg had moved to the West Coast from New York in 1955 with her then husband, the musician Hugh Anthony Cregg Jr. By the time she met Welch, however, she had divorced, and Magda was living with the younger of their two sons, Jeff, in Strawberry Point. As Jeff Cregg remembers it, he first met Welch early one morning at their home after Cregg and Welch had returned from a night at the bar. Despite Welch leaving an impression on the youngster, Jeff thought nothing more of it, because he was used to seeing poets and artists coming in and out of the house after raucous nights in Sausalito. To him Welch was like every other one of Magda's friends, and he expected nothing more from him than from any of the rest. However, before too long Welch was a regular visitor, and when Magda sold the house in Strawberry and moved to a new home up on the hill, on Buckelew Street in Marin City, she rented the extra room in that house to Welch, as he was needing a new place to stay. Having a beautiful view over the bay and a room in which he could write in relative quiet was

Figure 20.1: 52 Buckelew Street was the home in which Lew lived with Magda and Jeff from 1966 to late 1970. Courtesy of the Estate of Lew Welch.

a welcome bonus for Welch, and as Jeff remembers, he would sometimes "test out his stuff" on Magda and Jeff. While the relationship was initially one of friendship and convenience, Welch and Magda gradually became a couple, and the house on Buckelew Street became Welch's permanent home for much of the rest of his life. Were Marin City and Magda the solution he was seeking? If he had been truly lost, was this his salvation?

During the time between moving to San Francisco with Mary in late 1957 and moving in with Magda in 1964, Welch had been on an almost continuous tumultuous rollercoaster on which alcohol, poetry, depression, solitude, friendship, and uncertainty were only a few of the many turns. He had lived in countless apartments, cabins, shacks, and houses, held down a myriad of jobs, written lots of poems that he had allowed the world to see and many more that did not pass muster. He had gone from what he saw as a staid, suburban marriage to one passionate relationship after another. His mental state had swung from manic highs to the debilitating lows of ennui and inactivity. And now, in his room overlooking San Francisco Bay, Welch found something as close to "home" as he had ever had. His admission that he "wasn't even wandering" reads in retrospect like a clarion call. When Jeff Cregg says that Welch needed a place to stay, it seems more likely that he actually needed a great deal more than that.

Figure 20.2: Lew with Magda and Jeff, Christmas, 1967. Courtesy of the Estate of Lew Welch.

Suddenly Welch had found a certain modicum of stability. He had not only found a partner who could inspire him and deal with his mood swings and drinking but had also become a stepfather to Cregg's sons — a situation that propelled him into a world he had never previously known. With his elder brother Hugh at prep school on the East Coast, Jeff was largely alone in experiencing Welch as a surrogate father. Hugh would return during vacations, but before long he had started to forge a career for himself in the music industry, and his visits to Buckelew Street became fewer and fewer.[3]

Jeff remembers his stepfather partly as a "very entertaining and likeable guy" who could command attention immediately. Although he also witnessed Welch's gradual slide into irreversible alcoholism from close range, he has fond memories of the role that Welch played in his life: camping, fishing, hunting, singing, and writing poetry. Yet his initial impressions of Welch were largely the same as those he had about other boyfriends Magda had had. She would ask Jeff what he thought of them, while all he wanted was his own father.

On that Saturday morning in 1964, Jeff remembers that Welch "got my attention the very first time I laid eyes on him."[4] But despite the growing affection between Welch and Magda over the ensuing years, Jeff often felt, initially at least, that Welch was never being entirely honest. While Welch and Magda were clearly living as a couple before too long, Welch was apparently less enthusiastic about broadcasting this fact, preferring instead to say that he was merely renting a room at Magda's house. Indeed, in his correspondence, it takes Welch some considerable time to openly talk about Magda (and Jeff) in familial terms at all, and the apparently casual nature of their relationship at the time is highlighted by the fact that Welch even moved back to San Francisco for a six-month period at the beginning of the following year, renting a room at the now-famous Beaver Street apartment he shared at one time or another with Philip Whalen, Albert Saijo, Richard Brautigan, and John Montgomery. Only on departing Beaver Street for good in early October 1965 did he make the final and definitive move back to Marin City and Magda.

In time, however, Welch warmed to the role he played in Jeff's life. In addition to helping him name his first band—"God, Mother & Country"—Welch also bought Jeff his first gun (like father, like son), took him on trips into the mountains along the Oregon border, and, as Cregg's childhood friend Kremas Carrigg recalls, "was a pretty good role-model. It seemed that he did look over us as kids in a pretty good way and I think he did give good advice about life."[5]

Years later, when Welch planned to travel to Chile in the late 1960s to work as a lecturer or visiting "'Murcan poet," he was specific in stating that Magda and Jeff were to accompany him and that the terms would have to be sufficient to cover the costs of traveling and living as a family. Similarly, when he accepted the position of poet-in-residence at the University of Northern Colorado in 1970, he stipulated that his family included "me, my wife Magda, my son Jeff 15 yrs, [and] my dog" (IR2, 165). And with that, for the first time in more than a decade, Welch was clearly content in his role as a self-proclaimed family man.

PART III
The City

There will never be another poetry conference
in Berkeley; Berkeley is too bizarre.

—Robert Creeley, widely reported
(but unverifiable) quote from the
Berkeley Poetry Conference

21
Holding Court

Throughout the early 1960s, Welch gave increasingly more poetry readings as his reputation gained momentum. From those initial readings at Reed to Jack Spicer's intimate gatherings in San Francisco, one constant in his performances was his unique mix of poetry, anecdotes, and songs. An important element of theatricality set his readings apart from those of many of his contemporaries, resembling at times more the work of a cabaret artist than of a poet.

At Reed, Welch had first played with the notion of setting existing texts to music—his renditions of Shakespearean works being well-known among the pieces that he had composed. As part of the annual senior year arts festival, Reed had staged *Much Ado about Nothing*, and Welch was cast as Balthasar, the "roving minstrel with a real Elizabethan lute," thanks to his ability to play the guitar. He was also required to set music to one of the songs from the play, "Sigh No More, Ladies," which he duly did, to the melody of a West Texas blues ballad. Welch's approach was simple:

> I tried to go through the scholarship and find out how Elizabethans sang it, and I got so discouraged I couldn't do it, couldn't find it anywhere. . . . So finally just as a joke set it to, like, a West Texas ballad and it fit perfectly because, of course, you know all our hill-country ballads and everything had an Elizabethan origin, both from England and Scotland. (HWP, 70)

The song was an apparent hit, and Welch's delight at the reception it received is clear in a letter to Dorothy, in which he wrote,

> Everyone is simply wild about it. Everyone liked it so well that I have been used as publicity for the whole of the Reed summer

artistic activity . . . and have three engagements on the radio, my record, with my own voice singing to my own accompaniment on the guitar, and on every disc jockey's program, and numerous personal appearances lined up. Not the least of which is a tentative offer to sing in one of the larger of the local nightclubs. Even the jaded ears of the disc jockeys are turned by my tune. Every day more and more people get excited by it and it may just catch . . . and if that happens I may just end up with hundreds of thousands of dollars. (IR1, 14)

He somewhat ambitiously ended the letter with, "I am not going to be ruined by this. Fame only means that a lot of people whom you do not know, know you."

Recordings of the pieces, which were made for a radio show in Portland to advertise the festival, seem to have been lost, but Dorothy wrote after his death that they were "charming." Many typescripts of this and other passages and songs from plays including *Twelfth Night* and *The Tempest* have survived, and such is the careful notation of notes and keys that, when played, they sound as good today as they must surely have done then. Indeed, some ten years after he graduated, Welch returned to Reed and was amazed to hear that his rendition of "Sigh No More, Ladies" had become a sort of oral tradition and was still being recited by students.

Having studied music at Stockton College, and coming as he did from a musical family, Welch had a background and keen hearing for tone and pitch that provided him with the advantage of being able to produce pieces that, while often frivolous and comical, also came across as professional. His eye for detail and constant self-criticism and editing dictated that frivolity could never be mistaken for a lack of accuracy or seriousness of composition. It was, after all, an aspect of his writing that was invariably found in his poetry. And far from limiting himself to poetry, Welch also began to experiment with other literary forms. In addition to short stories and essays, some of his more eclectic works are the plays he wrote at Forks of Salmon, which offer a striking contrast to the poetry produced during that same period.

While many of the "Hermit Poems" are richly layered vignettes that muse on a multitude of Welch's immediate concerns while at the Forks, "Twins (A Play, A Dance With Words, A Play Who Songs)" is a humorous experiment into how words can be used to mimic movement and specific sounds can be used to make music. Played out in two short scenes with carefully constructed props, the one-man play is set in New York and features Alvin and Edgar Cone, fifty-year-old twins with a long-standing ambition to go to Los Angeles. Indeed, they have spent almost a third of their lives working and saving for the trip. Alvin is an elevator operator, Edgar a waiter. One goes up and down, the other back and forth. "That is the dance: Alvin up and down. Edgar back and forth. In a prominent place is a bucket or wash tub. It must be resonant since coins dropped into the tub provide the music for the first act."[1]

The dialogue between the twins is clipped and repetitive, mimicking the coins that are dropped into the washtub with a loud clank. This clank in turn seems to act as the beat for the first of several songs they sing, titled "We're Saving Our Money to Go to Los Angeles."

Their conversations are staccato. The sentences are short and often have single-syllable retorts that carry an overriding sense of lamentation. As such, "Twins" offers a bleak vision of the fruitlessness of such focused frugality in the face of ambitions that appear impossible because of factors that have not been previously considered (and that are ultimately beyond their/our control). It is a play about stagnation rather than possibility. About acceptance rather than adventure. In many ways, it represents an aspect of the philosophy that Welch was gradually developing about the necessity of realizing the limitations of our existence and nevertheless acting on them. Saving for fifteen or sixteen years to take a trip that was never likely to happen is akin to spending years in a job you despise while telling yourself otherwise.

> Alvin: I guess we're too old for Los Angeles.
> Edgar: Evidently.
> Alvin: There aren't many elevators here in Los Angeles.
> Edgar: All of the jobs for waiters are filled.
> Alvin: That is not it.
> Edgar: No, that is not it.
> Alvin: 50 is just too old for Los Angeles.
> Edgar: I should have realized that.
> Alvin: We both should have realized that.[2]

Much of this performance-oriented structure would eventually culminate in the stage acts Welch called his "One Man Plays," the cabaret-like pieces in which he endeavored to provide his audience with, as Welch's former Reed college housemate and friend Grover Sales wrote in his review of the show, a "total theatrical experience [in which] he whistles, chants, improvises, weeps, croons" and showcases not only "powerful and communicative verse," but also his "remarkable ear" and "resplendent signing voice." In addition, Sales wrote about Welch's poetry,

> Much of Welch's poetry can be viewed as modern sermons, Bob Dylan for mature adults, songs of protest and love that incorporate the sounds, the vulgarisms and the music of our time. . . . What should be clear is that an evening with Lew Welch is a far more rewarding experience than most of the so-called "legitimate" entertainment in town. (IR2, 142)

Welch's one-man plays were similar in structure and content to those given around the same time by members of the radical San Francisco theater company the Mime Troupe, formed in 1959 by the actor R. G. Davis with the intention of offering actors the opportunity to perform alternative forms of theater that diverged from the mainstream.

While still affiliated with the San Francisco Actor's Workshop, Davis began by producing silent pieces that were seen more as events than as performances. Having trained as a mime artist in Paris, Davis was aware of how powerful and effective silence could be in transmitting theatrical messages. But as the company grew and developed, so too did the ideology that acting could become an even more powerful medium in the transference of ever more important social messages. While the events themselves were often alternative renditions of classic works, the act of performing them in public spaces at no cost was a radical departure for theater in the city at that time. As the group evolved, this notion would turn out to be a key ingredient in what would later become the life-acting employed by members such as Emmett Grogan, Peter Berg, and others when, in 1966, they splintered into what would become the Diggers.

It was also around this time that Welch met the experimental film-maker Robert Nelson, who had also worked with the troupe, providing not only a short film (*Oh Dem Watermelons*) that was originally intended to be shown in the intermission of one of the troupe's most notorious productions, A *Minstrel Show, or Civil Rights in a Cracker Barrel*, but also using some of its members for several of his early works.

However, by the time Nelson met Welch, Nelson had firmly established himself within the avant-garde movement and had collaborated with a wide number of painters, actors, and composers, including William T. Wiley, William Allan, and Steve Reich. This relationship would not only give Welch the opportunity to broaden his creative horizons by acting in Nelson's film pastiche of the life of the legendary nineteenth-century tightrope artist Charles Blondin, *The Great Blondino*, but also offer him a chance to expound his views on national television as part of a spoof talk-show called "What Do You Talk About?" Directed by Wiley as part of the Dilexi Series and featuring Nelson as the host, the show featured Welch as one of three guests, alongside local Bay Area sculptor Bill Allan and an art student from the San Francisco Art Institute, Dominic Laducer. In the program notes for the show—which aired in May 1969 as part of a series that represented "a pioneering effort to present works created by artists specifically for broadcast" and featuring, among others, Robert Frank and Andy Warhol—the "ever-charismatic and highly talkative" Welch is described as stealing the show.

Welch, chain-smoking and holding court, creates his performance as a consummate talker. While the intention may have been to parody the talk-show format and the often-banal discussions that take place on them, Welch's charisma and ability to tell engaging stories actually makes the show an interesting spectacle and provides a fascinating insight into Welch's high school years and passion for football. Given that so little is known about this period of Welch's life, his anecdotes of friendships, quarterbacks, and spending time watching future NFL greats like Frankie Albert in action at Stanford add a more down-to-earth layer to his often complex and philosophical character.

In the second part of the show, Allan takes the stage and talks drily about various aspects of fishing, and although Welch is equally as animated, given his own vast experience in that field, this segment is little

more than a dreary tête-à-tête between two fishermen on a riverbank—
which was surely the whole idea in the first place.

In *The Great Blondino*, Welch plays the cop alongside Wiley's brother
Chuck, who plays the lead role. The film was shot without a script
(indeed it is almost entirely devoid of dialogue) or any clear plot structure
or narrative development, with Nelson and Wiley using various friends
and acquaintances to fill the roles as they materialized—the film was shot
"on weekends over an extended period, using whoever was available and
whatever story ideas came to mind at the time."[3] Welch was cast as the
"ever-present, though not very intimidating, plainclothes policeman,"[4]
and there is a certain irony in his playing a part almost entirely without
speech or structure. His appearances are fleeting, accompanied by bursts
of dramatic music; his role relies more on the power of observation and
expression, complementing the underlying message that Blondino was
walking a proverbial tightrope through his life and that, on such a perilous
journey, words are perhaps meaningless. What speech the film does have
is a veiled warning about how radical or off-beat life choices are often seen
as subversive, with Welch uttering such lines as, "When we first heard
about this fellow we were sure his operations were not in the national
interest" and "We feel that this man is going to cause certain harm." For
a poet whose primary concern was the accuracy of language in conveying
a message, his dialogue in *The Great Blondino* echoes what he had done
in "Din Poem."

Yet the film's disjointed and uncertain nature may just be what primarily appealed to Welch. As Blondino wanders, fantasizes, and dreams
his way through a quasi-tragic universe, so Welch often felt himself to be
on some tightrope between clear involvement with and disenfranchisement from the world around him. In some respects, therefore, *The Great
Blondino* was yet another vehicle with which Welch could experiment
and muse on the idea of distancing himself from perceived norms and
conventions while retaining a certain sense of inevitable symbiosis.

––––––

Despite his sense of disenfranchisement, Welch continued to immerse
himself in the practice of writing and reading. The mid-1960s saw not
only the publication of *Hermit Poems* but also the more expansive collection of poems that he had put together for Oyez Press, titled *On Out*.

Established in 1964 by Robert Hawley and Stevens Van Strum, Oyez was a Berkeley-based press born after conversations between the two founders at the Jabberwock café on Telegraph Avenue. As a former student of John Wieners, Robert Duncan, and Charles Olson at Black Mountain College, Hawley was a perfect foil for Van Strum, who at the time was working at Cody's Books—another pioneering literary institution in the Bay Area. Both had backgrounds in bookselling, and they were well-placed to start such a business because of Hawley's previous experience working for Olson at Black Mountain. Indeed, Olson had encouraged and helped Hawley develop his interest in American literature and start a career in books in the first place. Thus, for Hawley at least, Oyez was the logical continuation of a passion for poetry, fine literature, and printing that had blossomed more than a decade earlier.

For poets such as Welch, Oyez was an essential venture in a milieu that was fast becoming a minefield of power plays and in-fighting. It seems that, as poetry began to capture a wider audience and the interest in San Francisco–based poets grew, so the publishing houses—big and small—were, as Welch put it, "jockeying for power" (IR2, 133).

Still finding it difficult to attract larger publishers, Welch, like many other poets, was dependent on the willingness of small-scale presses like Oyez to publish his work, and his need to find a publisher at almost any cost is clear in his correspondence with Hawley. Although he would work with Dave Haselwood again, it was apparently Hawley's new press that provided him with the best chance at that time.[5] Having nothing in print and a limited amount of work in magazines, Welch somewhat desperately wrote, "I just want to get these poems printed, by anyone who can do it, and who wants to do it because they like the poems" (IR2, 133). Persuading Hawley was important for more than one reason, first because of Welch's need to have his poems in print and second because of his participation in the upcoming Berkeley Poetry Conference. To have a book to sell there was advantageous for both him and Oyez, given that Hawley and Van Strum could use that event as a means of recovering some of their outlay relatively quickly. Indeed, Welch makes a point of mentioning this to Hawley when submitting his poems.

In comparison to *Hermit Poems*, which Donald Allen managed to publish just after the conference, *On Out* was a much larger and more ambitious project, which would therefore take more time and effort.

Welch's manuscript contained poems taken from the prior fifteen years, going back as far as Reed College, and was clearly an attempt to make up for lost time. As always, Welch was concerned with coherence and included poems that would reflect this, including many new works that were put there to clarify the shape of the book. Initially, Welch envisaged that the collection would be split into three parts, namely "Early Poems and Finger Exercises, Portland Oregon 1948–1951," "'They Boarded Up the Fountain from the Cold' Chicago 1951–1957," and "'After Wasting Nearly a Fourth of My Life' Chicago 1/57–10/57, San Francisco 10/57–."[6] In the end, however, *On Out* was limited to works covering the decade between 1950 and 1960, and the section format was dropped in favor of a single selection. It was printed by Graham Mackintosh, who had by then established himself as a master printer in San Francisco working for White Rabbit Press and others.

The book's strength is the simplicity of its design, which in many ways contrasts with the poems it contains. Although published in August—thus missing the opportunity to profit directly from the conference—the book was a success nonetheless and gave Welch an initial platform from which to orate his message. Dedicated to Magda, the opening poem reflects Welch's ideology circa 1965 and is also a statement of intent in that it suggests that the poems within can (or should) be read from a particular perspective. In some ways it is a defense of poetry and the lifestyle of poets. It is also a rallying cry to those who lived that way at the time in order to pave a path for others to follow later. While the collection was a departure from the more focused poetry that Welch had written at the Forks, *On Out* went some way toward helping him reach a wider audience, and he found an unlikely supporter in the shape of Marianne Moore, who, upon receipt of the collection, commended Welch for writing "compactly some things I've always wanted to say" (IR2, 134).

By 1965, Moore was very much the elder stateswoman of American letters, having transcended poetry to become not only a best-selling and award-winning poet but also a society figure in New York who rubbed shoulders with the rich and famous. She even had the honor of throwing the season-opening pitch at Yankee Stadium in 1968, and as such represented the cult of celebrity in a way that Welch could have only dreamt. The two had been corresponding since 1963, when Welch had sent her a copy of *Wobbly Rock*. Her response at the time was both encouraging and inspirational, having commented on Welch's writing skill and the fact that it was partly his inclusion in *The New American Poetry* anthology that had made it possible for the then editor, Irita van Doren, to convince her to write the subsequent review in the *New York Herald Tribune Book Review* (IR2, 106). In that piece, titled "The Ways Our Poets Have Taken in Fifteen Years since the War," Moore wrote of Welch's natural quality and the success of his technical skill. She also commented on his being an observer, of how his observations on the state of the planet were also applicable to poetry, and how a line such as "the trouble is always and only with what we build on top of it" effectively creates a bridge between ecology and society and the (in)accuracy of language or (over)use of structural devices in composition. It was a review that must have appealed to both Welch's ego and his poetic sensibilities.

Moreover, in one of her later letters, Moore provides Welch with what could be argued was the best compliment he was ever paid, namely, to quote Goethe, "Genius is not superhuman exertion but naturalness" (IR2, 110). It was this "naturalness" that Welch had taken from Stein and Williams, and as such, this comment must have felt like a validation for his methods and his frequent unwillingness to compromise if the work did not flow as he would have liked. To successfully emphasize an image through the employment of enjambment was a means to an end—but for it to be praised was even greater. Rather than talk in generalisms, Moore specified particular lines from particular poems as being praiseworthy. From "Chicago Poem," she recognized (and liked "as much as ever"), "His color faded with his life. A small green fish." And, from "Wobbly Rock," the lines "even a good gust of wind will do it" and "I used to watch the Pelican," where Moore perhaps understood Welch's use of the pelican as a symbol of maternal self-sacrifice that could have referred to Dorothy and what she meant (and continued to mean) to her son (IR2, 134).

22
Was *Berkeley Too Bizarre?*

In the context of twentieth-century poetry, the Berkeley Poetry Conference must surely rank among the most important events in the history and development of American letters. In the decade since the Six Gallery reading, the underground had moved upstairs and now firmly occupied a place in the cultural mainstream. Jack Kerouac had become a figure in the shadows of American literature, and any remaining semblance of a Beat Generation had been lost in stereotypes, media hype, and the next new fad.

The idea for such a huge conference of "alternative" poets was not a particularly new one. A similar event had been organized in Vancouver by Warren Tallman and Robert Creeley in the summer of 1963. On that occasion, a large delegation of American poets traveled to Canada to join local up-and-coming writers in what was a gathering of many of the brightest and best North American poets at that time. Delegates from the United States included Ginsberg, Whalen, Duncan, and McClure. And although Welch, who was living at the Forks at that time, never participated in 1963, two years later, when Tallman organized the Fifth Annual Festival of Contemporary Arts at the University of British Columbia, Welch made the journey north and read alongside Jack Spicer.

The 1965 festival, which was curated by the renowned Canadian American architect and professor Abraham Rogatnik, took place in the first week of February, and Welch read on the first day. It appears, however, that having Spicer and Welch in attendance was something of a challenge for Tallman, given that they disliked each other's work. The situation was a far cry from the days when Spicer had offered Welch the opportunity to read in his parlor as an up-and-coming poet. By now Welch had become an established figure on the poetry circuit himself; however, Spicer's apparent desire to be the center of attention made Welch's presence in Vancouver and, more specifically, as a fellow guest lodging in

Tallman's house, part of the problem. Tallman's wife Ellen remembers it thus: "I think that Jack was annoyed by Lew being there. I don't think that Lew was exactly a favorite person [of Jack's]. But Jack was also a person who liked to be there by himself. He liked the attention."[1]

While the Tallmans were ultimately charmed by Spicer during his stay, and the audience was in raptures after his reading, Welch was not exempt from praise either, with the Canadian poet Maxine Gadd remembering how she and others had found Welch to be "far nicer and sweeter than Spicer" even if, on this occasion, his reading was perhaps less memorable.[2]

On his return from Vancouver, Welch immediately started a series of seminars at the UCSF Millberry Union, having been invited to give an eight-week poetry course on medicine and literature to interested medical students. In writing about what the course would offer them, Welch highlighted the fact that the two disciplines were often linked:

> The arts of Medicine and Literature have always been very close. Many of our best writers were also doctors: Rabelais, Maugham, William Carlos Williams, just to name a few, and Gertrude Stein's education was PreMed. Doctors have always used literature to gain a better understanding of the hearts and minds of men — Freud's use of Greek Drama and Dostoyevsky is, of course, a salient example. I want to offer members of the U.C. Medical Center a course that will face problems of compassion and human understanding shared by writers and doctors alike.[3]

His appointment at the Millberry Union was not the first time he had read there, having held a reading at the union the previous November. On that occasion he had continued to propagate the notion of the poet as a spokesperson and how poetry should not be

> an aloof, library art. I want my poetry to stand as the record of man's life: how he walked about the planet, weeping, looking, loving, crashing. Our main job as animals blessed with Human Being is to discover the earth — to live with all the wonder around us; to live, somehow, with joy.[4]

After the successes of the Auerhahn event and the Old Longshore-man's Hall Freeway reading, the Millberry reading was a return to the one-man shows that would begin to feature more and more in his sched-ule and that brought that year to an end on a completely different level from that of twelve months prior, when, exhausted and broke, he had returned from the Forks. Could the year to come signify a continuation of his progress? Or would his many demons resurface once again to under-mine the gains he appeared to have made?

It seems, however, that the seminars were also a success. At least enough of a success to warrant Welch being asked to give another series the following year. Indeed, it was to be the first of many such teaching appointments he was given at UC over the next five years. However, he had two other interesting and important events to attend to: first, a reading at the UCSF Medical Center in April, the subject of which was very close to Welch's heart, and second, the impending, and now famous, poetry conference to be held in UC Berkeley's California Hall in the summer.

———

Gertrude Stein had been one of the primary driving forces behind how and what Welch wrote since the day he had been forced to wait in James Wilson's office at Stockton College. She was a huge presence in his methodology and how he approached his craft. Therefore, when the opportunity arose to get involved in a large exhibition being held on Stein in San Francisco, Welch jumped at the chance. The exhibition, titled "Looking at Pictures with Gertrude Stein," was primarily organized by Elizabeth Coffelt and ran from April 6 to April 30 in the University Medi-cal Center. Coffelt was the program coordinator at the university, and, together with renowned designer Gordon Ashby, she conceived, edited, and coordinated the exhibition in anticipation of the fact that it would go on to travel the entire country over the next three years.

Comprising several hundred photographs of Stein during her time in Paris, the exhibit also featured copies of paintings by contemporary artists and friends as well as extracts from her writings, with particular attention being paid to "Pictures," the essay she chose to read in 1934 at the opening lecture of her triumphant return to the United States after thirty-one years.

The colloquium opened in San Francisco on April 6, and Welch was part of a three-man panel that read excerpts from Stein's poetry and plays two days later. Alongside him were Robert Duncan and Philip Whalen, both of whom had agreed to get involved at Welch's behest. Indeed, Welch had asked Duncan in a hastily scribbled note only the previous month, writing simply, "Dear Robert, Do you want to be on a G. Stein panel with Phil W. and me? . . . I thot each of us would give a 10 min. talk, then babble & answer questions."[5]

Welch's further involvement in the event consisted of his editing the exhibition catalogue, and it is tempting to think that he also wrote the biographical notes, which contained the closing remarks about Stein's presence creating "a star and sextant whose importance (for American letters) has yet to be fully realized."[6] If Stein's influence had waned in the minds of many, Welch for one was determined to keep flying her flag and to use her influence to reiterate his own message. After all, as Stein had once said, and Welch surely knew only too well, "there is no such thing as repetition . . . [but rather] insistence."[7]

———

Much has been written about the Berkeley Poetry Conference of July 1965: Charles Olson's long, drunken, and often incoherent ramblings (for some "an absolute travesty," for others "a tour de force"); Ginsberg's designation at the conference as "Secretary of State of Poetry"; Spicer's witty and cynical lecture on "Poetry & Politics"; and Michael McClure's decision to boycott the conference entirely because of what he felt was a clear act of neglect on the part of organizer Richard Baker, who had inexplicably failed to offer him (or Philip Whalen) the chance to participate.

For his part, Welch's involvement is partly infamous—not because of his reading but because he reportedly provided Olson with much of the liquor and for drunkenly heckling during Olson's performance. When Olson ran out of whisky during the intermission, Welch was quick to offer him a bottle of "lousy wine," and Olson carried on drinking, getting drunker and drunker until, after three hours, he had to be asked to vacate the stage. Indeed, although Robert Duncan walked out of the reading in protest at Olson's rambling and attacks on fellow poets such as Ginsberg and Welch, he conceded many years later that "I had been alienated by Olson's turning on Lew Welch—tho God knows what any one would have done, Welch was so out of order." [8] Welch's reputation as a self-proclaimed "lush" preceded him, and, given that, in addition to heckling, he was also prone to emotional outbursts, Welch often unsettled his audience.

Yet when his reading took place on the penultimate evening of the conference, and Welch had been introduced by Duncan as an "awkward fellow" and "another veteran of the Buddhist San Francisco universe,"[9] he spoke with a sense of purpose and passion to an audience that was clearly enjoying itself. Starting with a comment on George Stanley's statement about the war in Vietnam, which most of the participants signed in a collective declaration of their discontent at the situation there, Welch took a clear political stance in stating that people such as they have an obligation to speak out in the face of such atrocities, and that since his (albeit minor) involvement in World War II, America's position in the world had "exactly reversed itself" and that he, for one, was

> ashamed to be a member of such a nation. . . . We are now the aggressor and unless good men speak, it's going to be very bad. . . . Now maybe speaking doesn't count for very much but it happens to be my job and that's all I can do.[10]

Working through a mix of old and new poems, Welch used the conference to focus attention primarily on the musicality of his work. He led the audience on a journey through his methodology, discussing rhyme, structural choice, jazz, and his desire to give a platform to the voices of the nation.

Beginning with "A Round of English," which Welch himself described as "the literary biography of the author," he immediately sets the tone for what will be a performance that ranges from the serious to the whimsical. He sings of Shakespeare, Milton, and Shelley, and of graffiti on "shithouse walls." He talks of how when writing "Graffiti" he could not make the rhythm fit to a standard musical beat because he writes for human speech patterns that do not adhere to such rhythms. And of how the renowned jazz pianist, vocalist, and composer Bob Dorough had helped him with it.[11] He explains how his latest compositions—called "Leather Prunes"—are "a whole new take on language" and that they are intended to provide him with more space to use language in different forms so that he can express things he felt he had been unable to do because of a need to say things plainly such that they are understood.

The title of these new works was derived from a "spontaneous quip" that Richard Brautigan was prone to saying, namely, "Oh, Flap City. Oh, those leather wings. I didn't get your cherry, and I don't want your prune."[12] Welch was seemingly so enamored by this that he not only quoted it regularly himself but also, much to Brautigan's delight, used it to arrive at an appropriate title for these new works.

The "Leather Prunes" are Welch's ode to Gertrude Stein (he once called them "a kind of literary structure, a backwards Tender Button"[13]) because they also provided a certain ambiguity of intent as far as meaning was concerned. During his later poetry workshops at Berkeley Extension, Welch spoke of challenging students to extract themselves from the "word-boxes" that imposed limitations on both the language they used and its perceived meaning. His own definition of these works was that most of the "Leather Prunes" were

> one man plays. A man of many voices stands alone on a bare stage and makes a greaseless theater. Since there are no wings, in the set-lumber sense that is, there can be no useless hysteria behind them. Farewell Judy Garland.
>
> Titles, scene changes, set directions, speaker's names, and curtains are all part of the performed text. Otherwise the timing, the shape, is not revealed.[14]

Welch's subsequent performance of "Abner Won't Be Home for Dinner: A Play" is met with a mixture of stunned silence and howling laughter. It is a work that experiments with the notion that plays can be performed as one-man shows, unaccompanied by music or complicated by directors, actors, and sets. Welch wanted to have language stand alone. To "cut through the grease" of such accompaniments and distractions was similar to how his other experimental works, such as "Twins," also used language for specific linguistic purposes.

"Abner" is an apparently disjointed collection of words, phrases, and sounds that have little cohesion and even less meaning. The strength of the play lies in its absurdity and in how the words manage to somehow provide a clear image despite their obvious randomness. What the audience perceives the words to mean is entirely personal and thus perfectly illustrates what Welch wanted from his work at that time and what he had begun to write about some years earlier on his hermitage. For example, in Act I, Scene 3, Abner is involved in an act that creates repetitive "thumpings." These words do not relate to language but rather to action or sound, yet any possible meaning is tied in the minds of the audience to its relationship to the sentences that come before and after.

Abner: ThumpThumpThumpThumpThumpThumpThump
          ThumpThumpThumpThumpThumpThumpThump
Jessie: How do you know when you're through?
Abner: ThumpThumpThumpThumpThumpThumpThump
          Shut up
          ThumpThumpThumpThumpThumpThumpThump
Jessie: You never say you love me anymore.
Abner: Thump. ThumpThump
          Thump.[15]

What this thumping refers to is intentionally unclear. To one audience member it could relate to hammering. To another it could imply masturbation. To a third an increased heartrate. And while the sentences provide some sense of meaning, they too are limited in scope. Thus, Welch is creating nets in the minds of his audience in which they will catch these images before regurgitating them in some concrete sense later.

In all respects, both his one-man plays and other leather prunes can be seen in the context of Welch's ideas on perception and what happens when words are spoken. In a long and heavily edited unpublished poem from the Forks that begins "A perception, by definition, is discontinuous," Welch discusses the notion that when a word is spoken it immediately occupies a space that is connected to the individual's perception of what that word means. A picture is created in the mind that attaches several attributes to the word that, according to Welch (and Stein), limit understanding, inasmuch as the perception is "partial, arbitrary, and remembered." Welch contends that these perceptions, while true to each individual, are held "behind the eyeballs" and that the mind then builds nets in which to hold them for further analysis. And it is this analysis (or attachment of meaning) that creates the agony and suffering of interpretation. There is then the necessary continuation of thought to speech, or initial perception to the creation of an image and the utterance of the said image in speech. He goes on to suggest that

> All speech comes from behind the eyeballs.
> I open my eyes.
> Something comes in
> It whorls about in there and
> (very possibly)
> I cast out what I have made and
> Claim to see something.[16]

Welch offers a vision of how individuals create their own images that may or may not adhere to conventional and accepted constructs of what a word "means." He claims to see "something" that might be different from what that image conjures in the mind of another, and thus proposes that poetry is "never about anything more real than anything else you can say." This proposition in itself questions how poetry should be both created and read, and it posits the idea that there is a difference between "what goes on behind your eyeballs" and "what goes on in front of your eyeballs." As such, this too offers a clear insight into Welch's poetic processes and in particular sheds even more light on the thinking behind the leather prunes. Similarly, it draws a clear line back to Stein and the fact that, for Welch, her writing exists simply on the level of words and word relationships and that these are "constructions out of words that are suggested by an object" (HGS, 72) rather than defined by it.

Welch further develops this idea in an unpublished essay from the early 1960s in which he distinguishes between the "seer" and the "seen."[17] In doing so, he similarly proposes that images are not seen *by* the eye but rather *through* it, and paraphrases William Blake's proposition that "I do not distrust my vegetative eye, I see through it not with it" as the means to substantiate this.[18] Taking the works of Van Gogh, Renoir, and Cezanne, Welch asks us to see that the world they painted, while unique, is also, like the language of the poet, a means through which to provide perspective. If Van Gogh had not painted the world in a "curious yellow," would we be able to see it as such? And would its "meaning" be the same?[19]

   Welch's performance at Berkeley continued with a reading of "Leather Prune for J. Edgar Hoover," which, unlike "Abner," is less a play and more a combination of words/sounds and seemingly random sentences. Various works from *Hermit Poems* followed, in which he offers explanations where necessary but generally lets the poems do the talking. His final poem was a rendition of "What Strange Pleasures," which was met with rapturous applause at the close. There is no crying, no emotion other than excitement and laughter, nothing for audiences or fellow poets to be unsettled by. It is a consummate performance by a poet at the top of his game, comfortable on stage and content with his material.

23
A *Teacher of Truths*

The majority of my students trust me, sometimes to a degree almost
beyond my powers and my strength.
— Lew Welch to Elspeth C. Smith, correspondence, June 1969

Having taught the short poetry course at the Millberry Union in the
spring of 1965, Welch's next teaching opportunity was one that would
offer him a certain continuity—financially at least—for much of the rest
of his life. Shortly after the Berkeley conference, Welch was approached
by Elspeth Smith, who as director of the program at the University of
California Berkeley Extension was responsible for finding temporary staff
members who could facilitate short-term courses once or twice a year.
Thus, Welch began a ten-week poetry-writing course at the university in
September (Poetry Workshop 819), which ran on Thursday evenings and
cost $30. The flyers for the course were typically engaging, not only high-
lighting Welch's approach but also employing the rules such as outlined
in his next collection of poems, aptly titled *Courses*: "For those who want
to write and those who want to listen. Free-wheeling discussions of poetry.
Appearances by a guest poet from the current scene. Continued tutorial
services for the gifted. No credit. No blame. No balm."[1]

Over the years that followed, Welch's classes were often full, and his
teaching style was invariably what kept people coming back. His approach
was simple: give those in attendance a wide-ranging overview of poets and
poetic styles, then try to get across his ideology such that the students were
inspired to look for their own sense of what poetry is and how it might be
written. Yet his classes were also full of people looking for insight into the
city's poetry scene rather than having any serious poetic ambitions themselves. Thus, Welch would often fall back on his natural ability to hold
court and provide anecdotes to stimulate discussions among the students.

In addition, he would read his favorite works by contemporary poets to generate opinion or set assignments for students to read works from the *New Anthology* or the City Lights Pocket Poets Series.

Welch's notebooks contain various entries that relate to his teaching practices, with one carefully outlining not only the poets he was planning to include but also the titles of lectures he would be giving. The list is grounded in the classics, Romantics, and early modernists—a clear reflection of Welch's own schooling and his thoughts on where poetic education must begin:

> Rimbaud, Auden, Yeats, Crane, Keats, <u>Blake</u>, Dryden, Rabelais, Swift, Wyatt, Catullus, Dante, Petrarch, and "the bedrock" Greek anthology, China, the oracular voice and the white goddess. Song: Dylan, Dorough, Burns, Inventa Fugue, Blues, Calypso. Lectures: 1. The Work of Poetry 2. Language is Speech 3. Yeats' Tower 4. The Immediate Tradition.[2]

One of his former students, Richard Hughey, once wrote that Welch's lessons were "disorganized . . . and it appeared as though preparing them was not one of his staple pleasures."[3] That, of course, is not to say the lessons were not enjoyable. On the contrary, Hughey remembers Welch's classes as ranging from "interesting lecture-and-discussion" groups to "impressive and distressing"—impressive on account of Welch's continual desire to "divine a poet's intention and then relate to it on a personal level" and distressing because many of the poems he chose to recite were "truly awful." Yet this did not represent a drawback or a hinderance to Welch, who would shamelessly show emotion at reading these works, many of which were taken from local poets who were using the poetic milieu in the city at the time to render their feelings on all manner of political, social, and cultural issues. However, in Welch's expert hands these "raw and uncrafted" poems shone like "fireflies" with their "earnest cries of lamentation and despair."[4]

The inclusion of musicians and music styles in his classes is also an obvious choice, given his own musical background, with Robert Burns in particular occupying an important place in Welch's idea of what constituted literary tradition. For Welch, Burns represented the poet who was not only able to write poems that morphed effortlessly into song but also did so by bending language such that it sounded simple. Many years later, when looking back at the position of poetry and its public perception, Welch spoke of his desire to take poetry back "into the bar" and to give the public something of the apparent simplicity Burns employed in his poems and songs. Indeed, Welch even ventured to suggest to David Meltzer that "Burns wrote as fine a set of lyrics as any poet ever wrote," with particular attention paid to the traditional Scottish folk song "Loch Lomond," which Welch commended for both its lyricism and the musicality that belies its complexity.[5]

While Hughey and other students spoke glowingly of Welch as a teacher, others found his methods unconventional and upsetting. His eccentricities were not for everyone, and he was subjected to one particularly malicious onslaught by a student who attacked not only his methods but also his personal hygiene and sobriety.[6] At a time when Welch was balancing his poetry classes with his work as a ship's clerk on the docks, he would often go straight from the waterfront to the lesson in his work clothes, not having had the time or opportunity to change. This was, of course, a regular occurrence, and one not entirely unusual in a world where poets and artists often needed to hold down numerous jobs to make ends meet. However, it seems that the student in question took exception to Welch's teaching style and made an official complaint about his having given the students "a liberal education in 4-letter words and filth" and how his being an "outright revolutionist" who "lives in the black ghetto" was clearly having a negative influence on "our young people whose parents are paying their way to better things."[7] The letter was sent to the professor of education at Berkeley Extension, Morton Gordon, who had served in several capacities at the university and who would go on to have a distinguished career in education. It drew further attention to the fact that Welch "often came to class so dirty you would not want to sit in the front rows." However, it is the erroneous assertion that Welch had been in "a mental institution" and that his "ego should not be comforted at the expense of our young people's degeneration" that is most revealing of the complainer—a fact not lost on Elspeth Smith, who was charged with communicating the complaint to Welch. In her letter to him, she somewhat revealingly writes that she has "not solved the problem of designing a poetry class for a suburban real estate salesman" who studies poetry as a "hobby" and who was clearly alienated by Welch's approach from the start.[8]

Welch's response to the criticism was typically forthright, humorous, and achingly honest. He parries every criticism in clear rebuttals of truth. Yes, he drinks, but is rarely drunk, and never in class. No, he does not smell bad, but does, through necessity, come to class in work clothes. Yes, he swears, but only in his role as "poet/shamen/druid/priest," who also use curses in their poetic endeavors. Indeed, by doing so Welch challenged his students to see beyond what they know and understand. In teaching the "daemonic art" of poetry he was attempting to instill new courage in his listeners by assuaging their fears of what attaching meaning to words could imply where no meaning is necessary. By his own admission this had been a successful approach over the previous five years, and this class was no different, judging, as he did, by the "greatly improved work" and "unusually warm thank you's" he had received from the students on the final night.[9]

Welch's openness and honesty make it hard to understand, as Smith herself had written in her reply to the student who had complained, why students would not have approached him directly to discuss any issues they may have had. Indeed, Welch says such himself, welcoming criticism and regretting that he had "failed" this particular student. He also recognizes the difficulty the university has in creating a poetry course for such a wide-ranging audience, comprising "physics professors, computer programmers, nurses, doctors, teachers, housewives, whores, thieves, musicians, transvestites, nuns, the very young and the very old."[10] And yet, in teaching the course the way he wants to, Welch admits to not being able to pull any punches in his approach. If that rankles with some, so be it. But his punches are "ultimately very kind," like "shaking someone awake on the subway—[who] might otherwise have missed his stop."[11]

While Welch's response is full of truth and compassion, he is also clearly aware of his role and the possibilities that this affords him to effect change and influence people. When reading about how he has been able to win the absolute trust of suicidal students or others who have in one way or another been misled or mistreated, Welch reveals another side of himself: a teacher in the mold of James Wilson or Frank Jones, not the "parrots of murk" but the real teachers, committed to their task and driven by the innate desire never to fail those in genuine need. This little-known aspect of Welch's character is often lost amid the same hyperbolic language relating to his smelling bad, being doped up, and, as alleged by the complainant, leaving students "nauseated, mentally and emotionally."

In addition to the purely poetic perspective, Welch's preparatory notes for his workshops also include a more philosophical treatise on life in the States, the aforementioned essay, "How to Survive in the United States."[12] The essay seems to have been written as a means of providing advice and offering a foundation on which students could successfully write poetry while understanding the milieu in which they currently found themselves.

The essay, although unfinished, is a biting and humorous description of a country torn asunder by the contrasting images being broadcast nightly on the television. On the one hand, images from Hollywood and the silver screen, while on the other hand, the horrors of Vietnam on the news channels. Or the clear differences between the continuing sensibilities surrounding the Cold War and the rise of commercialism and disposable commodities that came to symbolize America in the 1950s and 1960s. He juxtaposes Pepsi and Mary Poppins against Russian aggression and political paranoia. He uses Rexroth as his spokesmen: "You can't terrify children with atom bomb drills and expect to calm them with Coca-Cola."[13] Yet despite the underlying sense of desperation in what he writes, he does so while offering hope and support to those "kids . . . who still read me and trust me."[14] More than a simple critique of living in the United States, the essay offers a point of view intended to guide rather than to preach. Welch never implores his readers to agree with his opinion but rather asks them to see it as a means for coming to their own conclusions: "I don't want you to believe what I say here, just because I say it. Go out and look for yourself, and then decide."[15]

In many ways this didactic approach empowers the students/readers, rather than asking them to merely appropriate a secondhand idea and make it their own. Indeed, this observation is often repeated in his former students' many testimonials about his classes. His honesty and desire to be truthful shine through. He was conscious of a need to tell the truth in the face of a government that avoided it, or a corporate machine that sustained itself through the fabrication of situations intended to mislead—an irony surely not entirely lost on a former advertising executive. It is the humble car salesman who Welch uses as the epitome of this corporate greed and exploitation, the salesman as a thief who will ultimately manipulate his customers such that they become disproportionately indebted to him:

He tells you he's going to save you time by giving you a car right now and letting you pay it off for the next 3 years. What this really means is that he's buying you, part of you, for 3 years. For 3 years you're a partial slave to him. You get a car which gradually decays, and he gets a piece of a boy, like a percentage of a fighter, who gradually gets tougher. The car will fall apart in less than 3 years, so unless you're a very good trader, you have a piece of junk and he has a part of a very real guy. You.[16]

In the same way that armies are made up of "kids who can't even drink or vote [and who] are given expensive weapons to kill other kids who cannot even drink or vote," so the second-hand car dealer creates a situation whereby he is just as selfish in his goals as the general or government are in theirs.[17] Therefore, none should be trusted at any cost.

Erroneously citing Mario Savio, one of the most prominent figures of the student rebellion that gripped UC Berkeley in late 1964 and spawned the "free speech movement," Welch suggested that the maxim "never trust anyone over the age of 30" was indeed true (the quote was actually originally said by Savio's FSM cohort Jack Weinberg, who, according to Savio, was the movement's primary tactician) and that anyone over that age was really only interested in buying, owning, manipulating, or controlling the youth of America for their own gains. Welch further implied that even decorated generals sending young men to their deaths in the jungles of Vietnam were somehow ignorant to the truths of war and the role of politics therein, being instead mere pawns in a system intent on making some point at the cost of individual lives (such as a "loss of prestige").

In the wider sense, "How to Survive in the United States" is a means of giving its readers a sense that they have a choice and that there is a need to come to the understanding that living in the United States is about choices that are often made by others. The youth of America need to be aware of exploitation and try to take control of their own choices through two simple maxims: "Stay out of debt" and "Right job. Right conduct." While this second maxim is a reference to the need to do a job with honesty and "total absorption" so that the natural result will be "something good and satisfying," the notion of "right conduct" is also clearly linked to the fourth leg of the Eightfold Path as prescribed by the Buddha. In this sense, "right conduct" means to live ethically and morally in "the absolute sense rather than by the standards of any particular time or society." Thus, when making decisions within the framework of a society that strives to impose choices upon us, right conduct is key to living by our own values—something that Welch had done and would continue to strive toward doing even as his writing and teaching imposed a certain construct within which his own life was now framed.

In another unpublished essay from 1966, Welch goes further in his description of "right work[,] right conduct" by specifying that the two are actually inseparable.[18] He suggests that the other aspects of the Eightfold Path are unimportant, given that these two "imply them all" and that to focus on work and conduct is fundamental to understanding what is "wrong" with America. He suggests that both ideas are mistaken by the public at large and that work/conduct is akin to what used to be known as "ethics" and "morality"—a process of understanding rather than a state. Or, more directly, "an idea-group which only has meaning as a process, or as the Buddhist puts it, a way . . . a series of actions."[19]

He then provides a concrete example of what work "means" in America and how it is now used simply as a way of specifying income and/or status. The notion of what the work *is* has been lost, and the idea of work is purely a means to an end in the minds of most people. We are what we earn, not what we do.

> The word "work" in America, has come to mean: "how I make my money." The work itself, that comes of it, how well it's done, whether or not you like doing it, and most especially, whether or not it needed to be done in the first place, is almost never considered in any way.[20]

The aforementioned construct of teaching and writing, both of which involved the emotion of performance, was driven not only by Welch's brutal self-criticism and censorship but also to some extent by his fear that no one was listening to him. In his preparation for a reading at the San Francisco Museum of Art, under the auspices of James Schevill, who was then the director of the San Francisco State Poetry Center, Welch wrote in his program notes that

"what was the 'Beat generation' is now down to a few survivors, each of whom went his own way. Most of us are gone (as so many makers go early) into prisons, loony bins, penthouses, graves, and the other silences of whatever desperation" (IR2, 139).

In some ways Welch is calling attention to the fact that this "generation" has now all but died out and with it the attention that the public once held for its members (although as Gary Snyder apparently once said, "There is no Beat Generation—it consisted of only three or four people, and four people don't make up a generation"[21]).

Figure 23.1: Lew giving classes in African American literature at the college in Marin in late 1968. Photographs © John Doss.

In writing to Schevill, Welch is keen to establish his credentials in numerical terms, challenging Schevill to a bet on how many people he can attract to his reading. However, this could equally be construed as a challenge to himself. After all, Welch was forty by this time and may have been trying to prove his continued relevance. He even goes as far as to specify the terms of the bet before Schevill has even had a chance to respond ("I bet you a fifth of good whiskey I can get more than 300 people to hear me" [IR2, 136]); while promising to cover any additional costs that Schevill may incur if the bet fails, Welch introduces the notion of what the poet needs to do in order to make his audience hear him. He does this by using Bertolt Brecht's poem "On Teaching without Pupils," which the German had written while in exile in Denmark during the 1930s. In the same way that Brecht is suggesting that his period in exile is like being "a man to whom no one is listening,"[22] so Welch suggests the same about his work and its performance. He tells Schevill that he has "learned to make them hear . . . what I make them hear is my poetry" (IR2, 138). And he has done that by being among those poets who, in his mind at least, have reshaped the notion of poetry readings into a new art form, removed from traditional ideas and now manifest as "a new theater far more flexible, challenging, and entertaining than what we had in the past" (IR2, 139).

This boast was far from being an empty one. On October 26, 1966, Welch read before an audience of several hundred at the San Francisco Museum of Art, doing so with an assuredness and poise that left its mark on audience members and critics alike. Stephen Suleyman Schwartz wrote of seeing Welch that night and of his "hilarious" rendering of Eliot's "The Waste Land" in the style of the jazz singer Billy Eckstine. It is a moment that is also captured in Grover Sales's review of the evening for the *San Francisco Chronicle* in January 1967:

> [Welch] composes tunes of disarming simplicity, has a resplendent singing voice well suited to the American ballad styles of Billy Eckstein [*sic*] and Herb Jeffries, both of whom he can imitate with uncanny precision. A superb mimic, his version of Eckstein singing Eliot's *The Waste Land* was one of the many delights he held for the museum audience, which seemed in a justifiable state of whooping hysteria much of the time. (IR2, 141)

Welch had an ability to engage with his audience—something he had developed and retained over many years—and he was as far removed from being the "man to whom no one was listening" as he was to "teaching without pupils."

24
Revealing the Mysteries

Despite the publication of *On Out* and the relative success of his readings and teaching at UC Berkeley Extension and Millberry Union, by the end of the year Welch had sunk into an all-too-familiar bout of acedia and self-doubt. Where other poets saw such successes as reason for added confidence, Welch's response was more than merely negative. In late 1965, Doug Palmer asked him to contribute to a new collection of works that Palmer was putting together titled *Poems Read in the Spirit of Peace and Gladness* as well as take part in a benefit reading for the San Francisco chapter of the IWW, which Palmer had helped establish. Despite the lofty intentions of creating "a socially anarchistic utopianism with poetry leading the way," which must surely have appealed to Welch's sensibilities at the time, his answer to Palmer's request is telling.[1]

Primarily a street poet writing under the pseudonym Facino Cane, Doug Palmer had started out producing mimeographed sheets of his poems and using them to make money on the streets of North Beach. By the end of 1965, however, he had read at the Berkeley Conference, was organizing readings himself, and was compiling what would become a comprehensive overview of Bay Area poetry. Palmer met Welch at the Jazz Workshop, which was one of San Francisco's premier nightclubs during the 1950s and 1960s, where both had gone to listen to a show by *Synapse* magazine editor Dave Hazelton's wife, Jeanne. The story goes that Palmer handed out his mimeos to the audience, and upon reception of one, Welch wrote a counter poem on the reverse and gave it back to him. As Palmer remembers, that poem became a talisman to encourage him in his poetic endeavors.[2] Yet by the time Palmer approached Welch about the anthology, it seems that the latter was the one most in need of encouragement. His response was nothing if not straightforward: "The only reason that I haven't answered you sooner is that I'm all published out (like fucked out) and/or don't have any confidence, delight, or NEED to make public all this bull-shit writ out over all these pages."[3]

While conceding that he may be prepared to participate in a future reading of some sort, the idea that he could muster what Palmer had asked for—namely ten pages of poetry—was simply beyond him. Welch wrote, "I don't have any 10 pages right now, or 10 poems, or 10 minutes, somehow."[4] And while Snyder (from Japan), James Koller, and many others set about arranging the particulars of their contributions to both a reading and/or the anthology, Welch simply offered his apologies by saying that at that moment he was "not a poet or person with head set to pick out 10 pages of anything."[5] It seems his job as poet was temporarily redundant.

Despite this, Welch did manage to write a short foreword for Palmer's own collection of poems, *Poems to the People*, in which he praised Palmer/ Facino for his courage in not only laying himself bare to the public but also for putting himself at the mercy of the authorities (Palmer had been arrested along with Dave Hazelton—aka Cinzano—for "begging" for food or small change when handing out mimeos). Is his work subversive? Does it upset the public interest? Why, asks Welch, are the police worried? The poems hold the key.

Not many poets will face the public bare, Facino does. The fuzz worry. Why? That's a real question. Figure it out. Read these poems.[6]

———

In terms of publishing his own work, it would be another three years before Welch had anything new in print. The ensuing time was largely spent reading, teaching, and being a partner and stepfather in Marin City. Indeed, when the time came to print a new chapbook, he was very much in teaching mode; having reestablished his working relationship with Dave Haselwood, the pair collaborated on the creation of what, in artistic and aesthetic terms at least, might have been their most ambitious project to date: *Courses*.

Firmly engaged in the education system, Welch set about writing a series of short statements that would encapsulate his more humorous takes on education and posit the idea that certain lessons can also be learned in a more frivolous fashion. Yet rather than being singularly humorous vignettes, they also offer pointed criticism of education in general.

As publishing houses, especially the smaller ones, jostled to create eye-catching works that might set them apart from the competition, *Courses* constituted something out of the ordinary. The relative sobriety of *Hermit Poems* and *On Out* was in stark contrast to what Welch and Haselwood now intended to publish. With a relatively small print run in mind, the pair set about hand-cutting a large piece of brown suede that would act as wrappers for the book. The slim volume had a cover stamp in gilt letters and Welch's autograph embossed on the rear.

The collection includes such subjects as "Geography" (which was based on a dream that Magda had about not being able to tell East from West[7]), "Aesthetics," "Theology," and "Math," as well as a credo ("We refuse the right to serve anybody"), an oath ("All persecutors shall be violated!"), a graduation address, and a poem that has gone on to become one of his most well-read: "The Basic Con."

The book is interesting for a number of reasons, not least because of Welch's mix of humor, historical awareness, and disguised criticism. During a reading of the entire book at the Renaissance Corner in San Francisco in the summer of 1969, Welch said that the collection was an alternative to having to go to college at all and that if people listened (or read) carefully they would be better equipped to deal with what they are required to face in the world. He sketched a vivid image of students having to walk through a gate over which the college motto is hung in big brass letters — "No Credit No Blame No Balm" — and that the courses on offer were there to help students understand and work with the other information they were given by (one assumes) less interesting lecturers. The motto clearly implies that students are solely responsible for their learning, and in the event of any feelings to the contrary, there would be nothing in the way of compensation. The truths being told are blunt and unremitting. Any proverbial medicines to soften the blow are not on offer.

"History," for example, offers a bleak vision of the cyclical nature of warfare and our part in sending young men and women to die in battle or return traumatized by what they have experienced. It is not only the story of two world wars but also, more directly, the reality of US involvement in Vietnam, where GIs returned to a cold shoulder from the very people who had sent them there in the first place and were left alone to deal with the debilitating aftereffects of what they had witnessed or experienced. There is a certain detached matter-of-factness when Welch states that

> Every 30 years or so, Elders arm Children
> with expensive weapons and send them away
> to kill other children similarly armed. (RB, 112)

There is also resignation in the fact that "nothing else changes" and that, despite the best intentions of governments to establish criteria with which to punish war criminals, these criteria are often disregarded as priorities change and attentions switch elsewhere. This is certainly the inference in the second half of "History," when Welch simply writes

> Mr Krupp got the whole works back
> By producing a single document
> From his briefcase. (RB, 112)

As a history lesson, the tale of Alfried Krupp is as enlightening as any in terms of how war retains the attention of those in power only until the next crisis comes along, when suddenly morality is forgotten and an unsettling lack of interest prevails. Krupp was head of a company that had supported the Nazis and helped to supply Germany with arms and materials during and after the rearmament of the 1920s and 1930s. When Hitler transferred sole propriety of the company to Krupp in 1943, it was seen as an act of gratitude for Krupp's continued support (alongside his being named minister for armament and war production), and when Krupp used prisoners from concentration camps as slave laborers in his factories, he did so in close consultation with the German SS. As a result, he stood trial after the war in Nuremberg, where he was sentenced to twelve years in Landsberg Prison (where Hitler had narrated the contents of *Mein Kampf* to Rudolf Hess in 1924).

However, Krupp served only three years, because his release was arranged by then US high commissioner for Germany, John J. McCloy, who in a strange twist of Welch-related fate was also instrumental in the decision to begin the internment of Japanese citizens such as Nyogen Senzaki after Pearl Harbor. In addition, McCloy also saw to it that all the property and holdings Krupp had previously held were returned to him, apparently on production of "a single document from a briefcase"—a metaphor for the ease with which such momentous crimes can be swept aside. The presiding judge at Nuremberg, William J. Wilkins, later wrote, "imagine my surprise one day in February 1951 to read in the newspaper that John J. McCloy, the high commissioner to Germany, had restored all the Krupp properties that had been ordered confiscated."[8]

This bleak history lesson is followed shortly after by "Botany," which immediately creates a huge contrast in terms of its frivolity. Welch lures the reader/listener in and then offers this contrast by way of highlighting the fact that education can be both serious and humorous:

> Consider the Passion Flower:
> Who'd ever think a plant would go to
> so much trouble
> just to get fucked
> by a Bee (RB, 115)

The choice of language, however, was not always met with the same enthusiasm as the message. During his poetry-in-residence at Greeley in the summer of 1970, Welch was invited to read at the Phi Delta Kappa fraternity banquet, only for the ladies in the audience to be shocked into stunned silence by the profanity in the poem's final line. In his remembrances of Welch's visit, former student George Redman recalls the evening and the reading of "Botany" as such:

> The audience beamed, even when he got to the "Consider the Passion Flower." . . . Lew read the first line, then stopped, just twisted his mustache ends, and asked, "Who knows what a Passion Flower is?" One of the administrators [sic] wives answered: "Yeah," says Lew, "It's got three of this for the Trinity, and twelve something or other for the apostles, it's full of symbolism but few know what it is, I guess I picked the wrong flower, you know." The audience beamed, then he smiled, and started over with the poem, and read straight through. Silence—except for the English majors in the audience. The chilled reaction was one of YOU CAN'T SAY THAT WORD IN FRONT OF LADIES!!!![9]

Courses continues with "Philosophy," then makes way for "The Basic Con," which Welch once said was the only compulsory course in the entire college. In it, Welch offers a vision of how people can be conned into following the ideologies of others (governments?) who ask allegiance for a cause based not on nobility but on lack of personal purpose (other than greed?). He sketches an image of a society in which "these, and an elite army of thousands" collect taxes to the detriment of the majority and then use that wealth not only for personal gain but also to wage war and claim the lives of those who have not been duped into believing their rhetoric:

> Finally, all this machinery
> tries to kill us,
> because we won't die for it, too (RB, 116)

It is the same critique of the ruling classes that found its form in many of his other works, but in the published version of *Courses* is confined to this single poem. However, in another poem that was likely intended for use as a later "postgraduate course," this theme of greed is also posited, albeit with slightly more humor:

CITY PLANNING

"Ribs, please Mr Gray,
And a taste of Muscatel."

In itself "The Basic Con" is arguably the most successful venture of all to come from *Courses*. Welch clearly saw it as the poem that best construed his opinions at that time, and when he was invited to give a reading at Jack Shoemaker's Unicorn Bookshop in Santa Barbara, Welch chose the poem to be made into a small broadside to accompany the event. Shoemaker, who had founded the press and bookstore with Alan Brilliant in 1966, had relocated there from the Bay Area some years earlier and had started holding occasional poetry readings to generate money not only for the store but also for his publishing endeavors at Unicorn Press. To that end, Welch was invited, and a run of 450 broadsides was printed.

The pair had never previously met, despite both being active in the Californian poetry scene, and Gary Snyder had prompted Shoemaker to approach Welch about reading. In a letter from early 1967, Shoemaker first outlines his plans for the reading series, listing previous poets who had read, such as Snyder, Basil Bunting, James Schevill, and Richard Brautigan, before trying to convince Welch that the trip would also be financially worth it. Indeed, he underlines his intention to "pay *all* expenses incurred and a $50 reading fee."[10]

The reading itself took place on April 22, 1967, but Welch's preparation was predictably besieged by calamitous problems. Living in Marin City meant that he had to travel a long way to get to Santa Barbara, and despite initially deciding against driving (knowing that his car was unlikely to make it all the way there), he did so anyway. Inevitably, his "pore Ol' Ford" managed to get as far as Goleta before breaking down and prompting the cost of a flight home—thereby negating any chance of a profit he may have made from the reading. Indeed, he even asked Shoemaker if there was any money available in his "poet kitty" to help cover his costs. Not only did he have to pay for a flight home ($35) but also his car had to be scrapped, and, to add insult to injury, he apparently contracted some gastro-related problem that sidelined him from work too.

In the end, it appears that Shoemaker was left to take care of the car and to somehow salvage anything he could from it ("I figure whatever the junk dealer gives for the car should go to whoever has to go to the trouble of getting that business done"[11]), and Welch instructed him to not only get as much for it as he could but to also "Kiss ol' Ford for me—he was a real good guy!"[12]

If, as suggested in "The Basic Con," Welch was unhappy with the ideologies of government and their apparent corruption, this was merely another example of a general discontent he had had with the United States for years. From his plans to flee to Chile to his opposition to the war in Vietnam and his opinions on the ruling classes, the poem was a political stance of sorts that found an even greater outlet in his essay-writing that, for a short period, looked like it might take precedence over his poetry. As the counterculture gathered pace, Welch used this as an opportunity to make his feelings clear in a more direct fashion than via poems. Aside from published essays such as "Greed" and "A Moving Target Is Hard to Hit," both of which discuss the situation in the United States without being directly critical of it by name, Welch wrote a long journal entry that is not only highly critical of the country but also challenges it directly to exert more influence at home and abroad. Unlike, for example, "How to Survive in the United States," which was more advisory in tone, Welch contends in this piece that the citizens of the United States live in dual states: the "Utopia," represented by natural beauty and technological advancements, and the "Bad Life," which is split into two, namely the personal and the national. He suggests that the personal aspects of American life amount to "man & wife working, moonlighting, madness, insecurity, drugs, tranquilizers, moral despair"[13] and that this happens in a country that is an "aggressive killer, paranoid, despised of all nations, morally bankrupt . . . a war economy with no war."[14] He challenges America to meet its responsibility of great wealth and power and to use that not only to benefit others but also to eradicate greed and alleviate conflicts. Welch then asks

> Can you become the great beneficent sharer of plenty? Or will you persist as the big frightened killer of the world? It is just not thinkable that a nation with this potential, with this achievable dream, act as we do now, in Vietnam, and everywhere else in the world. [15]

This is a call to arms. A challenge to halt capitalist-consumerist tendencies and make the American dream a reality through beneficence and a willingness to forgo profit for compassion. While there may be a naive idealism in what Welch is suggesting, there is nonetheless a contradictory belief in the country he is being so critical of.

> America faces a challenge today that is greater than any nation has ever faced in any place or time. We can remove all physical want from the world. We can actually do it. But we can only do it if we give it all away. If we ask for no profit. If we ask for no reward. . . . It would be easy to do this. We are, ourselves, so rich, we do not need more riches. We are, ourselves, so powerful, we do not need more power.[16]

Yet the Vietnam war raged on, domestic affairs continued to be fractious, and the "basic con" was still a reality. As such, Welch saw his job as a poet expand, and temporarily at least, he became ever more politicized,

where possible taking the opportunity to speak out at readings and during
his classes. His increased profile, despite having relatively little work in
print, was making it easier for him, and his list of performances grew
throughout the late 1960s. He participated in various benefits for ecology
groups, antiwar demonstrations, community-based activities, and peace
marches across the Bay Area and beyond. And as he continued to give
poetry workshops at the UC Extension, so Welch also continued to read
his work with the same exuberance and emotion as always. His apparent lack of published work did not seem to act as a deterrent, neither for
Welch nor for the organizers of these events. Yet far from being inactive
on the writing front, Welch was actually hard at work on not only social
manifestos but also the compilation of his collected poems and a further
collection that would be among his most poignant and insightful yet—
certainly when read in retrospect.

Since the publication of *Courses*, Welch had been in discussions with
Donald Allen about the possibility of bringing out a collected works, to be
titled *Ring of Bone*. Allen had mooted the idea to Grove Press and its then
editor-in-chief, Richard Seaver. Indeed, it was not the first time that Allen
had asked Seaver to publish Welch, having done so around the time of
Hermit Poems. Yet, unlike then, it seems that Grove was at least willing to
consider this collection, even if it would be a long-term project on account
of the restrictions they had on the publication of works of poetry. After consideration, Welch signed a contract and negotiations began as to how best
design and package the collection. Welch's idea was to issue the book as
a square-shaped hardcover with a seven-inch recording of Welch reading
the work tipped into the back. James Laughlin and New Directions had
done this with the publication of Lawrence Ferlinghetti's *Starting for San
Francisco* in 1960, and the format clearly appealed to Welch and Allen. In
a letter to Seaver from September 1969, Allen writes that

> Lew points out in his preface that his is a very oral poetry, and he
> does read very well. We could tape him in a studio here if you are
> interested in this. Then he suggests the record be sold separately
> for $1 when the book is brought out as a paperback.[17]

Seaver's response was apparently less enthusiastic, because in correspondence that follows there is no mention of even a publication date,

let alone a possible recording. However, it would be wrong to think (as Welch later did) that Grove had no real intention of publishing the collection at all. Indeed, while the situation dragged on for months, with Welch initially receiving no acknowledgment that his manuscript had even been received by Grove (despite having a contract), there is mention of the fact that Grove requested the original lion stamp that had featured in Welch's previous chapbooks and broadsides. Yet he still had little or no idea when the book would be released, and although optimistically writing that it would be out by spring 1970, a letter from Marilyn Meeker, his editor at Grove, fails to shed any light on this, stating only, "I cannot give you any definite information about the publishing schedule at this time" but that *Ring of Bone* would be published no later than "February 1962 [*sic*] in accordance with your contract."[18] This would surely have served only to compound Welch's frustration at the lack of progress, and indeed by midsummer he had given up all hope of the book ever seeing the light of day at all through Grove, writing to Clifford Burke that they "will dally, and when the date runs out they will try for an extension, and I won't give it to them. My contract has it that the book must come out on or before Jan. 30, 1972" (IR2, 171).

His alternative to Grove was the idea to print the entire collection with Burke so that he could simply present it to a new publisher already proofread and ready for publication. This way, he wrote to Burke, the book would be "beautiful beyond all believing" (IR2, 172).

The choice of Burke as Welch's new collaborator coincided with the fact that, around the same time, the pair was also collaborating on the design and printing of Welch's latest (and last) collection, *The Song Mt. Tamalpais Sings*. Although they had known each other for some years, this was the first opportunity they had had to work together on a book project, despite Burke's long-held appreciation for Welch's work. In his book *Printing Poetry*, Burke even goes so far as to say that the origins of that book had their beginnings at the Berkeley Poetry Conference, where he first heard Welch read his poems. He also left the hall that night with not only a copy of *On Out* (signed by Graham Mackintosh) but also a new understanding of the craft of reading poetry in the manner that Welch did.

In the years that followed, Burke and Welch became good friends (something that Welch was keen to maintain, especially after he had upset Burke by openly attacking T. S. Eliot in a class, a fact that Burke had found unacceptable and for which Welch wrote, "You were right to be angry. . . . I have no right to speak to these people so facetiously and crookedly"[19]), and he would visit Burke at his printshop on San Francisco's Schrader Street, discussing poetry and printing, and finally working together on a couple of free broadsides, including "Springtime in the Rockies, Lichen," which featured artwork by Magda and was handed out by Richard Brautigan at the Whole Earth Catalog Demise Party held at the Exploratorium in San Francisco a few weeks after Welch's disappearance in 1971. Of Welch, Burke later wrote,

> Lew was a great poetry enthusiast, his own as well as others,' and he mentored a number of younger poets, some of whom I published. I think it was that enthusiasm, along with his revelatory reading and his book, that invigorated my dedication to my craft.[20]

As a printer it must have been a joy to work with a poet who understood the fundamentals of printing and bookmaking and who believed in Lloyd Reynolds's dictum that "if you don't have ink ground into your fingertips, you don't love printing and you don't love books!"[21] For Burke, *Tamalpais* was the highlight of his collaborations with Welch. Because he had had the benefit of hearing Welch read, it was a book that allowed him to "come very close to reflecting his voice on the page," which is surely the ultimate challenge when putting the words of a poet like Welch on paper.[22] Indeed, Burke said of the project that it was one of the most *well-balanced* books that he has ever made, and in an epitaph to his friend wrote that "the only dead poet I've ever wanted to print is Lew Welch, who some of us claim still lives, and whose spirit is still alive in the memory of his being and teaching."[23]

The Song Mt. Tamalpais Sings was published as Maya Quatro Five—one of a series of thirteen chapbooks published by Jack Shoemaker featuring poets such as Robert Duncan, David Meltzer, and Philip Whalen. Welch's contribution was a small collection of his most ecologically aware poems. It also features "Song of the Turkey Buzzard," the poem that for many has come to represent his real suicide note—the one that sheds light on a decision he would not make for another two years but that foreshadows much of what many people think played out on that fateful day in May 1971 when he disappeared.

25
Creating the Condition You Describe

In the annals of counterculture folklore, the events and performers that have gone on to define "hippiedom" and the "Summer of Love" in the eyes of the mainstream were (and in many respects still are) Woodstock, Altamont, and the Isle of Wight festivals, along with the likes of Scott Mackenzie, the Mamas and the Papas, and Jimi Hendrix. Psychedelic campervans, rainbow-colored sarongs, marijuana, and free love add to that stereotype. And while all of these undoubtedly contributed to that image, the reality is of course considerably more complex. Although 1968 is often seen as the pivotal year in terms of the counterculture and the coming revolution, it is arguable that the previous year was actually far more important in establishing the parameters for the sweeping cultural upheaval that was to follow—at least in terms of notions relating to community-building and social revolution.

Much has been written about the events of January 14, 1967, in Golden Gate Park. The Human Be-In, the gathering of the tribes, the coming together of myriad like-minded souls on a wave of optimism and anger, the music, the poetry, the collectivity. What had started after a chance remark by one of the organizers of the Love Pageant Rally, Michael Bowen, after the event on October 6, 1966, ended with thousands of people immersed in arguably the first and most important multidisciplinary festival of its kind in the United States.

The Love Pageant had been organized by Bowen and the poet Allen Cohen to coincide with new legislation in California banning LSD, which had by that time become more or less the staple drug of an entire community. In passing this law, state authorities were essentially handing a sweeping mandate to law enforcement agencies to crack down on any and all distribution and/or use of that or any other drug in whichever way they saw fit. This had been their tactic for months, and passing a new law was merely a means of legitimizing their actions. The subsequent increase in drug busts led to widespread confrontations on the streets of Haight-Ashbury, resulting in even more protests, riots, and curfews, which had been an aspect of life in the city on and off since race riots had been sparked the previous September by the shooting of a young Black man by a white police officer at Hunter's Point.[1] As the atmosphere became more volatile, Bowen and Cohen called for a demonstration more in keeping with the pacifist sensibilities of the counterculture as a whole, and thus the Love Pageant was born.

Cohen summed up the idea by saying, "We wanted to create a celebration of innocence. We were not guilty of using illegal substances. We were celebrating transcendental consciousness. The beauty of the universe. The beauty of being."[2] As such, Bowen and Cohen were helping to establish the framework within which such an event could take place, as well as provide a template for what was to come in the future.

By the time of the Human Be-In, the volatility and violence in the city was at such a level that there was a serious feeling among large sections of the community that the only alternative was to keep on holding peaceful protests. As America seemed to explode in a nationwide revolution against any and all forms of authority and inequality, events such as Bowen's were seen as key in promoting togetherness and collaboration in the face of increasing and widespread fragmentation. As Richard Alpert had said on the evening of the Love Pageant, it is about "Humans being. Being Together." To which Bowen replied, "Yeah. . . . It's a Human be-in."[3]

The date of the event was determined by Gavin Arthur, the astrologer and sexologist who had published *The Circle of Sex* in 1962, in which he contended that sexuality and sexual orientation were linked with astrology and star signs. As well as befriending the likes of Neal Cassady and Allen Ginsberg, Arthur had also crossed paths with Welch and later drew up an astrological chart for him. The result, dated August 9, 1968, was specified by Arthur as a splash pattern—an even spread of the planets over the twelve points of the chart, suggesting both flexibility and a scattered range of interests. While there is certainly a positive interpretation of this chart, in Welch's case it is tempting to connect this with his regular periods of inertia, inasmuch as scattered interests may well result in scattered focus; Snyder once said that Welch often had difficulty finding his focus and using it to one end—in this case, poetry. Indeed, Snyder highlighted this problem in an anecdote from when the two roomed together at Reed:

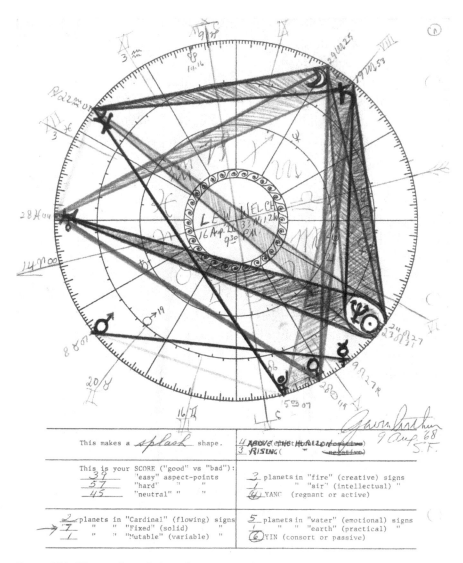

Figure 25.1: The astrological chart that Gavin Arthur made for Lew in August 1968. Courtesy of the Estate of Lew Welch and UCSD.

Welch would often wake early with all manner of grand plans for the day ahead but be subsequently unable to get out of bed. Snyder contended that this was an example of how his friend had many great ideas and interests (as confirmed by Arthur's chart) but was frequently unable to muster enough energy to see them through.[4] Indeed, Welch himself seems to underline this notion in a poem titled "Memo Satori":

> 2½ hours ago a bell rang
> I live the winter morning,
> half-clothed in a dark room,
> trying to plan the day (RB, 17)

Although "Memo Satori" was written in 1958, the primary image was originally contained in a poem that Welch wrote at Reed, and as such is just as likely to apply to the morning ritual that Snyder describes. The untitled poem, included in a letter to his mother, reads,

> In my middle youth I learned
> To leave each just-comfortable bed
> Before the sheets began to bind
> Before the visions in the mind turn. . . .
> I lived successive winter mornings
> Standing
> Never fully clothed
> Having not attended the alarm,
> Trying to plan the day.[5]

———

If a shortage of focus could be manifested as a lack of confidence, then the poem that Welch subsequently dedicated to Gavin Arthur titled "Sweet Death" exemplifies this perfectly. Indeed, the poem harks back to something that had troubled Welch for some time—his reaction to love and, more specifically, sex. In a letter to Kerouac from 1959, Welch uses Arthur as an example of someone who only succeeds in being attractive to the opposite sex after announcing celibacy: "I quit (hopefully Gavin Arthur's 60 year+ statement that only after announcing celibacy did the ladies really get down to business and lay him" (IR1, 169). And as such, Welch announces his celibacy too, admitting he had "never been a real cocksman, always fell in love." He may have said that it was because he had had nothing but "bad sex, worse love," but his own insecurities about his sexual prowess surely also played a role, as did his seemingly constant need for love despite his inability to readily accept it. In "Sweet Death," the poet describes himself as a "crab with lifted claw," suggesting that

Welch's defensive reaction to offerings of love is akin to developing a loathing of people who deprive themselves of death, as if suicide is in some way related to an inability to accept love.

————

The Human Be-In came to be with the usual technical malfunctions, as the poets and musicians on hand took the opportunity to pool their collective resources and highlight creativity as an ingredient for social cohesion. With music being the perfect medium to establish some sense of togetherness, the poetry readings and chanting of mantras added a spiritual dimension to this collectivity, and as Allen Ginsberg later said, the scene was about gathering people together who were "embued with a new consciousness and desiring of a new kind of society";[6] it was an event about love for one another, love for ourselves, love for the planet, the city, the park. In an act of what could be described as the ultimate social cohesion, the Hells Angels had been asked to provide security (a tactic that was both daring and obvious but that would have disastrous consequences the following year at Altamont), and by the time the sun had set and Snyder and Ginsberg led a final chant of "Om Sri Maitreya," the crowd drifted off, with Ginsberg asking them to pick up their trash on the way. The people left the field cleaner than they had found it.

While optimism about the event abounded, Emmett Grogan was scathing, seeing it as nothing more than a money-spinning opportunity for the organizers and "one great big fashion show" attended by a largely white suburban population in a "showcase for beaded hipsterism."[7] Among these hipsters were Welch and Magda. And although Welch never openly spoke of his thoughts on the event, his notions concerning the increasing influx of people into San Francisco—and the inevitability that some of these were jumping onto a bandwagon—were highlighted in an essay he wrote less than two months later.

Welch's short essay, "A Moving Target Is Hard to Hit," was "gestetnered"[8] by the Communications Company and distributed as a free pamphlet on the streets of Haight-Ashbury in the spring of 1967. In writing the essay, Welch inadvertently revealed his thoughts on the Human Be-In by highlighting what he felt were the potentially disastrous consequences of this influx of hippies and travelers as well as the need to circumvent these consequences through what he called "dispersal."

Welch clearly felt that the city was being overstretched and that nothing good would come of the burden being placed on it by these pleasure-seeking pseudo-revolutionaries. There was also the question of whether this type of manifestation had any effect, or whether such displays were merely an excuse to propagate good will and community in the face of far larger problems that continued unresolved. Indeed, Ginsberg himself had doubted the efficacy of the event, asking Ferlinghetti, "What if we are all wrong?"[9] Many people foresaw these consequences as catastrophic. And, as such, Welch meant his essay to be read as a warning or a challenge. Or perhaps it was merely advice to anybody willing to listen. There is, however, a peculiar and pointed resonance in the notion of dispersal as a form of personal—or indeed collective—revolution.

The precursor to this essay, however, was a longer and more autobiographical piece titled "Co-Lively," which was the basis for not only "A Moving Target Is Hard to Hit" but also the later essay, "Final City, Tap City." In "Co-Lively," Welch muses on the nature of codependence and what society needs to do or has done to achieve it. Harking back to his experiences at Big Sur in 1962, Welch uses the gathering of that group to highlight in some small way the logistics of dispersal. He wants people to live and move in small groups but feels that this notion of dispersal needs to be named. It needs to have a moniker with which people can identify and that avoids the stereotypes associated with "commune," which would encourage "every red-neck in Washington to gun us down as commies, which we most definitely aren't."[10] Thus "Co-lively."

Yet what is a "moving target" anyway? Welch had used the idea of a moving target some years earlier to refer to something that, even then, was particularly prevalent, namely a keen sense of antiauthoritarianism and the need to avoid any attempt at dictation by the state. Citizens and the groups they populate are by default targets. And moving targets are, of course, more easily missed, so citizens who want to avoid the restraints and obligations of society, government, and expectation are thus better served by "floating." It may seem strange in the subsequent context of politics and counterculture revolutions, but Welch's "moving target" was initially a reference to Nyogen Senzaki's floating zendos.

The situation at Marin-an a decade earlier ("It seems to me the 'Mur-can Zendo must float. *It is harder to hit a moving target*" [IR1, 164]) had forerun much of his current ideology concerning transience and the need

Figure 25.2: Lew
on the balcony at
his Buckelew Street
residence in Marin
City, California.
Photographer
unknown. Courtesy
of the Estate of
Lew Welch.

for people to move in order to escape, on the one hand, the ever-tightening
legislation at both city and state levels and, on the other hand, the increasing sense Welch had that what was happening in San Francisco at that
time was merely an example of a cultural sensation (or perhaps manifestation) overstretching itself. It is an ideology very much in the same vein as
the one that he largely practiced, and it is interwoven with his notions
concerning revolution and anarchism, as well as his desire to similarly
preempt vulnerability (the idea that overstretching the alternatives to
the status quo caused the newly developed culture either to implode or
become nothing more than a theory or paradigm). It is of course an ideology as old as time and one that had been consistent throughout his life,
although he was neither a revolutionary nor an anarchist in the truest
sense. He once wrote in a short treatise on the existence of the law within

society that "anarchy is true and impossible. It is impossible to live in a society (mindless conspiracy) without law (mindless conspiracy). Yet it is true: 'All is free where law is not.'"[11] Welch's penchant for solitude, his love of nature, and his unwillingness to bend to authority (which may well have started with the often uncomfortable and uncompromising relationship he had with Dorothy) were all strands of a belief system that, by extension, was also very closely linked to one of the fundaments of counterculture existence at that time; namely, Peter Berg's maxim that in order to become a "participant" in a society pushing for change, it was essential to "create the condition you describe."[12]

———

As a founding member of the Digger movement, Peter Berg was at the forefront of various counterculture happenings in San Francisco during the mid-1960s. A member of the Mime Troupe, Berg became one of the intellectual driving forces behind much of what subsequently occurred with the Diggers and their successors, the Free City Collective and Planet Drum. For their part, the Diggers were "community-anarchists" concerned with, among many other things, facilitating a viable alternative to the growing reliance on and coveting of material wealth. Taking their name from the seventeenth-century Digger (or True Leveller) Movement established by Gerrard Winstanley in northern England, the American Diggers molded many of Winstanley's groundbreaking Communist ideologies to suit a contemporary milieu. Not only anticapitalists, Berg and the other members of the group also grew to champion ecopolitics as well as notions of community-building and codependence. However, before any facilitation of alternatives could occur, people who "observed" needed to become people who "took part." In his essay "A Watch-Word to the Citie of London and the Armie," Winstanley states that "action is the life of all, and if thou dost not act, thou dost nothing." And although in Winstanley's case this relates more specifically to writing rather than acting, it is also true to suggest that this is a precursor to the idea of "creating the condition you describe" through actions rather than observations.

In the same way that the Mime Troupe was an uncompromising, truthful, and iconoclastic collaboration of like minds, so the Diggers would employ this same sense of radicalism (or indeed anarchism) to posit a well-constructed antidote to the trappings of capitalism, a shift into what Peter Coyote, another member of the group, called "a culture offering more enlightened possibilities for its members than the roles of *employee* or *victim*."[13] Yet as stage actors, Berg and the others in the troupe were just that—employees—and they were part of what was essentially a business that was there to serve the purposes of society's most dominant classes—a situation that, as mentioned above, Grogan also ascribed to the Human Be-In. So, despite its radical nature, the troupe was seemingly just another cog in the capitalist wheel. Thus, there was a clear need for a change of direction, given that the "condition" (as described by Berg) was still largely the "creation" of others.

And so the Mime Troupe necessarily morphed into the Diggers, and many of the members went from stage acting to what was known as "life acting"—a term that both politicized and justified the profession. Berg later described the initial idea as "taking theater off the stage and into people's hands. So I had evolved a concept of guerrilla theater, and guerrilla theater was to actively engage people in some action, or witness some event that would make them sort of a conspirator."[14]

Performances were then given that were totally devoid of structure or of financial obligation on the part of the audience. This was a notion of "free" being used in a premeditated and intellectual manner. And it was all part of a wider plan to propagate the notion of what the word "free" actually meant—not only in terms of liberation from convention but also in the manner of providing goods and services free of charge, which is a notion that itself generally runs contrary to human nature, at least in capitalist societies such in the United States, and is thus all the more powerful. In understanding this distinction and putting it into practice, the Diggers were drawing a proverbial line between what they felt could be co-opted, which was pretty much everything, and what could not. In doing so they created a cooperative sense of giving precisely because it was at odds with conventional capitalist ideologies. And that was also potentially beyond both the understanding and the influence of the authorities. They felt that "free" could not be co-opted. *Free* then became a byword for the Digger movement. And, as the ideology expanded, so did the supply. Free theater performances developed, as well as free food, free newspapers, free banks, and much more besides. This notion of "free" was crystallized by one of the Diggers' visionary voices, Billy Murcott, who saw the relationship between human desire for wealth and status and a near-pious obsession with property and capital once it has been gained, or indeed even more so while it is still coveted. So, one of the conditions that the Diggers created was called the "articulation of autonomy," whereby people (or rather participants) were to be forthright and take responsibility for what they felt ought to be done—the idea being, quite simply, that if we have fantasies, we should actualize them and try to build societies of like-minded people around them, thus distancing ourselves from conventionality and making it difficult for the mainstream to interfere or get involved—in one sense, thereby, creating a moving target.

For his part, Welch wrote three manifestos during the period in which he was associated with the Diggers—"Moving Target" followed by "Greed," which was published in the *San Francisco Oracle* in October 1967, and "Final City, Tap City," which featured in the final edition of the *Digger Papers*, published in conjunction with Paul Krassner's *The Realist* in September 1968.

Highlighting his thoughts on capitalism and the environment, and the influence of both on humanity at large, all three essays merge seamlessly with the Digger ideology of taking responsibility for social change and creating the condition you describe. Key figures within the Diggers felt that the freedom being assumed (and even appropriated by that time) by the gathering masses in Haight-Ashbury was freedom at a price. As hippiedom (which many felt was a media-driven phenomenon anyway) expanded, so the very freedom that was sought was automatically compromised. Grogan in particular railed against the increasing conventionality of the unconventional. If LSD was to be the means of escape from reality, then he felt it was a charade propagated for gain by the likes of Timothy Leary and Richard Alpert, who Grogan saw as bogus charlatans. Grogan felt that this was culture for gain, like the Human Be-In, and that the only true means of escape was to draw a clear line and live outside the profit, private property, and power premises of Western culture. In the eyes of Grogan and others, the Magic Bus, for example, was just another example of this property and power paradigm, despite it seemingly being used as a means of symbolizing the counterculture. Grogan felt that the effectiveness of rebellion from within society had been lost, and so took a clearly marginalized stance for the free and widespread distribution of their ideas on mimeographs. And so, in turn, the *Digger Papers* were born.

With the help of Chester Anderson and Claude Hayward, manifestos were printed and disseminated. Ideas were shared and seeds sown. The pamphlets warned, advised, and informed. They targeted and attacked the mainstream. They urged caution in the face of what was becoming an exploitative and overburdened scene. Anderson even went so far as to compare Haight-Ashbury to the war in Vietnam—without which the whole scene may never have actually materialized at all—saying that Haight had become a place where "minds and bodies were being maimed as we watch" by the increasing use and misuse of drugs and the overstretching of resources within the community.[15] Thus, the Diggers established their platform for politics amid the music, drugs, and spirituality, and in turn Welch's notion of dispersal was actualized.

Rather than offer advice or suggest a call to arms, however, Welch's manifestos simply described the world as he saw it and how he had become a "participant" within it. In this sense, Welch was both observing and participating. As the atmosphere in San Francisco changed—becoming more volatile, violent, and unpredictable—and the influx into Haight-Ashbury began to spiral out of control, so the stakes were raised. Not only were the political stakes raised after the events of the Chicago Democratic National Convention in 1968 but also the anticapitalist backlash was itself starting to have a severely negative impact on the environment. The thirst for power and wealth was being diluted with an equally voracious thirst for "freedom" as the hippies, Hells Angels, Black Panthers, poets, painters, and assorted assemblages of social detritus each fought for their own articulation of autonomy. For the Diggers, it was the substrata of society forming autonomous groups who could, in theory, act together to effect social change and propagate social cohesion. Yet the employees (or foot soldiers) in this battle were making a victim of not only the city but also the earth beneath its shiny asphalt streets. Peter Coyote wrote,

> All of a sudden, the game got real hard and real gritty and the city was overrun. That was coincident with the thinking that a lot of us had been doing about the planet: about what the eternal reality was under this thin sheet of asphalt. And we were sensitive to the growing ecological crisis and the fact that a culture that was pissing in its life-support system couldn't continue.[16]

This line of thought echoed Welch's, who for many years had been frequently removing himself from the urban environment and who had already written about the encroachment of concrete as a slow death knell for nature. Now this could be combined with an increasingly critical stance on the consequences of "greed" in any and all forms, whether capitalism or the rapacity for what is "free" and available. In Welch's mind, the targets were seemingly now moving less and blurring more.

For many, the solution lay in letting the planet reestablish itself and recover lost ground in its battle against exponential urban growth and population overload. This required dispersal: the moving target. Welch's idea (rather than concrete plan—as, in his mind, humanity is always defeated by the plans it makes) was to fragment into smaller nomadic groups that could move around the natural parks and forests of California unhindered while leaving a relatively small footprint. Welch wanted the escape—or the dispersal—to perpetuate a change rather than be a mode of merely surviving. Leaving the grid was saving the grid. Every potential tribesperson was one less taking their toll on Earth. One less hippie tapping into popular culture; one less employee chained to the pantheon of capital and property. It was something he had previously practiced *and* preached, and he understood that by removing yourself from the epicenter of any overpopulated scene, "the whole insane machine would go roaring by" (HWP, 7)—thus missing its target. He emphasized the freedom that nature had to offer, and the fact that national parks were abundant in California, and more important (and ironic given that they were maintained by a government agency, the National Park Service), they were *free*. He suggested that people play a waiting game. Beyond the invisible line that denotes the cultural divide, communities should gather in retreats. Work together. Codepend. They should populate the mountains, the beaches, the beautiful places. Practice vigilance. Turn their ears from the tumultuous din of America and look inward. In time, he wrote, "the Planet will germinate—underneath this thin skin of City, Green will come on to crack our sidewalks! Stinking air will blow away at last! The bays will flow clean!" (HWP, 20) and one day, when the foul dust settles, we can return to ghostly cities overrun with tangled foliage and try to start again. We can be participants in the reclamation of society rather than observers, employees, or victims of whatever cultural phenomenon is the flavor of that week. Welch ends his essay by writing, "Haight-Ashbury is

not where it's at. It's in your head and hands. Take it anywhere." However, the original ending of "Co-Lively" was more generic (Welch clearly
amended the essay to suit the current situation and appeal to that specific
demographic), and as such is more encompassing and interesting:

> Co-liveliness in thousands of buses and cars and camps and
> mountains and City-Pads. The already invented answer to Final
> City, Tap City. The way out. Out.
> Wheels. Mobility. Real families. Co-livelies. The use of *all* this
> generous Planet, at will, with loved ones, friends. It is all in our
> head and hands. Take it anywhere. And wait.[17]

Within a matter of months, the increasing violence was to become
a reality in his own life, as the social unrest plaguing the city spilled out
into the suburbs. Popular wisdom suggested that going across either the
Golden Gate or Bay Bridge was enough to escape and create a safe distance between what was happening in the city and in the outlying communities. However, even these communities were not spared the violence,
racial tension, and antiauthority activities. In August 1967, Welch wrote
to Snyder, who had returned to Japan in March, that

> the revolution is finally happening. Detroit and 40 other cities blew
> up in July—the 1967 total is 70 cities and towns. Marin City, my
> home town, blew up a weekend ago so bad we thought it prudent
> to evacuate ourselves. . . . It's not so much a racial revolution as
> a revolution of the poor. Detroit rooting was integrated—spade
> cats helping white cats into the high window. Not so much about
> colored skin as about colored T.V. (IR2, 144)

Welch's criticism of this "revolution" is clear, and in his mind at least, any
nobility and morality attached to the cause has been replaced by looting
and theft. He also tells Snyder that "it's pretty scary living in violence, I
really don't want Jeff to get hurt, or Magda."

Furthermore, the consequences were not limited to "rifle fire." The
rapid increase in drug users and deterioration of any sense of control
meant that pushers were being ousted by organized crime syndicates,
who had moved into "Hashbury." Welch writes of people being sold acid

laced with STP (which was the street name for the drug 2,5-dimethoxy-4-methylamphetamine, or DOM, developed by the renowned pharmacologist and "godfather of psychedelics," Alexander Shulgin) and LSD spiked with methadone, which resulted in "people, good ones, blowing their minds irreversibly. Like, gone. Away" (IR2, 145)—this constituting an extreme form of dispersal: the chemical way.

26
Peonies and Columbine

> Dear Richard, Here we are in Greeley looking at TV in a motel.
> Greeley has no booze inside city limits. There are big trees here.
> Elms. Lew & Magda
>
> —Lew Welch (IR2, 173)

Welch's penultimate teaching position was in Colorado. Having worked
at several University of California campuses for much of the previous five
years, the five-week residency at the University of Northern Colorado at
Greeley in the summer of 1970 was an opportunity not only to broaden
his horizons but also to extract himself from the Bay Area for a while,
where he was finding it increasingly difficult to reconcile himself to his
life. He wrote to his old friend at the Forks, Katherine George, that while
it may be "better than any urban area in the world . . . it just may not be
good enough for those, like us, who are blessed with the choice of moving
away" (IR2, 166). Welch felt that events in the city were going downhill
fast, and he hankered after the relative solitude of the Forks, writing that
George's annual newsletter about life there had gotten him "messed up"
about whether to join her and her family or stay in Marin City with Magda
and Jeff. It was a dilemma that may well have helped him choose to take
the job in Greeley in the first place, given that Colorado represented one
of the "big Western States" he loved so much—states in which he could
indulge in the passions that really excited him, such as hiking and fishing.
Indeed, these were the very passions that he had so joyously indulged in
at the Forks. Yet it is also unavoidable to see the ever-more-destructive
nature of race relations, the heightened problem concerning drug abuse
among the influx of hippies, the looming specter of Vietnam, and his own
alcoholism as mitigating factors in his need to vacate the city.

The previous summer he had secured a post at the Urban School of
San Francisco, teaching high school freshmen in what he called "literary

get-togethers" that he hoped would help him snap out of his most recent downer. Bemoaning his mood and the changes that were taking place in his life at the time, Welch wrote to his friend Terence Cuddy that he felt "old and feeble" and that it was as always just a "matter of stopping the booze absolutely" (IR2, 158). Indeed, he admitted to it being his fifth attempt at going dry, and that it was a "drag."

The Urban School was—and still is, in some respects—a radical alternative to conventional schools in San Francisco, and hiring Welch was merely one example of this. In his acceptance letter to the first headmaster of the school, Robert Wilder, Welch wrote indirectly of his joy at not being asked to sign a contract that forbade him from burning "flags, politicians, draft cards, or money" (IR2, 155) but rather one that seemed to allow him to focus on teaching his students to observe and listen as writers do, without any obstacles or limitations. However, before the academic year had ended, circumstances once again intervened and Welch felt compelled to discontinue his work there in the face of what he called "our frightening situation"—namely, the present crises facing the United States both at home and abroad. Rather than turn his attention to using these crises in his lessons—as he would in Greeley later that year—he instead admitted defeat and wrote that his voice had been "silenced" by "those I call my enemies" (IR2, 169). Exactly who these enemies were remains unclear, other than their general description as having "fascistic" minds and desiring to silence "normal, life-loving teachers" such as he.

During that same summer, having been invited by Whalen to join him in Japan, Welch was presented with the opportunity to remove himself from the situation at home and relieve the frustration that he was feeling. It had been Welch's long-held desire to go to Japan—indeed, since his return to the West Coast and immersion in the world of American Zen—and with two of his closest friends there, Whalen's suggestion not only constituted the perfect remedy for his discontent at the state of the nation but also offered him the chance to get back on the wagon again—and stay there.

Having heard that his old friend was "all nervous again," Whalen encouragingly wrote "SHAKE NOT, NEITHER SHALT THOU SHUDDER NOR TREMBLE QUIVER THROB THRONG, for LO! the world & the Flesh & the Devil ain't worth it, Selah!"[1]

The gravity of this nervousness is then put into stark perspective through the description of a recent earthquake that Whalen had experienced in Kyoto and that, like Welch's emotional state, was only temporary. He goes on to ask Welch,

> Have you considered the advantages of a sea voyage? I doubt that it would hurt you and Magda any at all to take a slow boat out of San Francisco, to land at Kobe, where I could meet & bear you all away to Kyoto for a while. . . .
>
> Traveling is, in itself, so occupying & maddening & exciting that it gives one a whole new view of the general self, at least for a while. "Look at all that beauty out there—the sails of the herring fleet." . . . Please try to get better. We all still love you.[2]

In his reply, Welch candidly writes that he has "finally really hit bottom on the alky trip" and that he had ended up in hospital "flipped out, body screaming for peace and mind on bad death/suicide trips."[3] He admits to being unaccustomed to sobriety and that he needs time to make peace with his new self, having lived almost continually of late in a state of overexcitement and nervousness. As he had written to Cuddy previously, however, it was a serious attempt at "stopping the booze absolutely," and as such was an admission that the search for peace was better than the addled mind.

———

Welch and Terry Cuddy had been friends for a number of years and had read together in March 1968 for a live broadcast on KPFA radio. Titled *New Writer's Forum,* the show was recorded at the Straight Theater in San Francisco, and also featured Philip Whalen, David Meltzer, and Charles Upton, in addition to Welch and Cuddy. Welch's performance is typically assured, being in many ways the now tried and tested material he had been reading since the Berkeley bash. As well as reading a selection of poems and "Leather Prunes," Welch also sang a song written by Magda called the "Hippy Chick's Lament," in which he displays his natural talent for singing unusual rhythms. While the others also read their poems, Cuddy sang three folk songs about his experiences as an inmate in San Quentin. Indeed, Cuddy was partly instrumental in organizing a reading Welch gave with Richard Brautigan at the prison in August 1969.

Brautigan and Welch's reading at the prison was one of many performances that musicians and poets, including Frank Sinatra, the Grateful

Dead, and Country Joe and the Fish had made there over the previous decade. Indeed, performances in or at prisons had become something of a mainstay in raising awareness of what many believed to be the harsh and inhumane conditions in which prisoners had to live. Welch's motivation may have simply been based on doing a favor for his friend, but there may also been an element of protest. Either way, the reading left an indelible impression not only on Welch but also on the inmates—who Welch later described as most appreciative. The subsequent newspaper article covering the event wrote of it being "casual, almost intimate" and that the inmates in attendance were "engrossed" by the poetry being read.[4] The visit, which was held in the prison's Garden Chapel, lasted an hour and was a mix of poetry and "informative and spontaneous" discussion. Indeed, such was the success of the reading that, when Cuddy wrote to Welch on August 31 to offer his thanks, he suggested that a repeat performance might be possible:

> I really want to thank you and Brautigan for coming and reading. It just knocked everyone out. I talked with Chaplain Tolson [who had helped set up the reading along with Intern Chaplain Harold Dodd] and he said that if you at some future date would like to repeat it, he'd be anxious to arrange it.[5]

Cuddy, who Welch dubbed the "poet-folksinger," had already made a name for himself earlier that year by somewhat audaciously handing a song he had written to Johnny Cash before the country star performed for the inmates in the follow-up to his highly successful concert at Folsom Prison the previous year. During the show at San Quentin, which was also recorded and released later that year as a live album, Cash briefly mentions Cuddy before launching into a typically Cashesque rendition of the song, titled "I Don't Know Where I'm Bound."

Such was the immediacy of their friendship that when Cuddy became eligible for parole in September, Welch wrote an impassioned plea to the parole officer, lauding Cuddy as one of the finest poet-singers in America. Welch offered this opinion on the basis of his being "an expert on human behavior and value" and on account of his having spent a lifetime around people from all walks of life—including, apparently, convicted felons. To his mind, Cuddy was now a model citizen who was not only reformed but was needed outside of prison in order that he might use his gifts to help "ease" the troubles of the world. Indeed, Welch concludes his letter by contending that "to keep a man of such gifts confined in jail is a shame. We need him out here. We will, and he has many friends, do everything we can to befriend this outstanding man" (IR2, 160).

In a letter shortly before his release, Cuddy's mother, Geraldine, also wrote to thank Welch for being a friend to Terry and for the faith he had shown in her son before asking if he could put in a good word with the parole board. Whether Welch's plea made any difference is unknown, but Cuddy was released later that year. And by the time Welch and Magda left for those "big Western States," such was Welch's friendship with Cuddy that not only had Welch written a song (including lyrics allegedly written by Brautigan) called "The Trailer Is Leaky" that he asked Cuddy to put to music, but the couple had even rented their house on Buckelew to him too—at which address he remained in one capacity or another until after Magda and Welch split up in early 1971.

The job in Greeley was only one of a number of engagements that Welch had lined up for the first half of 1970, engagements that, as he was quick to tell George, went about lifting any financial woes he had. In addition to Greeley, Welch was due to give readings on campuses in Vancouver, Salt Lake City, Pocatello, and Logan, all of which would bank him in excess of $6,000. Add to that the hope he still had for the successful publication of his "big book" by Grove Press and his contentment—financially at least—is clear.

Before heading north however, Welch and Magda paid a visit to Santa Fe, where James Koller and Drummond Hadley had made their homes in recent years and where, as consequence would have it, Gregory Corso was currently also staying (records suggest that he had just spent some time in Santa Fe jail). Welch wrote of his being impressed by the city and expressed his admiration for what he conceived to be a certain authenticity and expansiveness rarely found in other cities. Indeed, compared to his thoughts on Greeley, Santa Fe seems like a more likely destination for a poet-in-residence than northern Colorado. Magda also wrote of his having enjoyed the visit, telling Donald Allen that they had had a great time in what was a great place. And, as if to reiterate the difference, her letter, like

Figure 26.1: Lew and
Magda in the garden
of their Marin home.
Photographer unknown.
Courtesy of the Estate of
Lew Welch.

those of her partner, also bemoaned the situation in Greeley as "a mixed
bag of revelation and boredom" and related how the barrenness of life in
small-town America was causing her to realize what she was missing back
on the West Coast.[6] For Welch, the lack of winos, skid row, and racial
diversity, as well as having to "drink in miserable Oakie bars or plastic
motels" (IR2, 172) clearly constituted a marked contrast from what he was
used to. Coupled with the perpetual scent of manure from what were then
the largest cattle feeding pens in the world, the picture is complete. Unlike
the couple's descriptions of the town however, their descriptions of the
university campus are positively glowing. He describes the beauty of the
campus, the quaint Englishness of the gardens, and the amenities he had
in "the perfect Norman building" he rented with Magda. She in turn wrote
to Allen about the "aura of concentration, openness, and friendliness."[7]

The poet-in-residency itself began on June 14 with a reading at
McKee Hall. Welch had been invited to participate as part of the university's centennial celebrations, and he was being paid $2,000 for the
privilege. Preparations for the position were put largely in the hands

of Professor Neal Cross, who when charged with the task had written to Welch explaining his dismay and total confusion about what such a residency would entail. His initial correspondence begins by addressing Welch as "Mrs." and proceeds, quite comically, to wonder just what a poet-in-residence might actually do besides "reside in Greeley."

Welch had been given the task of teaching the Forms and Meaning of Poetry class—about which Cross knew equally little—and was asked if he might consider for himself what the course would look like. Surely there would be more to it than Welch merely chatting to "a Creative Writing Club of little old ladies in tennis shoes" or being "viewed" by the local Rotary Club, as if being a poet was something of an attraction or novelty.[8]

Cross appears to have been put at ease by Welch's reply, writing in return that the poet's "cordial letter" had reassured Cross that its "content and spirit" would make it easy to "invent" the duties Welch was to carry out. However, any jokes quickly disappeared as Cross realized the enormity of Welch's task:

> I've just looked up your Forms and Meanings class. You have a real chore there. It meets from 11:15 to 1.10 on four days a week, Monday, Tuesday, Thursday and Friday. So you've got 40 hours of teaching—which is a hell of a lot for one class.[9]

Undaunted, Welch set about creating the terms and conditions for his classes, ever conscious that the poetry be at the forefront, even in the face of administrative requirements such as the grading of writing and the evaluation of critical ability.

The local media also ran several short pieces on Welch's appearance, making reference largely to the McKee Hall reading while offering some background information on his philosophy regarding poetry and on his continued commitment to environmental issues and his support for the occupation of Alcatraz by Indigenous Americans—a situation that had been highlighted in the mainstream American media since the prior year. Indeed, Welch had attended and read alongside Snyder, Corso, Kyger, and others at a benefit reading held the previous December at San Francisco's Glide Memorial Church to help raise awareness and funds for not only the island's occupants but also for the city's American Indian Center, which had burned down some months earlier.

Welch had long had an interest in the plight of Indigenous Americans, and his previous references to how society should fragment and live in tribes is all the more poignant given that, despite the support of Hollywood celebrities and local musicians and poets, the occupancy of Alcatraz ended with the forcible removal of the final protesters after the government cut off the electricity to the island and all but starved the protesters of water and other essential supplies. If anything, this merely served to bolster Welch's idea that self-sufficiency was the only way forward. While reliance on government-provided amenities and services was far from being the only reason that the prolonged success of the occupation failed, it was certainly a mitigating factor.

Welch himself never mentioned his alliance to the Indigenous American cause in the interviews he gave for the *Greeley Tribune*, preferring instead to focus on his approach to the residency and how he planned to perform and assist wherever necessary. He highlighted his own gregarious nature and flexible repertoire, which included being the master of ceremonies at "fashion and flower shows," singing the "Star-Spangled Banner" in a "strong ballpark baritone," and generally carrying on in "whatever way is wanted."[10] His self-promotion was as tongue-in-cheek as ever, and rather than use the platform for political ends, he simply focused on the poetry and how he felt that, if poems could not be extracted from the "din around us," then he had failed in his job. His commitment to poetry was as pure and important as it always had been.

In addition to what was presumably a similar reading list to his previous classes, Welch added an essay that helped underpin the message he had propagated in *Courses*—namely, that the university in its current state was archaic and in need of a radical overhaul. The work, titled "Don't Send Johnny to College" and written by the eminent scholar Hugh Kenner, argued that students who are merely at college to kill time before entering the employment market are clogging up the system for students who genuinely desire a higher education. While different in some aspects to his own thoughts on the issue of education, it is interesting to see the passages that he underlined and to imagine him challenging those students who had enrolled not to be the "Johnnies and Jonnies" in his class but to excel on Welch's terms by focusing on the essentials of learning, such as those outlined in *Courses*.

While Kenner laid the blame largely at the feet of uninspired and ambivalent students and their parents, Welch railed against the teachers and the system in which they worked (were these then the "enemies" at the Urban School of which he had written?), arguing that

> problems in schools exist because bad teachers and stupid people run them. Students want to learn. Angry students are the result of teacher failures. For me, the college experience was perfect. For them it is really not satisfying, it is terrible. I honestly believe what you have is a bunch of lazy teachers.[11]

His approach was one of stealth and rebellion. By being the opposite of the "lazy" teacher and applying himself with as much energy and enthusiasm as he could muster, Welch set about trying to inspire and excite—a formula that, given the student testimonials provided throughout his teaching career, was more often a success than not. And the same was certainly true in Greeley. As he said in his closing presentation at the University of Northern Colorado, the key to successful education and to slowly changing the system for the better was, as Gary Snyder once said, to "live in their house and every night go down into the basement and nibble at the foundation."[12]

———

Welch's poetic output appears to have waned during this period, with his poems being largely short and humorous or pointedly dedicated to particular people. Having completed what would be his final collection of poems the previous year, and with Grove Press still largely noncommittal about when *Ring of Bone* would be published, Welch's work from the turn of the decade displays a certain acceptance that perhaps his time was running out. Many of the later works are directed at friends and fellow poets. They are concise and at times melancholic snapshots of Welch's mindset, and most revealing of all was the fact that he had already thought about how any new book of poems would be different from his previous ones. Indeed, the planning he so meticulously insisted on before was now replaced by something not only far more chaotic but also revelatory.

His new book was to be called *Cement,* and a draft preface he wrote on the Salt Lake City stop of his university tour of 1970 outlined the intended structure: "All of my previous books were structured so that each poem nourished, and was nourished by, the others. This book is not like that. It is a wasted field in which, like blocks of cement, the wreckage of my mind is scattered" (RB, 208). This image is the most powerful of any he left us in the final year of his life. Far more than the planned structure of any future collection, this was a moment of self-revelation. Much has been written about his disappearance and the perceived romanticism of offering yourself to nature and alighting on the wing of a turkey buzzard, but there is surely nothing akin to the honesty and despondency that it takes to write of "the wreckage of my mind" and to see it, in many ways, as similar to the urban spaces that he felt were lying in ruin—concrete jungles devoid of soul or meaning, blocks of cement in a wasted field.

Alongside short and melancholic vignettes such as "Small Sentence to Drive Yourself Sane" and "Getting Bald," Welch had planned to write many poems that have either been lost or were never written. Titles, all revealing in themselves, include "40th Birthday," "Dragon Fly [Birth]," and "The Last Mussel Feast—A Curse."[13] Yet two of the longer surviving poems are most indicative of his state of mind—resigned and accepting of its fate yet thankful and honest in a manner more veiled than in his planned preface.

The first poem is "Inflation." Dedicated to Neil Davis, the owner of the No Name Bar in Sausalito, the poem provides humorous commentary on the notion of inflation and the negative effects it has on the drinker. Tapping into two of the most prevalent aspects of Welch's life—alcohol and finances—its humor masks the seriousness of the message, namely that inflation will not stand in the way of Welch's drinking. As the years pass, so prices increase, but the alcoholic will continue to drink until "penniless and drunk." However, there is no sense of anger or dissatisfaction, merely a shrug of drunken shoulders and another few dollars into Davis's cash register.

The second of the two is "The Wanderer," which draws inspiration not only from William Carlos Williams once again but, more interestingly, from the I-Ching, or the ancient Chinese *Book of Changes*. While the poem is a hopeful reminiscence on Welch's life, ending with his admittance that Magda's love has saved him from eternal wandering, it could also be argued (as he stated in an earlier poem called "How to Give Yourself Away") that love in itself "can never free us" (RB, 199). The act of being found, and accepting the consequences, is of course a form of achievement, but total freedom from suffering cannot be found so easily. Indeed, had Welch, as he contends, chosen to follow the sun and move in a westerly direction as "all Men" do, his entire life as we know it would have been eradicated in a "single step," and the abundance he now felt would have been achieved almost instantly. Yet such a reading, while understandable when facing the "wreckage" of his mind, would surely have rendered his life worthless—which in itself is surely the greatest and most tragic admission of them all.

———

As the final poem in *Ring of Bone*, "The Wanderer" provides a certain finality, but in some respects it could be argued that the prelude to this can actually be found in the aforementioned poem, "How to Give Yourself Away," which Welch had intended to deliver as the Sunday morning sermon at Richard Brautigan's Invisible Circus extravaganza at the Glide Church in February 1967.[14] Although very much akin to his spoken-word performances, the poem also draws considerably from Welch's wealth of philosophical and literary knowledge in order to provide a "sermon of gladness" that posits the idea that the "uninvention" of the self should be our ultimate goal as humans.

While Welch had long proposed that "reinvention" was key to becoming our ultimate selves, this changed in the later part of the 1960s to the more abstract and philosophical notion of uninvention. Welch argued that there were several aspects of human nature that, rather than used to effect change, need to be tempered or eliminated in order for us to arrive at a truer sense of ourselves. He begins by stating that temptation should be avoided and that compassion was, as Freud contended, merely an extension of the ego. Therefore, this idea that we need to somehow eradicate this ego is also reflected in earlier notions of *anatta* that Welch introduced in *Hermit Poems*. However, the "He-persona" that he adopted at Forks of Salmon is now replaced by the inclusivity of "we," as if Welch now sees himself as a spokesman for humanity at large. Yet is it not also the case that Welch had once called on Avalokiteshvara for guidance—the Bodhisattva of Compassion, embodying everything that was good about egolessness? And now even "we" has given way, in order to find our true selves? It may be, however, that Welch is indeed contesting the idea that compassion is selfish at all. Or at the very least that it is something we cannot separate ourselves from so easily. Despite such contentions, and the sense that compassion as a positive human attribute had "withered in dry fields of SELF-suspicion" (albeit, as history has told us, temporarily), the human need to bloom was greater: even if it naturally resulted in

> strange blossoms.
> Burnt-out brains.
> Endless tapestries of
> What was seen at Cactus-time,
> Methedrine,
> Mad houses,
> Movies cast on walls being looked upon by
> movie-makers casting dreams upon the walls . . . (RB, 200)

These movies are then illustrated by means of Plato's "Allegory of the Cave," in which the philosopher claims that humanity is trapped in an illusion of appearance versus reality and that language is only partially capable of rendering the truth. Welch embedded an explanatory sonnet within "How to Give Yourself Away" that summarizes Plato's idea before suggesting that Welch is one of the few who has grasped the notion of what is happening inside the cave and is apparently equipped to deal with it. Indeed, he even offers himself as a guide, leading "Barbara" toward the light and presumably to a mutual understanding of the glare outside.

Welch writes that, rather than see the shadows of ourselves cast on the "blackened wall," we need to turn our backs and walk into the glare, the self. Gradually uninvented. He further suggests that

> We have looked too long upon reflections of the
> Light.
> No, not even reflections! More nearly shadows of
> our Selves. The light BEHIND us. The image cast
> upon the Wall is still ourSELF. (RB, 201)

So, rather than simply accept these mere reflections, we need to make our way to the opening into which the light shines. And the opening in any of the forms that allow for departure—"the open mouth, the door, the way, the ragged circle, Light"—will then lead us to the goal of uninvention, namely

A Gladness as
remote from Ecstasy
as it is from
        Fear. (RB, 204)

Only then will we have given ourselves away.

———

In some ways, Welch's poems serve as a constant reminder of his philosophies and the incongruities that can be found there. Although "How to Give Yourself Away" was not written until 1967, his notion of (re)invention—and thus much of what followed in his work—was rarely more clearly stated than in a poem from 1960: "Invention Against Invention."

Very much a continuation of "Din Poem" or, as Philip Whalen later suggested, an obvious conclusion to it, "Invention Against Invention" leads us through the streets of San Francisco as Welch enumerates the details of what he sees and experiences there. The poem begins by asking, "How can we dismiss the needs of others / Having invented them?" (RB, 165).

Welch then states that feelings such as fear, guilt, compassion, and shame are inventions internal to each of us and that we will not recognize them unless they are released through certain situations. He then suggests that "these inventions are the enemy" and that the situations, feelings, and experiences that follow are all somehow linked to this list of inventions—which we are asked to add to—with sight being the primary means of legitimizing them.

In a similar occurrence to that which Welch had experienced in New York years before and in Bixby Canyon in 1962, he then witnesses an epiphany of sorts, whereby there is a realization that we need to exist outside the parameters of invention. Welch describes this epiphany:

Here, a giant crack occurs—both audible and blinding! (RB, 168)

In a lecture at Naropa University in 1980, Whalen further suggested that "Invention Against Invention" was very much a contradiction to what Welch would write in "Ring of Bone" in that here, as he opened himself up to everything around him, it prompted anxieties and fears rather than the state of enlightened awareness that resulted in Bixby Canyon.[15] In that sense, the line above is thus the moment when the poem flips and reveals itself.

As such, a sudden fissure appears into which a certain consciousness seeps and Welch's vision of humanity begins to further crystallize. This then is a more sweeping vision—a vision less personal that may well have acted as a catalyst for the more introspective searches that were to come. There is no let-up in the images that Welch witnesses on the city streets, yet now there is a sense that, having opened himself so purely to everything around him ("I decide to let it all come in!"), he is incapable of processing it without succumbing to tears. It is too much for him, and he tries to regain his composure, tries to become something—"a shape determined by things beyond invention"—that will be capable of understanding the key questions of life: "How did it come to this?" and "Can we do anything about it?"

Welch's answer, initially at least, seems to be escape. His vision of a weeping figure in the street, a "34 year old Negro (my age)" with a shoeshine box at his feet, is the final straw. He contends that to continue to live, he must remove himself from this environment but realizes that it will continue irrespective of whether he looks at it or not; plants will continue to grow, just as people will continue to be downtrodden by society and circumstance. Although he would physically remove himself on many occasions and in many different ways in the years that followed, it is the poem's final image that best sets the scene for what was to come and displays Welch's recognition of the internalized compassion and despair that such a realization engendered in him:

> I, Leo,
> Oak-leaf on my back,
> Striving to uninvent myself,
> Wish, tonight,
> On my 34 candles,
>   For a world where never,
>   Through man's invention,
>   Can a man be made to cry like that. (RB, 171–172)

27
Seeking Perfect Total Enlightenment

DIFFICULTY ALONG THE WAY
Seeking perfect total enlightenment
is looking for a flashlight
when all you need the flashlight for
is to find your flashlight.

(RB, 139)

Lew Welch experienced many difficulties during his life. His was a path fraught with obstacles and problems from as far back as he could remember. From the nomadic experience of his youth to the dysfunctionality of his parents; from the psychological demands of self-criticism and frequent mental instability to his use of alcohol as an antidote. Throughout his life Welch seemed to be continually looking for a flashlight to bring some sense of enlightenment to what was often clearly a troubled existence. He sought it in words, friends, women, work, religion, philosophy, forests, mountains, rooming houses, fishing trawlers, psychiatry couches, taxi cabs, alcohol, and Dorothy. Yet, despite all his efforts, he was ultimately unable to find it. It was almost always just out of reach. Either extinguished, out of batteries, or so faint as to be completely useless. And thus, in the end, he simply stopped looking for it and left. He was forty-four years old.

———

In the early morning of January 19, 1971, two Standard Oil Company tankers collided in dense fog near the Golden Gate Bridge, spilling 800,000 gallons of oil into the San Francisco Bay. The resulting slick quickly spread on the tide along the coast and into lagoons and marshes, causing widespread damage to marine and bird life and prompting a huge community effort the like of which had not been seen in the area since the earthquake of 1906. The public response drew together what the local media called an eclectic mix of "longhairs and hardhats" in what was later described as "the unifying potential of environmentalism, a consensual cause that could bring together hard hats and hippies."[1]

As Standard Oil and Chevron workers organized the rescue efforts and distributed supplies, so families, schoolchildren, office workers, public servants, and surfers from all over the area helped to recover and treat oil-soaked birds, spread protective straw on the beaches, and administer medical supplies to people injured by panic-stricken wildlife. The number of animals affected by the spill ran to more than seven thousand, most of which died. Estimates put the number of volunteers at forty thousand, but those involved suggested it may have been even higher. One among them was Lew Welch.

In a letter to James Koller a few days after the disaster, Welch wrote of "the panic of those who finally realize it's all over, that all that ecology stuff was true. . . . Quite likely every lagoon and marsh in the Bay Area will be sterile forever" (IR2, 176). John Doss, who had been a close friend of Welch for many years and with whom he regularly organized mussel feasts on the Marin County beaches, said that the spill "was like an ecological bomb went off under the Golden Gate Bridge and this is the Dunkirk of ecology."[2] It was, quite rightly, a pessimistic view and one that symbolized what the spill meant to everyone.

For the best part of a decade Welch had become increasingly more conscious of, and vocal about, the relationship between man and nature, and the situation hit him hard. What he had once told Bill Yardas while fishing out on the Duxberry Reef still rang true: "I would have liked to be an ecologist." So, the oil spill coming as it did at the same moment that he and Magda decided to break up was an alignment of circumstances that may have contributed more markedly to the events that transpired four months later than anyone could have imagined at the time. When Welch wrote to Koller of "those who realize it's finally over," could he have also been including himself? Was he one of "those"? Whatever he meant by his words, it is certainly true to say that the spill left its mark. When asked about the possible reasons for his disappearance, Doss said that Welch had been affected by the spill and that "he said that it was part of 'the coastal die-back.' That the U.S. was beginning to die on its coasts, and that he was heading for the hills."[3] Indeed, the poet John Montague went even further, saying that Welch "had become depressed by a calamitous oil spill in the Bay, and by how little was being done for such an ecological disaster."[4]

In the months between returning from Greeley and the oil spill, Welch had been working with Jack Shoemaker on producing a reprint of *The Song Mt. Tamalpais Sings* with additional poems and the same artwork that had been used for the *Springtime in the Rockies* broadside earlier that year. He had also been drinking as heavily as ever. Such was his addiction now that he was regularly required to admit himself into the hospital to relieve the pain and anxiety of his alcoholism. Every time he went in, he promised it would be the last, but it never was. He spent a drunken Thanksgiving with Dorothy (for which he characteristically apologized afterward, calling himself "such a bad alcoholic I barely function") and enjoyed a visit from Snyder and his wife Masa in early January. Indeed, Snyder's journal entries of that visit are among the most descriptive that remain of Welch at that time, revealing a side to his friend that had never been specified quite so graphically before but that might well add fuel to the fire of suggestions that his eventual disappearance was almost inevitable.

During their three-day visit to Mill Valley, Snyder and Masa went to Welch's house on Buckelew Street for a party. Also in attendance were Donald Allen, Locke McCorkle, Robert Creeley and his wife Bobby, Jack Boyce, and many others. Amid the conversation, music was provided by Jeff Cregg and friends playing bossa nova, and by the time most people had left Snyder noticed that Welch was getting increasingly drunk:

> I notice for the first time Lew's face when drunk. It's very different. A demon is in it—glad to be drunk—pleased—eyes gleaming— dishonest—A spirit still in him this morning when I take him to the plane, shit-eating smile as he promises not to drink—and while in his soused world Lew talks as much as he always does, but an exaggeration is thru it; suffused with a kind of dishonesty. Another being is driving.[5]

This other "being" was someone that Snyder had clearly never seen before. He had witnessed his friend sink into alcoholism from a distance, tried on occasions to help him through it, and, with the offer for Welch to join him on the San Juan Ridge, Snyder took the last throw of the dice.

However, he was quite rightly not all compassion and understanding. And it seems that others were also losing patience with Welch.

Notable by her absence that evening in early January was Magda Cregg. Snyder notes that he had visited her house two days prior to the party at Welch's, the implication being that the couple had already separated. For his part, Welch simply wrote to Koller, "Magda and I are breaking up. . . . It's so hard for me now, but I'll make it somehow, tho I know I'll never be the same" (IR2, 176).

The reasons for them parting are surely as complex as any breakup, but it is hard to see past the idea that Welch was becoming impossible to live with. Indeed, Jeff Cregg said that Magda had had enough of the hospitalizations, the delirium tremens, or finding Welch passed out on the floor.[6] She wanted more from life than she was getting. Or that he was able to give her. And so she left.

By late February, Magda had flown to Colombia for a six-month tour of South America, and Welch was about to head north for an extended stay with Bill Yardas, give a reading at Reed, and then return to finalize his move to North San Juan. His intentions are clear: build a cabin, change his lifestyle, and stop drinking. Although Welch admitted that this would not be easy, a visit to his old fishing buddy was a start.

Yardas and his family were now living on a stump ranch in Woodland, Washington, about an hour north of Portland. Yardas had exchanged the fishing grounds off the coast of California for Oregon waters, sailing out of Portland and working as a longshoreman on the waterfront at Longview. It was the perfect place for Welch to dry out, given that there was little alcohol and nowhere to go. He could focus on working and trying to regain some of the characteristics he and others felt had been lost through drink. In a letter to Magda in Bogota, Welch writes, "I've been straight over a month now & some of it is coming back. For one thing I can finally sleep. For another, the appetite is good. And the depression is now only a boredom."[7]

He also describes his plans for the cabin and how, when finished, it may result in a complete return to his former self: "Maybe by then it will all come back & I'll be something like the old Lewie!" (IR2, 180). The question, however, is what was it that Welch had lost? What part of him had been compromised? Aside from the physical and emotional aspects of being an alcoholic, was there an integral part of the poet now gone too?

Figure 27.1: Lew entertaining friends at Buckelew Street in Marin. Photographer unknown. Courtesy of the Estate of Lew Welch.

Welch had struggled with his demons for years. When he and Magda first met, she was already aware of a certain aura of inevitability, a certain sense that his life and eventual death were inextricably tied not only to drinking but also to Dorothy and his childhood. In some ways, all his retreats were temporary injunctions to the final sentence. With each return he would be reincarnated, and another period of optimism would begin, only for those demons to slowly eat away at his rebuilt foundations. And this inevitability was always there—under the surface, unspoken but present. In an interview with Stephen Fox for his 1976 book *An Appreciation of Lew Welch*, Magda even went so far as to say that, from the moment they met, she saw and understood this. Welch had told her in 1964 that he was ready to "cash in his chips" but that, because of her, he would "stick around for a while." For Magda, it was not the moment that Welch swapped his shotgun for a .22 Smith &Weston revolver but rather the constant battle over many years, the lack of control or will power. He was always "promising never to drink again & going on the wagon & then falling off the wagon & getting drunk again & all the guilt & all the destructiveness that goes with it."[8]

Yet for others who observed Welch from a safer distance, his alcoholism resulted in something more fundamental, something that ate at his very existence as a poet, namely his ability to be the teller of truths, the adviser, the voice.

For as long as he could remember, Welch had seen himself as a spokesman for his tribe. His job was to observe and warn, to watch and advise, to make the link between actions and words. Suddenly he had a vision of the future that could not be reconciled—and the Standard Oil spill was surely part of that. If the job of the poet was indeed "to tell the tribe certain things," then in his own mind Welch had failed. To have achieved some sense of success—what might be called the realization of the "Bread vs. Mozart's Watch" dilemma—for the sake of success only, was never what he had wanted. And he became, as Magda said, "distasteful to himself." So he simply stopped writing it down. Not stopped as in *quit*, but stopped as in having *nothing left to say*.

———

Welch's probable last poem was written on March 28, 1971, and in it he seems to underline this notion of his being, as Donald Allen once said to Lewis MacAdams, "finished" with writing. Welch further suggests that his "work" as a poet is done, and that unless the inspiration comes freely and easily, there is nothing more to say. Given that nothing ever came free or easily to Welch, this is a tacit admission of his decision to stop. To finish. His current relationship with the muse was no longer tenable.

> I gave that damned muse
> all the hard work she's gonna get
> From now on I get it all
> free or not at all.[9]

———

Final Interlude: "Hear My Last Will & Testament"

> I never could make anything work out right and now I'm betraying
> my friends. I can't make anything out of it—never could. I had great
> visions but never could bring them together with reality. I used it all
> up. It's all gone. Don Allen is to be my literary executor—use MSS
> at Gary's and at Grove Press. I have $2,000 in Nevada City Bank of
> America—use it to cover my affairs and debts. I don't owe Allen G.
> anything yet nor my Mother. I went Southwest. Goodbye. Lew Welch.
>
> (IR2, 187)

Welch left this note in his car on the day he disappeared. Despite all the
plans he had made to build a cabin, and all the work that had been done
by his friends to help him, his depression seemed to have reached such
a heightened state that this was his only conceivable way out. The note
itself is both honest and pragmatic, but there is little of the poet in it. It is
a measure of the man that, as he bowed out, he left his flair for language
and allusion aside and spoke as plainly as he ever had. Exhausted. No
debts. A simple goodbye.

Yet to say that he had left nothing to the imagination is a fallacy.
Indeed, he had already given the world a suicide note of such propor-
tions that what he left in his car was merely an administrative footnote.
When Jack Shoemaker published *The Song Mt. Tamalpais Sings* in the
fall of 1969, the collection ended with the poem "Song of the Turkey
Buzzard," which, in the years after Welch's death, has often been seen as
a premonitory poem. A poem that somewhat romantically outlines the
perfect death. The perfect reentry into the food chain. The transference
of the human to the bird. Welch soaring in the belly of a turkey buzzard.

"Song of the Turkey Buzzard," dedicated to Grateful Dead manager
and drinking buddy at the No Name Bar, Rock Scully, recounts incidents
both real and imaginary in Welch's life, when he had encountered both
the cougar ("how desperately I wanted to ride Cougar") and the turkey
buzzard. He writes of once hitting a buzzard with a .22, and of hearing the
"flak" as the bullet nestled into the bronze wing. He includes visions of
hatching and wounded birds, and of how they, like him, were "exhausted
. . . too weak to move their shriveled wings" and that one "wanted to die
alone." Welch had been drawn to the turkey buzzard in "a trance, a coma,

half in sleep and half in fever-mind," and he further acknowledges in his vision that the bird is

> The very opposite of
> death
> Bird of re-birth
> Buzzard
> Meat is rotten meat made
> sweet again . . . (RB, 135)

And that his body is reinvigorated through its reincarnation in this new form.

Through this final vision, in "Song of the Turkey Buzzard," Welch was able to create his "last Will & Testament," in which he elucidates the idea of his continuance: "with proper ceremony disembowel what I no longer need, that it might more quickly rot and tempt my new form." This reliance on a "new form" is clearly a notion that underpins Welch's poetic perspective in "Song of the Turkey Buzzard," and his life in general toward its end. He has become the "hatching bird" on the "fierce, sun-heat, sand," but rather than gasping for air, Welch awaits disembowelment so that he might become the meat on the talons of the soaring buzzards.

While the poem has since become his symbolic suicide note, Welch's thoughts on what his ideal totem animal might be can be found in earlier writings that establish a sense of the importance he attached to the animal world and how he might eventually become a part of it. In a letter to Whalen, he wrote of his desire to transcend human life by becoming something that he saw as existing "above" mankind, writing,

> I had my most beautiful death yet, a 3-month puss-bomb which popped about June 1 and blew me A (hyphen) WAY. I was escorted by Milarepa and Tamalpais on wings of Turkey Buzzard which, by the way, I find I ride. (IR2, 156)

The reference to both Milarepa and Mount Tamalpais here may suggest that he had had a particular vision of what that (or any future) death must consist of: namely neither one thing nor the other in isolation but a merging of two streams, or necessary interdependence. Indeed, since his

attention had long been focused on nature and the growing disassocia-
tion and disrespect of humanity toward it, so Mount Tamalpais became
something of a metaphorical *stupa*[10] to Welch, signifying not only something "perfect in Wisdom and Beauty" on Earth but also a final burial
site, where he could accept his death in this life and, more specifically,
celebrate his reentry into the next as integral to the continuance of the
natural order there. In this respect, Tamalpais serves as a symbol for all
mountains, and his final choice of the Sierra Nevada, while born of convenience, is merely the replacement of one mountain mother for another.

Some years earlier, Welch had written a poem titled "What the Turkey
Buzzard Said," which in many ways not only anticipated his later notion
of the relationship between humanity and the mountains but also crystallized much of his previous experiences and thoughts on how important
the mountain is or can be in our lives. The poem posits the idea that, as
the mountain assumes the role of a god(dess) in the minds of humans,
then, and only then, will we begin to learn and progress beyond what we
already are. He writes of how there is a mutual give-and-take in the form
of offerings and that, as humans accept those given by the mountain, so
the mountain rejoices. As humans offer "joyfull song" that has "the sound
of Awakening"[11] so the mountain in turn accepts and reciprocates that joy.
In addition, totem animals are also offered by the mountain, and Welch's
later relationship with the turkey buzzard may find its root in the fact that
"any sentient being" can be seen as a totem. Given the importance of
Mount Tamalpais to him, the mountain as a goddess may have "given"
the buzzard to symbolize his future state.

In another poem, "The Rider Riddle," Welch even suggests that if we
spend enough time on the mountain, "she will always give you a Sentient
Being to ride." He then asks the unanswerable question "What do *you*
ride?" (RB, 128), knowing that he has no choice in the matter but rather
that we are assigned a totem instead. He may have wanted the cougar, but
in the end he got the buzzard.

———

Welch's relationship with Mount Tamalpais began almost as soon as he
arrived back on the West Coast in 1957. From his early sojourns with Gary
Snyder to the later circumambulations and *kinhin* sessions, Tamalpais
had long represented a figure of extreme importance to Welch. He spent

a lot of time on the mountain, especially in the late 1950s, describing it to Dorothy as "very beautiful and quiet [with] many eucalyptus groves and a redwood forest."[12] Indeed, it was a mountain that also occupied a key place in the lives of many of Welch's friends. In the book *Tamalpais Walking*, cowritten by Snyder and master woodcarver Tom Killion, the mountain is simply described as a "beautiful coastal mountain, rising from the Pacific waves."[13] Although Welch saw something of the goddess in Tamalpais, Snyder was more modest in not describing it as either "sacred" or "spiritual" but rather celebrating what Killion calls "the mountain's qualities as grounded in the material world."[14]

Snyder's relationship with Tamalpais dates as far back as 1939 when, as a nine-year-old, he went on his first hikes there with his aunt. These and many later trips instigated a fascination that would last for many years and have an influence not only on Welch but also on many others. Indeed, Snyder's role in the mountain becoming a key site for Welch and others cannot be underestimated: it was he, along with Ginsberg and Whalen, who organized its first circumambulation in 1965—an event that established the mountain as something to be venerated and something that has gone on to become a regular fixture on the annual hiking calendar in Marin County.

In themselves, circumambulations have long stood as an essential practice in various religions or faith systems such as Hinduism and Buddhism. Known as *pradakshina* in Sanskrit, the practice has various purposes, one of which is as a ritual form of walking meditation that is intended to "produce good fortune by imitating the auspicious journey of the sun."[15] In reenacting the Tibetan Buddhist practice of following the direction of the sun around stupas, Tamalpais's future spiritual significance was ensured. Yet while Snyder wrote widely on the ecology, geology, and geography of Tamalpais and its surroundings, Welch's short collection is more of an ode to the mountain's symbolic nature as giver and sustainer than to its incarnation as a metaphorical stupa and, thus, perfectly captures Welch's reverence for it in a handful of short poems.

In one of his most beautiful and poignant odes to nature, "Prayer to a Mountain Spring," Welch describes his feelings with such simplicity that the message is as clear as anything he ever wrote:

Gentle Goddess
Who never asks for anything at all,

and gives us everything we have,
thank you for this sweet water
and your fragrance.

However, rather than having purely spiritual importance for Welch, Tamalpais also acted as a place where he could go when he needed an escape from one of the ever-more-prevalent demons he faced throughout the 1960s. One such example is recounted by Snyder who, increasingly concerned about his friend's alcoholism, decided that, in keeping with contemporary notions about getting to the root of problems through dropping acid, the two should head off along the Bolinas Ridge Road to a meadow that Snyder called "a special spot on Tamalpais . . . the very literal edge of the coast" to induce such "psychological explanations" that might uncover the causes of his friend's dependence on alcohol. Snyder describes the experience of how Welch connected with the buzzards circling overhead, of how he talked to them and said that he was not yet ready to be eaten, having filled himself with "too much poison" to be edible. He was experiencing his "flavor-sense" — the element of himself that would bridge the gap to his "buzzard angel." They dropped 300mg of acid between them, which induced an awareness in Welch that prompted him to transmit the "radiance and brilliance of the face of Krishna, the great vajra jewel," so much so that Snyder wept at its beauty. [16]

Welch's explanation for drinking excessively was that he had not been nursed enough as a baby — an argument he often mentioned, and one that indirectly places the blame for his alcoholism on his mother.[17] Snyder remembers responding by saying that he could become edible again and that humanity's responsibility to the world is "to be good to eat." Yet while the cleansing vision of becoming "tasty" through self-development helped Snyder quit smoking, Welch's addiction was clearly more deep-seated, and despite some immediate respite, the trip had little lasting effect on Welch other than cementing his relationship with the hawks and vultures circling high above throughout day. Indeed, as he sat with Snyder and watched them soar against the curved horizon of the Pacific, it is easy to see how this final collection of poems are rooted in such melancholic understanding that, just as San Francisco signified the final frontier in the western expansion of the continent, so Tamalpais signified a similar terminal station for Welch. And although his final resting place would be elsewhere, this understanding was a reiteration that mountains are the perfect places to offer yourself to the cycle of life and death.

The title poem of *The Song Mt. Tamalpais Sings*, while celebrating the mountain and the beaches that wash under it, is also a reflection on humanity; it is suggestive of Welch's reconciliation to the fact that, now more than ever, humanity was facing some measure of extinction and that Tamalpais symbolized, on the one hand, the furthest reaches of western expansion (both literally and metaphorically) and, on the other hand, the antidote to that. The poem's italicized repetition of *"This is the last place. There is nowhere else to go"* after every stanza leaves us in no doubt that Welch felt a certain sense of finality. Indeed, the closing lines offer a juxtaposition of resignation and hope in suggesting that, while humanity can go no further in its "wanderings," Tamalpais is the only place to which people need go. This in turn suggests the power of the mountain to provide us with everything we require, without the need to reciprocate this time. In addition to the herbs that she offers so that we might begin to "know her," there are also the mountain springs that replenish and reinvigorate. In many ways Tamalpais is the perfect home. She has food aplenty, a comfortable climate, and a rich tradition of sustaining inhabitants. While the West Coast may symbolize the end of the pioneer spirit and a final frontier, it need not result in what Whalen called the backwash of people moving inland again, nor signify the disillusionment that entered the American psyche when the frontier was reached.[18] Those "jeweled beaches" can just as easily be seen as a paradise or a resting place that leaves us wanting for nothing. Or, as Welch so poignantly put it in those prophetic closing lines of *The Song Mt. Tamalpais Sings*,

> This is the last place.
> There is nowhere else we need to go. (RB, 122)

———

While the final days of Lew Welch's life played out on the North San Juan Ridge, the story of that day begins five years earlier, in the summer of 1966, when Gary Snyder, together with Allen Ginsberg, Roshi Richard Baker, and the Swami Kriyananda, purchased a hundred acres of abandoned land there. During the mid-1800s, the San Juan Ridge—like much of the American West—had been subjected to deforestation, exploitation, and systematic hydraulic gold mining at the hands of frontiersmen and gold diggers. As had been the case all over the American West, the white man had not come to adapt to the land or its native inhabitants but instead "to make all else adapt to him" [19] by felling trees and clearing scrublands in their search for an American El Dorado. This search had been all-consuming, and when the land proved fruitless or the gold ran out, the prospectors moved on, leaving behind a state of utter ruination. Although the land once again belonged to the indigenous flora and fauna—including ponderosa pine, manzanita, and mountain yew, as well as deer, coyote, bobcat, and bear—prospectors had now stripped much of what had made it nourishing for so many indigenous species in the first place.

When Snyder decided to permanently settle on the San Juan Ridge in 1969, he did so with the clear intention of reinhabiting a place where his presence would be in harmony with its primary inhabitants. He sought to escape industrial civilization and reintegrate with nature by adapting to it and creating a mutually beneficial synergy through living wisely and carefully in nature. He replanted trees, dechannelized streambeds, and broke up the crude asphalt roads. However, his work was more than mere reinhabitation; the reasons for establishing an alternative community along the ridge were manifold. The poet and fellow San Juan Ridge resident Steve Sanfield later described his decision to join Snyder: "There is madness afoot in the world, and if not for our own sake, at least for that of our own children, we have to preserve certain sacred places, where we can see the moon and stars every night, and the loudest sound is the quail at dawn."[20]

Snyder, who had returned from Japan with his third wife Masa Uehara earlier that year, found a number of volunteers to help him build a home that was to be part Japanese farmhouse and part Native American homestead. The architecture fused elements of both, and the isolation ensured the lack of electricity and running water. This was not the typical "back-to-the-land, countercultural, utopian image of living outside of society [but rather] more like what the farm and the ranch in the West is."[21] In such a place, where wildcats kill the poultry and bears plunder the fruit trees, it is paramount that the inhabitants already have (or quickly develop) an intimate understanding of how such an ecosystem functions—and, more important, that they accept and embrace it.

As might be expected, construction of the building was slow and methodical, but by the summer of 1970 the single-story house was habitable. Local materials like "ponderosa pines from within three hundred feet of the site to frame, and local incense cedar for siding . . . [and] foundation stones [that] came from the middle fork of the Yuba River" were essential in the building process.[22] Snyder's intention was to make this house—named Kitkitdizze, after the Wintu word for the perennial low-growing, spicy-odored shrub also known as "mountain misery"—his long-term home. And such was the foundation, and the success of living there, that Gary Snyder still resides in the house as of this writing, in 2023.

Welch was among the many visitors to the house, and he would soon consider moving out to the San Juan Ridge himself. By late 1970, Welch had inquired about the possibility of building a small cabin there with Magda. As always, money was tight, and Snyder warned him that the Sacramento-based realtor who held the deed to the adjacent land was charging too much for it. However, this issue was quickly resolved by Allen Ginsberg, who offered Welch a portion of his land on which to build. In a letter to Snyder in January 1971, Ginsberg wrote that he had imagined sharing the land with Welch and fellow poet Philip Whalen when he bought it. He envisioned giving one-third of his land to each of them with a lifetime occupancy guarantee that would safeguard the land and any cabins that were built on it.[23]

Welch soon drew up plans to build a cabin and submitted them in early May to the group charged with their approval, the Bald Mountain Land Association. This association had been established by Snyder and company to tightly control the building of houses and cabins so that the area would not become too densely populated. One rule stipulated that when a home was built, it could not be visible to its nearest neighbor. As a result, Welch planned to build a small twenty-foot-by-twelve-foot structure "on a N.E. slope in a grove of oaks just at the meeting of the oaks and the pines. The site is not visible from the parking lot, from Gary's site, nor from the side, nearer the knoll. . . . neither is it visible from any of the present trails to the meadow" (IR2, 86). He then sought people to help him with construction work and finally seemed to be on the verge of reinhabiting a place of his own that might offer him the peace of mind and quiet contentedness he needed.

Welch started to put together a crew of people to oversee the build, including Jeff Gold, Zac Reisner, and Steve Sanfield. Many other friends and acquaintances were waiting to help, and most of them surely understood the importance of what Welch was trying to achieve. Although money was tight, and his fears of failure were never far from the forefront of his thoughts, Welch's confidence in the success of his plan was tangible (he even whimsically suggested his cabin be known as "Welch Gulch"). Detailed blueprints were drawn up, materials inventoried, and contracts sent. However, by mid-January 1971, Cregg and Welch had split up, and Welch's already heavy drinking had increased to such an extent that it was creating all manner of other problems—not least how he could build the cabin. Could this project, therefore, be a final solution to these problems? Or was it simply one venture too many?

In many ways Ginsberg's offer also presented Welch with the ultimate opportunity to put his words into action. For much of the last decade Welch had been advocating (and practicing) a return to nature. Echoing Snyder's philosophy that humanity needed to return to "a sense of self-contemplation, community and embeddedness in nature," Welch had already extricated himself from what he called "the world that is Man" on several occasions.[24] However, he had long struggled to reconcile himself not only to the civilized world but, more important, to his own shortcomings and idiosyncrasies. Previous periods of reclusion in remote cabins, on fishing vessels, in nighttime cab rides, or during solitary mountain hikes had only ever resulted in a temporary respite from this struggle that never seemed to last very long. Many of those closest to Welch, including Snyder and Whalen, felt that this was his final chance at attaining something close to his own ideals. Surrounded by close friends and immersed in a community of like-minded poets, artists, and woodsmen, the San Juan Ridge was to be Welch's long-term home. A place where he could be cleansed and inspired. Whalen later wrote, "I wish we could help him to live & not want to sap himself headlong into boring grave & tedious tomb &c."[25]

His preparations for building a cabin were at an advanced stage, and on the face of it he was excited—perhaps even overly so—by the prospect. The correspondence that remains between Welch and those who were going to help him indicates that nothing was wrong. His questions regarding materials, foundations, and dimensions were those of someone looking forward to the future. Indeed, at one point the architect in charge of the build, Bruce Boyd, even told Welch to slow down and that more patience was required to make sure that the plot on which Welch was due to build was properly prepared. Welch's patience, however, was running out.

In one final desperate attempt to get on the wagon, Welch had again turned to Antabuse, the drug widely prescribed to alcoholics to reduce their desire to drink—or, more accurately, to create such a reaction to alcohol when ingested that the prospect of drinking is enough to make someone think twice. Yet despite providing Welch with temporary respite, it also induced paranoia and depression—a deadly combination for a man partially crippled by self-doubt and disappointment already. He wrote of depressions and bewilderment and of how he had not realized quite how far removed from himself he was. This was in every sense his final shot at survival. However, in the days immediately before his disappearance, he was faced with not only self-criticism and doubt but also with the fact that others were also beginning to openly question if he was ever likely to change. In one of his journals, Snyder, for example, wrote of a wildflower walk in those final weeks during which Welch had shot a lizard. Afterward he said to his friend "you're not going to change,"[26] and while the admonition was perfectly understandable, given all the help and support Welch had received from Snyder and others, in some senses it may also have been too much for Welch in his fragile state. Indeed, Whalen later wondered if some of the criticism had not been too harsh.[27] Yet Snyder's continued compassion for his friend was now manifest in a reproof that was intended to brace Welch for the mountain he still had to climb: the path to his recovery. His friend and compatriot was being cruel to be kind. Snyder later told Whalen that, despite all his problems "the 'poet-light' was still shining even when Lewy was at his sickest."[28] But it was no longer enough. It is more likely that Welch had no energy for that climb, and having long seen the ease with which the vultures circled overhead, he had made up his mind that that was the path of least resistance. Whalen, for one, although angry and upset at what he saw as Welch's "final revenge upon his mama, and a complete rejection of all the rest of us"[29] also realized that if "he was sick & crazy & in pain, why not commit suicide if the pain is really too great & if it is never to end?"[30]

On the day of his disappearance, Welch and others, including Ginsberg and Jack Hogan, were staying at Kitkitdizze. In the morning Snyder read Samuel Johnson's pamphlet *The Vulture* to the assembled group. Published in 1758, Johnson's essay is a short fable about the relationship between men and vultures, and of how men are an essential part of the vulture's sustenance. Through their hunger for power and wealth, men wage wars that allow vultures to simply wait and feast on what remains after the battle is over. After listening to Snyder, Welch took his revolver and left for his camp. He failed to show up for dinner that evening, and the note he left was found the following afternoon by Zac Reisner.

Lew Welch's life had been consumed by an inner battle—his was a war to regain power over himself, and he realized it was one he couldn't win. By giving up the fight, he offered himself as sustenance—as he had proclaimed he would—and in his own way Welch became, as Johnson wrote, "a friend to the vultures."[31]

Epilogue

What actually happened that day will likely remain forever unsolved. Despite the extensive searches, no trace of Welch has ever been found, and the Nevada City Sheriff's Department consigned missing persons file no.121571 to a filing cabinet—where it remains to this day. Newspapers across the country ran stories about the disappearance and wrote follow-ups about the futility of the subsequent searches. The headlines were short, the column inches limited. In many ways the coverage was a bleak reflection of his position in the literary world at the time.

Since his disappearance, Lew Welch has been spotted on various occasions and in various locations. At one point he was reported found by the Deputy Sheriff Stan Kramer. Or seen in a Nevada City bank withdrawing money. Or as a fluttering vision around a candle in a mushroom-haze. Many of his friends initially refused to accept that he had taken his own life and held out the hope that he had simply extricated himself once again and was hiding out in a cabin somewhere in the mountains he so loved. Indeed, there are those who believe that he lived on under an alias or that his remains were found and never revealed so that his final wishes be fulfilled. However, as weeks turned to months, and months to years, so his death became an inevitable fact. His presence waned, or as Clifford Burke wrote, "my friend and teacher just faded away."[1]

However, his legacy lives on along the San Juan Ridge in the form of the Ring of Bone Zendo. He has appeared in literary magazines and anthologies across the world, and with Donald Allen acting as his literary executor, *Ring of Bone* finally saw the light of day two years after Welch's disappearance, and the collection was reissued by City Lights in 2012. Educational institutes across the United States and beyond used his texts in their courses then and still do so now. Whatever else he left behind, Welch *deserves* a biography because his life is a life worth telling.

And so, as this work started with Dorothy Brownfield, so it will end
with her. In providing Donald Allen with a biography for *Ring of Bone*,
she also included something akin to a eulogy that is the most fitting for a
life such as Lew's: matter-of-fact, resigned, and straight to the point:

> Why didn't Lew want to continue living? It was because he was ill,
> very ill physically. His twenty years of hard drinking had taken a
> terrific toll, apparently his liver was almost destroyed and his kidneys
> were not functioning properly. For the last two years of his life Lew
> felt badly all the time, and it was hard for him to enjoy anything
> (not even fishing which was his favorite form of recreation). Also
> the constant intake of alcohol and non-intake of proper food
> and vitamins had seriously damaged his central nervous system,
> including his brain. Lew was smart enough to realize all of this. He
> did try to stop drinking in 1970 and 1971 but the pain was too great
> and the depression too deep. Lew waited too long and had "used
> it all up." Physically, mentally, and emotionally Lew had used it
> all up in 44 years. He apparently felt ready to leave this not too
> pleasant earthly existence. So we who miss him should "not grieve"
> even though it is difficult.[2]

———

"THE WAY OUT, IS OUT" (RB, 199)

Notes

INTRODUCTION

1 Michael Davidson, *The San Francisco Renaissance: Poetics and Community at Mid-century* (Cambridge, UK: Cambridge University Press, 1989), 11.

2 Never particularly outspoken politically, Welch was nonetheless aligned with the sensibilities of the IWW and briefly flirted with the Portland branch of the John Reed Club. His unpublished essay "How to Survive in the United States" is as close to an overtly political outpouring as he ever wrote.

CHAPTER 1

1 Typescript of "The Mindless Conspiracy," Lew Welch, n.d., box 3, folder 8, Lew Welch Papers, MSS 13, Mandeville Special Collections Library, UCSD.

2 David Meltzer, *The San Francisco Poets* (New York: Ballantine, 1971), 192.

3 "Miss Brownfield Will Be Bride of Tomorrow at Quiet Home Wedding," *Arizona Republican*, May 25, 1924, p. 4.

4 "Miss Brownfield Will Be Bride," *Arizona Republican*.

5 Meltzer, *San Francisco Poets*, 194.

6 Meltzer, *San Francisco Poets*, 194.

7 Transcript of an unpublished draft for *The San Francisco Poets*, David Meltzer, box 10, folders 43–45, David Meltzer Papers, 1955–1973, Washington University Libraries, Department of Special Collections.

8 *The 1920 Phoenician*, Yearbook (Phoenix, AZ: Phoenix Union High School, 1920), 20.

9 Meltzer, *San Francisco Poets*, 190.

10 *Arizona Republican*, "Society," September 25, 1921, p. 18.

11 Meltzer, *San Francisco Poets*, 193.

12 "Brownfield Estate Valued at $15,000," *Arizona Republican*, May 12, 1921, p. 6.

13 "Brownfield Policy Paid," *Arizona Republican*, May 17, 1921, p. 6.

14 John B. Irwin and Donald C. Mackay, *Stanford Quad* (Palo Alto, CA: Stanford University Press, 1924), 283.

15 Kathy Johnston, email to the author, September 2, 2016.

16 Aram Saroyan, *Genesis Angels: Lew Welch and the Saga of the Beat Generation* (New York: Morrow Quill Paperbooks, 1976), 14; and Meltzer, *San Francisco Poets*, 193.

17 "Mammoth Auction," *Arizona Republican*, Sunday August 25, 1929, p. 19.

18 Meltzer, *San Francisco Poets*, 194.

19 *Fort Scott Daily Tribune*, short notice on back page, March 4, 1911, p. 8.

20 Meltzer, *San Francisco Poets*, 192.

21 Meltzer, *San Francisco Poets*, 192.

22 Meltzer, *San Francisco Poets*, 192.

23 Typescript of poem "A Poem for the Fathers of My Blood: the Druids, the Bards, of Wales and Ireland," by Lew Welch, n.d., box 3, folder 1, Lew Welch Papers, MSS 13, Mandeville Special Collections Library, UCSD.

24 Jeff Cregg, interview with the author, October 16, 2017, San Rafael, California.

25 Saroyan, *Genesis Angels*, 33.

26 David Schneider, *Crowded by Beauty* (Oakland: University of California Press, 2005), 178.

CHAPTER 2

1 Promotional material for Casa de Manana, University of California, San Diego Digital Collections, accessed October 30, 2021, https://library.ucsd.edu/dc/object/bb20846899/_1.pdf.

2 David Meltzer, *The San Francisco Poets* (New York: Ballantine, 1971), 193–194.

3 Meltzer, *San Francisco Poets*, 193–194.

4 Transcript of an unpublished draft for *The San Francisco Poets* by David Meltzer, box 10, folder 43–45, David Meltzer Papers, 1955–1973, Washington University Libraries, Department of Special Collections.

5 Transcript of unpublished draft for *The San Francisco Poets*.

6 Meltzer, *San Francisco Poets*, 194.

7 Welch's reading from the Teach-In can be found at https://archive.org/details/canhpra_000033.

8 Meltzer, *San Francisco Poets*, 223.

9 Transcript of unpublished draft for *The San Francisco Poets*.

10 Robert Service, "On the Wire," *The Complete Poems of Robert Service* (New York: Dodd, Mead & Company, 1945), 336.

CHAPTER 3

1 David Meltzer, *The San Francisco Poets* (New York: Ballantine, 1971), 195.

2 Welch gives a humorous description of this and other sport-related anecdotes in William T. Wiley's spoof talk show *What Do You Talk About*, which was part of the Dilexi Series aired on KQED in 1969. See chapter 20 in this book for more on Welch's performance on the show.

3 To put his high school efforts into clear perspective, in Welch's sophomore year the NCAA Track and Field Championships was won by future Olympic gold medalist Cliff Bourland in a time of 48.5 seconds.

4 Meltzer, *San Francisco Poets*, 202.

5 William G. Cutler, *History of the State of Kansas, Part 16, Bourbon County* (1886), https://www.kancoll.org/books/cutler/bourbon/bourbon-co-p16.html#BIOGRAPHICAL_SKETCHES_VAN_FOSSEN-YORK.

6 "In the Service," *Stanford Daily* 104 (46): 6 (October 27, 1943), https://archives.stanforddaily.com/1943/10/27?page=6&section=MODSMD_ARTICLE50#article.

7 Correspondence from Lew Welch to Dorothy Brownfield, May 4, 1945, box 1, folder 5, Lew Welch Papers, MSS 13, Mandeville Special Collections Library, UCSD.

8 Correspondence from Lew Welch to Dorothy Brownfield, March 11, 1945, box 1, folder 5, Lew Welch Papers, MSS 13, Mandeville Special Collections Library, UCSD.

9 Aram Saroyan, *Genesis Angels: Lew Welch and the Saga of the Beat Generation* (New York: Morrow Quill Paperbacks, 1976), 33–34.

10 Correspondence from Lew Welch to Dorothy Brownfield, April 26, 1945, box 1, folder 5, Lew Welch Papers, MSS 13, Mandeville Special Collections Library, UCSD.

11 Correspondence from Lew Welch to Dorothy Brownfield, May 28, 1945, box 1, folder 5, Lew Welch Papers, MSS 13, Mandeville Special Collections Library, UCSD.

12 Correspondence from Lew Welch to Dorothy Brownfield, November 7, 1945, box 1, folder 5, Lew Welch Papers, MSS 13, Mandeville Special Collections Library, UCSD.

13 Correspondence from Lew Welch to Dorothy Brownfield, May 4, 1945, box 1, folder 5, Lew Welch Papers, MSS 13, Mandeville Special Collections Library, UCSD.

14 Correspondence from Lew Welch to Dorothy Brownfield, July 6, 1945, box 1, folder 5, Lew Welch Papers, MSS 13, Mandeville Special Collections Library, UCSD.

CHAPTER 4

1 David Meltzer, *The San Francisco Poets* (New York: Ballantine, 1971), 217.

2 Meltzer, *San Francisco Poets*, 218.

3 Meltzer, *San Francisco Poets*, 218.

4 James R. Wilson, "Lew Welch," in *Hey Lew*, ed. Magda Cregg (Bolinas, CA: Magda Cregg, 1997), 26.

5 Wilson, "Lew Welch," in *Hey Lew*, 26.

6 Howard Greene and Matthew Greene, *The Hidden Ivies* (New York: HarperCollins, 2009), 275.

7 Greene and Greene, *Hidden Ivies*, 289.

8 Meltzer, *San Francisco Poets*, 201.

9 For a full explanation of these concepts, see Eric Paul Schaffer's introduction to *How I Read Gertrude Stein* (HGS, xvii–xxiii).

10 David Meltzer, *San Francisco Beat: Talking with the Poets* (San Francisco: City Lights Books, 2001), 339.

11 In *Hearings before the Committee of Un-American Activities, House of Representatives*, June 14 & 15, 1954 (Washington, DC: US GPO, 1954), 6614–6615.

12 Michael Munk, "McCarthyism Laid to Rest?" *Endpaper*, Spring 2006, https://www.reed.edu/reed_magazine/spring2006/index.html.

13 Munk, "McCarthyism Laid to Rest?"

14 Gary Snyder, interview with the author, September 23, 2019.

15 James Brown, "The Zen of Anarchy: Japanese Exceptionalism and the Anarchist Roots of the San Francisco Poetry Renaissance," *Religion and American Culture: A Journal of Interpretation* 19, no. 2 (2009): 209.

16 On July 1, 1950, a small number of US troops engaged the North Korean army at the Battle of Osan. It was the first battle to involve a collaboration between US forces and their South Korean counterparts. Although it ended in defeat, the battle signified the beginning of a more concerted effort by the United States to lend significant air and ground support to the South Korean army. Indeed, the initial four-hundred-strong US force grew to a little under 1.8 million servicemen seeing service in Korea between 1950 and 1953.

17 Correspondence from Lew Welch to Dorothy Brownfield, October 3, 1950, box 1, folder 10, Lew Welch Papers, MSS 13, Mandeville Special Collections Library, UCSD.

18 Gay Walker, "Lloyd Reynolds, Robert Palladino, and Calligraphy at Reed College," *The Calligraphy Heritage at Reed*, https://www.reed.edu/calligraphy/history.html.

19 Walker, "Lloyd Reynolds, Robert Palladino, and Calligraphy."

20 Stephen Prothero, *Big Sky Mind: Buddhism and the Beat Generation* (New York: Riverhead Books, 1995), 15.

21 Jackie Svaren, on Reynolds's philosophy, https://www.reed.edu/calligraphy/notable_quotes.html.

22 Philip Whalen et al., *Festschrift for Lloyd Reynolds* (Portland: Reed College, 1966), 5.

23 Philip Whalen, "'Goldberry Is Waiting'; or, P.W., His Magic Education as a Poet," *The Poetics of the New American Poetry* (New York: Grove Press, 1973), 455.

24 "Reynolds and the Writing Out of Texts," in *Excerpts from the Cooley Art Gallery Exhibition*, http://www.reed.edu/calligraphy/reynolds_cooley_show.html.

25 Walker, "Lloyd Reynolds, Robert Palladino, and Calligraphy."

26 Gary Snyder, *Mountains and Rivers Without End* (Berkeley: Counterpoint Press, 1996), 155.

27 David Schneider, *Crowded by Beauty* (Oakland: University of California Press, 2005), 57.

28 Schneider, *Crowded by Beauty*, 57.

29 Don Berry, *Moontrap* (Corvallis: Oregon State University Press, 1962), viii.

30 Lew Welch, "Poems and Commentaries," *Janus No.3* (Portland: Reed College, March 1950), 5.

31 Karen Alkalay-Gut, "The Dying of Adelaide Crapsey," *Journal of Modern Literature* 13, no. 2 (1986): 225–250, http://www.jstor.org/stable/3831493.

32 Susan Sutton Smith, ed., *The Complete Poems and Letters of Adelaide Crapsey* (New York: State University of New York Press, 1977), 31.

33 Unpublished drafts of various poems by the members of the Adelaide Crapsey-Oswald Spengler Mutual Admiration Poetasters Society can be found in the archive at Reed College: box WHA-002, folder 3, Miscellaneous: manuscript of "Adelaide Crapsey-Oswald Spengler Admiration Poetaster's Society"; correspondence about L. J. Reynolds Festschrift, Philip Whalen Papers, Reed College Special Collections and Archives.

34 Philip Whalen, *Off the Wall Interviews with Philip Whalen* (San Francisco: Four Seasons Foundation, 1978), 34.

35 Handwritten poem by Lew Welch, box WHA-002, folder 3, Miscellaneous: manuscript of "Adelaide Crapsey-Oswald Spengler Admiration Poetaster's Society"; correspondence about L. J. Reynolds Festschrift. Philip Whalen Papers, Reed College Special Collections and Archives.

36 Whalen, *Off the Wall Interviews*, 34.

37 Correspondence from Lew Welch to Dorothy Brownfield, January 10, 1949, box 1, folder 8, Lew Welch Papers, MSS 13, Mandeville Special Collections Library, UCSD.

38 Correspondence from Lew Welch to Dorothy Brownfield, January 10, 1949.

39 Correspondence from Lew Welch to Dorothy Brownfield, January 10, 1949.

40 Frank Jones, "Scenes from the Life of Antigone," *Yale French Studies* 6 (1950): 91–100.

41 Frank Jones, "Et Ego in Arcadia," in *Hey Lew*, ed. Magda Cregg (Bolinas, CA: Magda Cregg, 1997), 17.

CHAPTER 5

1 Barry Ahearn, ed., *The Correspondence of William Carlos Williams and Louis Zukofsky* (Middleton: Wesleyan University Press), 433.

2 Alan Soldofsky, "'Those to Whom Interesting Things Happen': William Carlos Williams, Kenneth Rexroth, Lew Welch, and Joanne Kyger, and the Genome of San Francisco Renaissance Poetry," *William Carlos Williams Review* 35, no. 2 (2018): 164–195, https://doi.org/10.5325/willcarlwillrevi.35.2.0164.

3 Whalen, "'Goldberry Is Waiting'; or, P.W., His Magic Education as a Poet," *The Poetics of the New American Poetry* (New York: Grove Press, 1973), 455.

4   William Carlos Williams, *The Autobiography of William Carlos Williams* (New York: New Directions, 1967), 377.

5   Jaap van der Bent, "A Hunger to Participate: The Work of John Clellon Holmes," PhD dissertation (Nijmegan: Radboud University, 1989), 12.

6   Michael Schumacher, *Dharma Lion: A Biography of Allen Ginsberg* (New York: St. Martin's Press, 1992), 122.

7   David Meltzer, *The San Francisco Poets* (New York: Ballantine, 1971), 204.

8   Meltzer, *San Francisco Poets*, 205.

9   Williams, *Autobiography of William Carlos Williams*, 374.

10 Gary Snyder, *Earth House Hold* (New York: New Directions, 1969), 113.

11 Snyder, *Earth House Hold*, 113.

12 Meltzer, *San Francisco Poets*, 205.

13 Meltzer, *San Francisco Poets*, 212.

14 James R. Sledd, *Language* 40, no. 3 (July–September 1964): 465–483, doi: 10.2307/411516.

15 L. Ron Hubbard, *Dianetics: The Modern Science of Mental Health* (Los Angeles: The Church of Scientology of California, 1950), 9.

16 Hubbard, *Dianetics*, 7.

17 Hubbard, *Dianetics*, 118.

18 On September 9, 1950, the American Psychological Association amended its Council Policy Manual to include the following:

    APA response to Dianetics
    In view of the sweeping generalizations and claims regarding psychology and psychotherapy made by L. Ron Hubbard in his recent book, "Dianetics," the American Psychological Association adopts the following resolution: While suspending judgement concerning the eventual validity of the claims made by the author of Dianetics, the American Psychological Association calls attention to the fact that these claims are not supported by empirical evidence of the sort required for the establishment of scientific generalizations. In the public interest, the Association, in the absence of such evidence, recommends to its members that the use of the techniques peculiar to Dianetics be limited to scientific investigations designed to test the validity of its claims.

19 Correspondence from Lew Welch to Dorothy Brownfield, April 30, 1951, box 1, folder 13, Lew Welch Papers, MSS 13, Mandeville Special Collections Library, UCSD.

20 Patanjali's yogic philosophy centers on the notion that in order to achieve *kaivalya*, or emancipation from *duhkha* (misery), one has to train the mind and body through a series of stages. These stages begin with the Samadhi Pada, in which Patanjali outlines the concepts and process necessary for "absorption into Spirit" or "communion with one's spiritual essence." This is followed by an explanation of the practice of yoga (Sadhana Pada) with practical advice and the introduction of the eightfold path. The third set of sutras is known as the Vibhuti Pada, intended as a means of gaining a "continuous stream of mindfulness ranging from concentration to meditation to absorption of an object's spiritual essence." Patanjali outlines these practices in 196 sutras, highlighting that the attainment of self-realization can be gained only through the employment of the Eight Limbs of Yoga (otherwise known as "solution-creators" and closely related to the Eightfold Path in Buddhism). In the end, these practices will thus help yoga practitioners tackle the "afflictions" that block the path to an understanding of their true selves.

21 Peter Manseau, *One Nation, Under Gods: A New American History* (New York: Little, Brown and Company, 2005), 381–382.

22 Manseau, *One Nation, Under Gods*, 385.

23 Correspondence from Lew Welch to Dorothy Brownfield, November 1, 1951, box 1, folder 14, Lew Welch Papers, MSS 13, Mandeville Special Collections Library, UCSD.

24 Dell Hymes, "A Coyote Who Can Sing," in *Gary Snyder: Dimensions of a Life*, ed. Jon Halper (San Francisco: Sierra Club Books, 1991), 392–404.

25 Correspondence from Lew Welch to Dorothy Brownfield, November 25, 1951, box 1, folder 14, Lew Welch Papers, MSS 13, Mandeville Special Collections Library, UCSD.

26 Meltzer, *San Francisco Beat: Talking with the Poets* (San Francisco: City Lights Books, 2001), 339.

27 Typewritten journal entry by Gary Snyder, February 6, 1952, box I 84, folder 1, Gary Snyder Papers, D-050, Special Collections, UC Davis Library, University of California, Davis.

28 Welch discusses this during his reading class at the Poetry Center in San Francisco in November, 1959. A recording of the event can be found at https://diva.sfsu.edu/collections/poetrycenter/bundles/191200; 03:34.

29 Welch Poetry Center reading, 1959, https://diva.sfsu.edu/collections/poetrycenter/bundles/191200; 04:10.

30 Otto Jespersen, *Essentials of English Grammar* (London: George Allen & Unwin, 1933), 17.

31 Legend has it that while working in advertising Welch penned the slogan "Raid Kills Bugs Dead" to accompany the bug designs that had been created by the advertising agency Foote, Cone & Belding for S.C. Johnson's Raid insecticide. The slogan was first used in 1956 and continues to be used today.

32 Typewritten journal entry by Gary Snyder, November 3, 1952, box I 84, folder 1, Gary Snyder Papers, D-050, Special Collections, UC Davis Library, University of California, Davis.

33 In *Genesis Angels* (New York: Morrow Quill Paperbooks, 1976), Aram Saroyan describes Welch suffering from an "uncontrollable spasm of the mad." His mind had become agitated by city life and the "loneliness and disconnection" he felt in Chicago. Saroyan's description of Welch talking incoherently or uncontrollably to a random neighbor is indicative of his having something akin to a manic episode.

34 Brownfield wrote this note regarding the payment for Donald Allen on the envelope of the letter she had received from Welch on July 10, 1952, box 1, folder 15, Lew Welch Papers, MSS 13, Mandeville Special Collections Library, UCSD.

35 Saroyan, *Genesis Angels*, 81.

36 Correspondence from Lew Welch to Dorothy Brownfield, March 28, 1955, box 1, folder 15, Lew Welch Papers, MSS 13, Mandeville Special Collections Library, UCSD.

37 Correspondence from Lew Welch to Dorothy Brownfield, January 16, 1954, box 1, folder 15, Lew Welch Papers, MSS 13, Mandeville Special Collections Library, UCSD.

38 Correspondence from Lew Welch to Dorothy Brownfield, January 16, 1954, box 1, folder 15, Lew Welch Papers, MSS 13, Mandeville Special Collections Library, UCSD.

39 Correspondence from Lew Welch to Dorothy Brownfield, January 16, 1954.

CHAPTER 6

1 There are a number of different interpretations of Verlaine's ballad, but the gist points at a translation along the lines of *Why would you have me remember that?*

2 "Emil Garber at Age 84, Newsman for 50 Years," *Chicago Tribune*, April 27, 1985, https://www.chicagotribune.com/news/ct-xpm-1985-04-27-8501250901-story.html.

3 Aram Saroyan, *Genesis Angels: Lew Welch and the Saga of the Beat Generation* (New York: Morrow Quill Paperbooks, 1976), 88.

4 This reading is printed as "Final City / Tap City," in *How I Work as a Poet* (HWP, 14–21).

5 "Emil Garber at Age 84," *Chicago Tribune*.

6 Saroyan, *Genesis Angels*, 88.

7 Correspondence from Lew Welch to Dorothy Brownfield, July 21, 1957, box 1, folder 16, Lew Welch Papers, MSS 13, Mandeville Special Collections Library, UCSD.

8 Correspondence from Lew Welch to Dorothy Brownfield, May 14, 1958, box 1, folder 16, Lew Welch Papers, MSS 13, Mandeville Special Collections Library, UCSD.

9 Dorothy Brownfield, "Brief Biography of Lew Welch," Donald Allen Papers, MSS 3, box 72, folder 3, Mandeville Special Collections Library, UCSD.

10 Brownfield, "Brief Biography of Lew Welch.".

11 The older of the two Keck brothers, George, had been a teacher at the Institute of Design, and Welch living in the Pioneer Co-Op meant that he and Mary were able to live in a house designed with the philosophy of the New Bauhaus firmly in mind.

12 "Pioneer Coop a Year Old," *Hyde Park Herald*, Wednesday, August 11, 1954, p. 3.

13 Jack Kerouac and Steve Allen, *Poetry for a Beat Generation*, LP record (New York: Hanover-Signature Records, 1959).

14 Transcript of an unpublished draft for *The San Francisco Poets* by David Meltzer, series 2, box 1-4, folder 13-45, David Meltzer Papers, 1955–1973, Washington University Libraries, Department of Special Collections.

15 Jack Kerouac, "Essentials of Spontaneous Prose," *Evergreen Review* 2, no. 5 (Summer 1958): 72–73.

16 Eberhart's article, titled "West Coast Rhythms," featured in the *New York Times Book Review* on September 2, 1956, and Lipton wrote a piece titled "Youth Will Serve Itself" for *The Nation* on November 10. Eberhart's article in particular was key because not only did it provide Welch with information about the San Francisco poetry scene in general but it also introduced him to "Howl," which Eberhart said was the "most remarkable" poem of any written in recent times. Eberhart had been invited to San Francisco by Donald Allen to attend a reenactment of the Six Gallery reading.

17 Richard Eberhart, "West Coast Rhythms," 18.

18 Carl Sandburg, "Chicago," *Poetry* 3, no. 6 (March 1914): 191.

19 Lew Welch at the Magic Lantern, April 22, 1967, "Introduction to 'Chicago Poem,'" recording, https://media.sas.upenn.edu/pennsound/authors/Welch/Magic-Lantern_04-22-67/Welch-Lew_05_Intro-to-Chicago-Poem_Magic-Lantern_Santa-Barbara_04-22-67.mp3, 01:09.

20 Arthur Waley, *The Nō Plays of Japan*, https://www.gutenberg.org/files/43304/43304-h/43304-h.htm.

21 Sandburg, "Chicago," 191.

22 Samuel Charters, "Lew Welch," in *The Beats: Literary Bohemians in Postwar America, Dictionary of Literary Biography 16*, ed. Ann Charters (Detroit: Gale, 1983), 65.

23 Sandburg, "Chicago," 191.

24 Typescript of poem "Brought" by Lew Welch, December 12, 1957, Donald Allen Papers, MSS 3, box 72, folder 3, Mandeville Special Collections Library, UCSD.

25 Joan Retallack, ed., *Gertrude Stein: Selections* (Berkeley: University of California Press, 2008), 49.

26 Jack Kerouac, *The Subterraneans* (New York: Grove Press, 1958), 68.

27 Douglas Brinkley, ed., *Windblown World : The Journals of Jack Kerouac 1947–1954* (New York: Viking, 2004), 215.

28 Lew Welch at the Magic Lantern, April 22, 1967, "Introduction and Opening Comments," recording, https://media.sas.upenn.edu/pennsound/authors/Welch/Magic-Lantern_04-22-67/Welch-Lew_01_Introduction_Magic-Lantern_Santa-Barbara_04-22-67.mp3, 03:02.

29 Garret Caples, "Two Chicagos," Poetry Foundation *Open Door*, May 8, 2014, https://www.poetryfoundation.org/harriet-books/2014/05/two-chicagos.

30 Chidi Ikonné, *From DuBois to Van Vechten: The Early New Negro Literature, 1903–1926* (Westport, CT: Greenwood Press, 1981), 14.

31 Langston Hughes, *The Big Sea* (New York, Hill and Wang, 1964), 29.

32 Correspondence from Donald Allen to Lew Welch, July 10, 1963, box 70, folder 19, Donald Allen Collection, MSS 3, Special Collections and Archives, UC San Diego.

CHAPTER 7

1 Correspondence from Lew Welch to Dorothy Brownfield, October 17, 1957, box 1, folder 16, Lew Welch Papers, MSS 13, Mandeville Special Collections Library, UCSD.

2 Correspondence from Mary Welch to Dorothy Brownfield, December 15, 1957, box 2, folder 5, Lew Welch Papers, MSS 13, Mandeville Special Collections Library, UCSD.

3 Typescript of essay "How to Survive in the United States" by Lew Welch, n.d., box 4, folder 5, Lew Welch Papers, MSS 13, Mandeville Special Collections Library, UCSD.

4 Correspondence from Lew Welch to Dorothy Brownfield, October 17, 1957.

5 Typescript of "Summer Goes Down the Other Side of the Mountain," by Lew Welch, box VI 146, folder 19, Gary Snyder Papers, D-050, Special Collections, UC Davis Library, University of California, Davis.

6 Typescript of "Summer Goes Down the Other Side of the Mountain," by Lew Welch.

7 Typescript of "Summer Goes Down the Other Side of the Mountain," by Lew Welch.

8 Typescript of "Summer Goes Down the Other Side of the Mountain," by Lew Welch.

9 Typescript of poem "For Mary" by Lew Welch, July 2, 1957, box 3, folder 20, Lew Welch Papers, MSS 13, Mandeville Special Collections Library, UCSD.

10 Correspondence from Lew Welch to Dorothy Brownfield, May 14, 1958, box 1, folder 16, Lew Welch Papers, MSS 13, Mandeville Special Collections Library, UCSD.

11 Handwritten draft of letter from Lew Welch to Philip Whalen, May 14, 1958, box 2, folder 33, Lew Welch Papers, MSS 13, Mandeville Special Collections Library, UCSD.

12 Many years later, he also attributed a debt of gratitude to the doctor who he felt had saved his life, the eminent physician Sandor Burstein, in a copy of *Hermit Poems* dated May 27, 1964, writing "Without his art, none of these poems."

13 Typewritten journal entry by Gary Snyder, July 7, 1958, box I 1, folder 6, Gary Snyder Papers, D-050, Special Collections, UC Davis Library, University of California, Davis.

14 William Hjortsberg, *Jubilee Hitchhiker: The Life and Times of Richard Brautigan* (Berkeley, CA: Counterpoint Press, 2012), 148.

CHAPTER 8

1 Alan Watts, *In My Own Way* (Novato: New World Library, 1972), 252.

2 Interview with Claude Dalenberg, by DC, October 7, 1994, http://www.cuke.com/Cucumber%20Project/interviews/ananda.html.

3 Interview with Claude Dalenberg, by DC.

4 David Schneider, *Crowded by Beauty* (Oakland: University of California Press, 2005), 89.

5 Joanne Kyger in an interview with David Chadwick, September 29, 1995, reproduced in *Beat Scene* 77 (Early Summer 2015): 39.

6 Ivan Morris, ed., *Madly Singing in the Mountains: An Appreciation and Anthology of Arthur Waley* (London: George Allen & Unwin, 1970), 67.

7 Donald Allen, ed., *The New American Poetry* (New York: Grove Press, 1960), x.

8 James Dickey, *The Death and Keys of the Censor* [Review of *The New American Poetry 1945–1960*, by D. M. Allen] *Sewanee Review* 69, no. 2 (1961): 318–332, http://www.jstor.org/stable/27540675.

9 Cecil Hemley, *Within a Budding Grove* [Review of *The New American Poetry 1945–1960*, by D. M. Allen] *Hudson Review* 13, no. 4 (1960): 626–630, https://doi.org/10.2307/3848017.

10 There is a certain symbolic significance in the choice of "stonefield" as the surname, given what Welch later wrote about the wreckage of his mind being similar to an expanse of cement blocks in a field. See chapter 25 in this book.

11 Michael Davidson, *The San Francisco Renaissance: Poetics and Community at Mid-century* (Cambridge, UK: Cambridge University Press, 1989), 150.

12 Schneider, *Crowded by Beauty*, 116.

13 William Hjortsberg, *Jubilee Hitchhiker: The Life and Times of Richard Brautigan* (Berkeley, CA: Counterpoint Press, 2012), 115.

14 Davidson, *San Francisco Renaissance*, 150.

15 Joanne Kyger, "Anne Waldman: The Early Years… 1965–1970," *Jacket* 27, April 2005, http://jacketmagazine.com/27/w-kyge.html.

16 Rick Fields, *How the Swans Came to the Lake: A Narrative History of Buddhism in America* (Boulder, CO: Shambhala, 1981), 255.

17 "Twenty-Sixth Annual Walk into History," Mill Valley Historical Society Guidebook, May 25, 2003, https://mvhistory.org/wp-content/uploads/2012/08/plugin-hist-walk-guidebook-2003.pdf.

18 Gary Snyder, in *A Zen Life: DT Suzuki, the Man Who Introduced Zen Buddhism to the West*, June 13, 2008, YouTube video, 1:14, https://www.youtube.com/watch?v=RVp9i4QIUUU.

19 Senzaki saw his zendos as floating because of their having no fixed headquarters.

20 *Zazenkai* means "coming together for zazen" and is most often interpreted as a Zen Buddhist retreat that is less intensive and shorter in duration than a traditional sesshin, while usually retaining the essential elements such as zazen (sitting meditation), kin-hin (walking meditation), Dharma talk, and chanting.

21 Carole Tonkinson, ed., *Big Sky Mind: Buddhism and the Beat Generation* (New York: Riverhead Books, 1995), 244.

22 Tonkinson, *Big Sky Mind*, 244.

23 Frank Chin, "A Buddhist Ideal," Frank Chin blogsite, February 7, 2017, https://chin-talks.blogspot.com/2017/02/a-buddhist-ideal.html.

24 Chin, "A Buddhist Ideal.".

25 Chin, "A Buddhist Ideal."

26 It is important to state that anything positive that Saijo experienced in the camp must not detract from the fact that most Japanese Americans experienced considerable hardships while interned, as well as discrimination and prejudice when they were finally released under Executive Order 9742 in June 1946. The vast majority lost their jobs, belongings, and homes as a result of EO 9066, and despite formal apologies and restitution under the Reagan administration in 1988, the order was a highly damaging one for more than one hundred thousand people of Japanese descent.

27 Chin, "A Buddhist Ideal."

28 Going for Broke: The 442nd Regimental Combat Team, September 24, 2020, https://www.nationalww2museum.org/war/articles/442nd-regimental-combat-team.

29 A common translation of these vows is

>   Beings are numberless, I vow to save them
>   Desires are inexhaustible, I vow to end them
>   Dharma gates are boundless, I vow to enter them
>   Buddha's way is unsurpassable, I vow to become it.

30 Tonkinson, *Big Sky Mind*, 245.

31 Correspondence from Gary Snyder to Will Petersen, January 20, 1959, box 1, folder 27, Gary Snyder Papers, Special Collections and Archives, Kent State University.

32 Welch wrote about this in the poem "He Writes to the Donor of His Bowl."

33 Saijo transferred this information in a handwritten note to Welch that included formal practices and structures for an entire zazen session.

34 Handwritten notes by Albert Saijo, n.d., box 2, folder 22, Lew Welch Papers, MSS 13, Mandeville Special Collections Library, UCSD.

35 Typescript for grant application to the Saxton Trust by Lew Welch, April 1960, box 2, folder 1, Lew Welch Papers, MSS 13, Mandeville Special Collections Library, UCSD.

36 Tom Killion and Gary Snyder, *Tamalpais Walking* (Berkeley, CA: Heyday, 2010), 14.

37 Correspondence from Lew Welch to Dorothy Brownfield, May 14, 1958, box 1, folder 16, Lew Welch Papers, MSS 13, Mandeville Special Collections Library, UCSD.

38 Ven. Shikai Zuiko O-sensei, "Kinhin: The Dignity of the Buddha," White Wind Zen Community, June 7, 1996, http://wwzc.org/dhathetext/kinhin-dignity-buddha.

39 Bob Greensfelder, "A Landscape Full of Stories," North Columbia Schoolhouse Cultural Center, http://www.northcolumbiaschoolhouse.org/landscapeofstories/bob-greensfelder/.

40 Chuck Oldenburg, "Krag's Ship," Mill Valley Historical Society, May 2013, https://www.mvhistory.org/history-of/history-of-homestead-valley/krags-ship/.

41 "Twenty-Sixth Annual Walk into History," Mill Valley Historical Society Guidebook, May 25, 2003, https://mvhistory.org/wp-content/uploads/2012/08/plugin-hist-walk-guidebook-2003.pdf.

42 The SFSC reading can be found at https://diva.sfsu.edu/collections/poetrycenter/bundles/191200.

43 Correspondence from Lew Welch to Philip Whalen, n.d., box 27, folder 7, Philip Whalen papers, BANC MSS 2000/93 p, The Bancroft Library, University of California, Berkeley.

44 Correspondence from Lew Welch to Philip Whalen, n.d.

45 Welch specifies that the sentence in question is on "page 208." However, when the novel was published by Coyote Press in 1967, it appeared on page 142.

46 Program notes for the San Francisco State College reading on November 8, 1959, box 8, San Francisco State College Poetry Center records, BANC MSS 78/161 c, The Bancroft Library, University of California, Berkeley.

CHAPTER 9

1 Ann Charters, *Kerouac: Selected Letters: Volume 2: 1957–1969* (New York: Penguin Books, 2000), 254.
2 Charters, *Kerouac: Selected Letters: Volume 2*, 224.
3 Steve Allen, *The Jack Kerouac Collection* (Santa Monica: Rhino Records, 1990), 16.
4 Blaine Allan, "The Making (and Unmaking) of PULL MY DAISY," *Film History* 2, no. 3 (1988): 185–205.
5 Correspondence from Joanne Kyger to Stan Perksy, November 30, 1959, box 8, folder 25, Joanne Kyger Correspondence, MSS 8, Special Collections & Archives, UC San Diego.
6 Paul Maher, *Kerouac: The Definitive Biography* (Lanham, MD: Taylor Trade Publishing, 2004), 397.
7 Albert Saijo, "A Recollection," in *Trip Trap: Haiku on the Road* (Bolinas, CA: Grey Fox Press, 1973), 5.
8 Typescript of *Bemisbag: A Movie*, by Lew Welch, n.d., box 4, folder 9, Lew Welch Papers, MSS 13, Mandeville Special Collections Library, UCSD.
9 Charters, *Kerouac: Selected Letters: Volume 2*, 258.
10 Charters, *Kerouac: Selected Letters: Volume 2*, 225.
11 Charters, *Kerouac: Selected Letters: Volume 2*, 225.
12 Charters, *Kerouac: Selected Letters, Volume 2*, 243.
13 Michael Schumacher, *Dharma Lion: A Biography of Allen Ginsberg* (New York: St. Martin's Press, 1992), 319.
14 Schumacher, *Dharma Lion*, 10.
15 Schumacher, *Dharma Lion*, 10.
16 Correspondence from Lew Welch to Ruth Witt-Diamant, December 14, 1959, box 8, San Francisco State College Poetry Center records, BANC MSS 78/161 c, The Bancroft Library, University of California, Berkeley.
17 Correspondence from Lew Welch to Ruth Witt-Diamant, December 14, 1959.
18 William T. Lawlor, ed., *Beat Culture: Lifestyles, Icons & Impact* (Santa Barbara, CA: ABC-Clio, 2005), 160.
19 Typescript for an unpublished draft of "Hard Start" by Lew Welch, n.d., box 7, Lew Welch Papers, MSS 13, Mandeville Special Collections Library, UCSD.

CHAPTER 10

1 Welch describes Ryoanji in *How I Work As A Poet* as "one of the most famous gardens in Japan" (HWP, 78).
2 Peterson's essay featured in *Evergreen Review* 1, no. 4.
3 Koan is a Japanese word that comes from the Chinese *kung-an*, which means public dictate. It is a reference to examples that are meant to guide life; or, in the case of Zen Buddhism, dictates are meant to be catalysts for awakening one's true/deep/pure nature. There is never an answer to a koan; any response will be rejected. Active thought is futile in trying to understand the riddle; instead the koan must be allowed to work on the mind during meditation and heighten intuitive understanding.
4 Rod Phillips, *"Forest Beatniks" and "Urban Thoreaus"* (New York: Peter Lang, 2000), 83.
5 Phillips, *"Forest Beatniks" and "Urban Thoreaus,"* 85.
6 Daisetsu Teitaro Suzuki, *An Introduction to Zen* (New York: Grove Press, 1964), 88.

7 Robert Sohl and Audrey Carr, *The Gospel According to Zen* (New York: Signet, 1970), 38.

8 Suzuki, *Introduction to Zen*, 92.

9 Welch had been hospitalized in 1949 after an outbreak of mononucleosis at Reed had affected a significant number of students.

10 Correspondence from Lew Welch to Dorothy Brownfield, July 15, 1950, box 1, folder 10, Lew Welch Papers, MSS 13, Mandeville Special Collections Library, UCSD.

11 Typescript of unpublished prose piece by Lew Welch, n.d., box 4, folder 9, Lew Welch Papers, MSS 13, Mandeville Special Collections Library, UCSD.

12 Typescript of unpublished prose piece by Lew Welch, n.d.

13 Typescript of unpublished prose piece by Lew Welch, n.d.

14 Jeff Cregg, interview with the author, October 16, 2017.

15 "Summary and Report," *English Journal* 34, no. 1 (1945): 47–52, http://www.jstor.org/stable/807000.

16 Typescript for grant application to the Saxton Trust by Lew Welch, April 1960, box 2, folder 1, Lew Welch Papers, MSS 13, Mandeville Special Collections Library, UCSD.

17 Typescript for grant application to the Saxton Trust by Lew Welch, April 1960.

18 Typescript for grant application to the Saxton Trust by Lew Welch, April 1960.

19 Typescript for grant application to the Saxton Trust by Lew Welch, April 1960.

CHAPTER 11

1 Correspondence from Gary Snyder to the author, November 11, 2021.

2 Typescript for grant application to the Saxton Trust by Lew Welch, April 1960, box 2, folder 1, Lew Welch Papers, MSS 13, Mandeville Special Collections Library, UCSD.

3 Handwritten journal entry by Lew Welch, n.d., box 6, folder 1, Lew Welch Papers, MSS 13, Mandeville Special Collections Library, UCSD.

4 Correspondence from Lew Welch to Dorothy Brownfield, March 26, 1950, box 1, folder 10, Lew Welch Papers, MSS 13, Mandeville Special Collections Library, UCSD.

5 Typewritten journal entry by Gary Snyder, n.d., box I 84, folder 1, Gary Snyder Papers, D-050, Special Collections, UC Davis Library, University of California, Davis.

6 Typewritten journal entry by Gary Snyder, January 21, 1950, box I 84, folder 1, Gary Snyder Papers, D-050, Special Collections, UC Davis Library, University of California, Davis.

7 Typewritten journal entry by Gary Snyder, January 12, 1950, box I 84, folder 1, Gary Snyder Papers, D-050, Special Collections, UC Davis Library, University of California, Davis.

8 Typewritten journal entry by Gary Snyder, January 12, 1950.

9 Both quotes from Whalen to Donald Allen, September 25, 1976, Donald Allen Papers, MSS 3, box 48, folder 11, Mandeville Special Collections Library, UCSD.

10 Albert Saijo, "Untitled," review of *I, Leo*, by Lew Welch, *Western American Literature* 13, no. 3 (Fall 1978): 277–278, University of Nebraska Press.

11 Kerouac, "Essentials of Spontaneous Prose," *Evergreen Review* 2, no. 5 (Summer 1958): 72–73.

12 Kerouac, "Essentials of Spontaneous Prose.".

13 "The Man Who Played Himself," *Evergreen Review* 5, no. 17 (March-April 1961): 97–105.

CHAPTER 12

1 Ann Charters, *Kerouac: Selected Letters: Volume 2: 1957–1969* (New York: Penguin Books, 2000), 256.

2 Charters, *Kerouac: Selected Letters: Volume 2*, 256.

3 Charters, *Kerouac: Selected Letters: Volume 2*, 256.

4 Ann Charters, *Kerouac: A Biography* (New York: St. Martin's Press, 1994), 310.

5 Charters, *Kerouac: A Biography*, 311.

6 Kerouac, *Big Sur*, 59.

7 Carolyn Cassady, *Off the Road: My Life with Cassady, Kerouac, & Ginsberg* (New York: William Morrow, 1990), 345.

8 Cassady, *Off the Road*, 346.

9 Cassady, *Off the Road*, 346.

10 Correspondence from Lew Welch to Dorothy Brownfield, August 24, 1960, box 2, folder 1, Lew Welch Papers, MSS 13, Mandeville Special Collections Library, UCSD.

11 Charters, *Kerouac: Selected Letters, Volume 2*, 309.

12 Kerouac, *Big Sur*, 180.

CHAPTER 13

1 Ronna C. Johnson, "Lenore Kandel," in *The Encyclopaedia of Beat Literature*, ed. Kurt Hemmer (New York: Facts on File, 2007), 170–172.

2 Donna Nance, "Lenore Kandel," *The Beats: Literary Bohemians in Postwar America: Dictionary of Literary Biography 16*, ed. Ann Charters (Detroit: Gale, 1983), 271.

3 Lenore Kandel, "People," *San Francisco Examiner*, October 21, 1979, 475.

4 Robert Ross, letter to the author, November 12, 2019.

5 Manjusri is the Bodhisattva of Wisdom and, like Avalokiteshvara, features in both Welch's poetry and prose. Indeed, Welch may have intended the poster to reflect not only his search for wisdom but also his desire to affiliate himself more openly with the lion—or "Leo" character—as suggested in his circus role as the "Magick Lion."

6 Robert Ross, letter to the author, November 12, 2019.

7 Polina Olsen, *Portland in the 1960s: Stories from the Counterculture* (Charleston, SC: The History Press, 2012), 36.

8 Olsen, *Portland in the 1960s: Stories from the Counterculture*, 38.

9 Handwritten journal entry by Philip Whalen, August 20, 1962, box 1, folder 3, Philip Whalen papers, BANC MSS 2000/93 p, The Bancroft Library, University of California, Berkeley.

10 Handwritten journal entry by Philip Whalen, Kyoto, May 1971, box 3, folder 6, Philip Whalen papers, BANC MSS 2000/93 p, The Bancroft Library, University of California, Berkeley.

11 Correspondence from Lew Welch to Ruth Witt-Diamant, July 29, 1960, box 8, San Francisco State College Poetry Center records, BANC MSS 78/161 c, The Bancroft Library, University of California, Berkeley.

12 This line is taken from a draft of a section in the poem that was eventually cut from the final version; box 3, folder 20, Lew Welch Papers, MSS 13, Mandeville Special Collections Library, UCSD.

13 Lew Welch at the Magic Lantern, April 22, 1967, "Introduction to 'Din Poem,'" Pennsound recording, https://media.sas.upenn.edu/pennsound/authors/Welch/Magic-Lantern_04-22-67/Welch-Lew_11_Intro-to-Din-Poem_Magic-Lantern_Santa-Barbara_04-22-67.mp3.

14 Lew Welch at the Magic Lantern, April 22, 1967.
15 Gertrude Stein, *Four in America* (New Haven, CT: Yale University Press, 1947), 167.
16 Stein, *Four in America*.
17 Duncan described "Taxi Suite" thus in his introduction to Welch's reading at the Berkeley Poetry Conference in 1965.
18 David Meltzer, *The San Francisco Poets* (New York: Ballantine, 1971), 209–210.
19 Correspondence from Lew Welch to Dorothy Brownfield, June 29, 1961, box 2, folder 1, Lew Welch Papers, MSS 13, Mandeville Special Collections Library, UCSD.
20 Correspondence from Lew Welch to Dorothy Brownfield, June 29, 1961.
21 Correspondence from Lew Welch to Dorothy Brownfield, June 29, 1961.
22 Correspondence from Dorothy Brownfield to Emil Garber, August 15, 1961, box 2, folder 1, Lew Welch Papers, MSS 13, Mandeville Special Collections Library, UCSD.
23 Meltzer, *San Francisco Poets*, 225.
24 Jennie Skerl, "Introduction," in *Reconstructing the Beats*, ed. Jennie Skerl (New York: Palgrave Macmillan, 2004), 4.
25 Transcript of an unpublished draft for *The San Francisco Poets* by David Meltzer, series 2, box 1-4, folder 13-45, David Meltzer Papers, 1955–1973, Washington University Libraries, Department of Special Collections.
26 Meltzer, *San Francisco Poets*, 225.
27 Correspondence from Lenore Kandel to Don Crowe, July 1962, box II 215, folder 86, Gary Snyder Papers, D-050, Special Collections, UC Davis Library, University of California, Davis.

CHAPTER 14

1 Correspondence from Lew Welch to Robert Duncan, May 18, 1960, box 67, PCMS-0110, the Robert Duncan Collection, circa 1900–1996, the Poetry Collection of the University Libraries, University at Buffalo, The State University of New York.
2 Correspondence from Lew Welch to Robert Duncan, May 19, 1960, box 67, PCMS-0110, the Robert Duncan Collection, circa 1900–1996, the Poetry Collection of the University Libraries, University at Buffalo, The State University of New York.
3 Correspondence from Lew Welch to Robert Duncan, May 19, 1960.
4 Correspondence from Lew Welch to Robert Duncan, May 19, 1960.
5 T. S. Eliot, "Baudelaire," in *Selected Essays* (New York: Harcourt, Brace and Company, 1932), 335–346.
6 R. H. Blyth, "Instant Zen," in *The Gospel According to Zen* (New York: Signet, 1970), 105.

CHAPTER 15

1 Philip Whalen, notebook entry, August 20, 1962, box 1, folder 3, Philip Whalen papers, BANC MSS 2000/93 p, The Bancroft Library, University of California, Berkeley.
2 Typescript of untitled poem by Lew Welch, box, 13, folder 9, Lew Welch Papers, MSS 13, Mandeville Special Collections Library, UCSD.
3 *Mountain Home: The Wilderness Poetry of Ancient China*, trans. David Hinton (New York: New Directions, 2005), xv.
4 T'ao Ch'ien, *The Selected Poems of T'ao Ch'ien*, trans. David Hinton (Port Townsend, WA: Copper Canyon Press, 1993), 8.
5 Industrial Union Manifesto, https://archive.iww.org/history/library/iww/industrial_union_manifesto/.

6 Stephen Schwartz, "A Recollection of Lew Welch," in *Hey Lew*, ed. Magda Cregg (Bolinas, CA: Magda Cregg, 1997), 56.

7 Stephen Schwartz, "A Recollection of Lew Welch," in *Hey Lew*, ed. Magda Cregg (Bolinas, CA: Magda Cregg, 1997), 55.

8 T'ao Chi'en, *Selected Poems*, 9.

9 T'ao Chi'en, *Selected Poems*, 9.

10 Edward Conze, *"The Perfection of Wisdom in Eight Thousand Lines & Its Verse Summary* (Bolinas, CA: Four Seasons Foundation, 1973), 52.

11 Typescript of essay "Commentary on the Prajñāpāramitā" by Lew Welch, n.d., box 3, folder 8, Lew Welch Papers, MSS 13, Mandeville Special Collections Library, UCSD.

12 Robert Creeley, *The Collected Essays of Robert Creeley* (Berkeley: University of California Press, 1989), 293–294.

CHAPTER 16

1 "Poem: Xuixi Yin," August 29, 2006, Politics Forum, http://www.politicsforum.org/forum/viewtopic.php?t=63305.

2 Interview with Donald Allen, by DC, ca. 1995, cuke.com, http://www.cuke.com/Cucumber%20Project/interviews/allen.html.

3 Lenore Kandel, *The Collected Poems of Lenore Kandel* (Berkeley, CA: North Atlantic Press, 2012), 154.

4 Handwritten ledger entry by Lew Welch, 1961, box 5, folder 6, Lew Welch Papers, MSS 13, Mandeville Special Collections Library, UCSD.

5 The Batman Gallery archive states that Welch consigned three of his works for resale on February 27, 1962. This date coincides with the Master Bat Group show at which Conner exhibited work alongside, among others, Herms, Lavigne, and Batman owner William Jahrmarkt.

6 Correspondence from Lew Welch to Don Crowe, January 6, 1961, box II 216, folder 77, Gary Snyder Papers, D-050, Special Collections, UC Davis Library, University of California, Davis.

7 Jack London, *Sea Wolf* (London: MacMillan, 1904), 57.

CHAPTER 17

1 Michael McClure, "Kirby Doyle and the Snows of Yesteryear," Empty Mirror, https://www.emptymirrorbooks.com/beat/kirby-doyle.

2 Thomas F. Merrill, *The Collected Prose of Charles Olson: A Primer* (Newark: University of Delaware Press, 1982), 138.

3 Correspondence from Lew Welch to Dorothy Brownfield, April 22, 1963, box 2, folder 2–3, Lew Welch Papers, MSS 13, Mandeville Special Collections Library, UCSD.

4 Untitled essay by Lew Welch, 1963, box 3, folder 8, Lew Welch Papers, MSS 13, Mandeville Special Collections Library, UCSD.

5 "Standing in the Soul," Steve Silberman interviewed by Robert Hunter, December 1992, http://artsites.ucsc.edu/GDead/agdl/silber.html.

6 This is a reference to the Sung landscape scroll paintings that Welch had once gone to see at the Seattle Art Museum with Gary Snyder.

CHAPTER 18

1 Transcript of an unpublished draft for *The San Francisco Poets* by David Meltzer, series 2, box 1-4, folder 13-45, David Meltzer Papers, 1955–1973, Washington University Libraries, Department of Special Collections.

2 The collection would eventually become *Hermit Poems*. Rago published a small selection in *Poetry* 104, no. 4.

CHAPTER 19

1 Don Carpenter, "Poetry at the Longshoreman's Hall," Literary Kicks, September 20, 1996, https://www.litkicks.com/Longshoreman.

2 Carpenter, "Poetry at the Longshoreman's Hall."

3 Piero Melograni, *Wolfgang Amadeus Mozart: A Biography* (Chicago: University of Chicago Press, 2007), 80.

4 Melograni, *Wolfgang Amadeus Mozart*.

5 Correspondence from Lew Welch to Robert Duncan, March 7, 1963, box 67, PCMS-0110, the Robert Duncan Collection, circa 1900–1996, the Poetry Collection of the University Libraries, University at Buffalo, The State University of New York.

6 Correspondence from Lew Welch to Robert Duncan, March 7, 1963.

7 Transcript of an unpublished draft for *The San Francisco Poets* by David Meltzer, series 2, box 1-4, folder 13-45, David Meltzer Papers, 1955–1973, Washington University Libraries, Department of Special Collections.

8 Transcript of an unpublished draft for *The San Francisco Poets* by David Meltzer.

9 Correspondence from Lew Welch to Robert Duncan, March 7, 1963, box 67, PCMS-0110, the Robert Duncan Collection, circa 1900–1996, the Poetry Collection of the University Libraries, University at Buffalo, The State University of New York.

10 An entry in one of Philip Whalen's notebooks suggests that he intended to contribute an article titled "I Was Born by Mistake, among Poor People" to *Bread*.

11 Correspondence from Lew Welch to Robert Duncan, March 7, 1963, box 67, PCMS-0110, the Robert Duncan Collection, circa 1900–1996, the Poetry Collection of the University Libraries, University at Buffalo, The State University of New York.

12 Don Carpenter, "Poetry at the Old Longshoreman's Hall," Literary Kicks, September 9, 1996, https://litkicks.com/Longshoreman/.

13 "Whoregon," *Open Space* 6, ed. Stan Persky, San Francisco: June 1964.

14 Stephen Vincent, "Poems in Street, Coffeehouse, and Print—The Mid-1960s," FoundSF, http://www.foundsf.org/index.php?title=Poems_in_Street,_Coffeehouse,_and_Print%E2%80%94The_Mid-1960s.

15 Typewritten journal entry by Gary Snyder, January 12, 1950, box I 85, folder 2, Gary Snyder Papers, D-050, Special Collections, UC Davis Library, University of California, Davis.

16 Stephen Vincent, "Poems in Street, Coffeehouse, and Print—The Mid-1960s," FoundSF, http://www.foundsf.org/index.php?title=Poems_in_Street,_Coffeehouse,_and_Print%E2%80%94The_Mid-1960s.

17 Handwritten notebook entry by Philip Whalen, July 1964, box 1, folder 6, Philip Whalen papers, BANC MSS 2000/93 p, The Bancroft Library, University of California, Berkeley.

18 The discussion was published in its entirety by Donald Allen's Grey Fox Press in 1977.

CHAPTER 20

1 Jeff Cregg, interview with the author, October 16, 2017, San Rafael, California.

2 Jeff Cregg, interview with the author, October 16, 2017.

3 After initially working as a harmonica player and backing singer in bands including Clover and Thin Lizzy, Hugh Cregg eventually found worldwide fame as a founding member of Huey Lewis and the News. There has also been a suggestion that he

adopted his stage name in honor of his stepfather, and although Lewis has acknowledged Welch's influence when he was teenager, there is apparently no truth in this.

4 Jeff Cregg, interview with the author, October 16, 2017.

5 Kremis Carrigg, interview with the author, October 16, 2017, San Rafael, California.

CHAPTER 21

1 Typescript of play *Twins: A Play in Two Acts* by Lew Welch, box 4, folder 12, Lew Welch Papers, MSS 13, Mandeville Special Collections Library, UCSD.

2 Typescript of play *Twins: A Play in Two Acts* by Lew Welch.

3 James Peterson, *Dreams of Chaos, Visions of Order: Understanding the American Avant-Garde Cinema* (Detroit: Wayne State University Press, 1994), 171.

4 Peterson, *Dreams of Chaos*, 170.

5 In addition to the establishment of Oyez, 1965 was also the year in which Dave Haselwood opened his independent publishing house, having left Auerhahn in the hands of his former business partner Andrew Hoyem at the end of the previous year.

6 Typescript of notebook entry by Lew Welch, n.d., box 3, folder 20, Lew Welch Papers, MSS 13, Mandeville Special Collections Library, UCSD.

CHAPTER 22

1 Lewis Ellingham and Kevin Killian, *Poet Be Like God: Jack Spicer and the San Francisco Renaissance* (Hanover, NH: Wesleyan University Press, 1988), 319.

2 Ellingham and Killian, *Poet Be Like God*.

3 *Synapse—The UCSF Student Newspaper* 9, no. 7 (February 10, 1965): 4

4 *Synapse—The UCSF Student Newspaper* 9, no. 4 (November 2, 1964): 3

5 Correspondence from Lew Welch to Robert Duncan, March 17, 1965, box 67, PCMS-0110, the Robert Duncan Collection, circa 1900–1996, the Poetry Collection of the University Libraries, University at Buffalo, The State University of New York.

6 This statement was included in a biographical note in the exhibition catalogue titled *Looking at Pictures with Gertrude Stein* (San Francisco: University of California, San Francisco Medical Center, 1965).

7 Gertrude Stein, *Lectures in America* (New York: Random House, 1935), 166.

8 Amy Evans and Shamoon Zamir, eds., *The Unruly Garden: Robert Duncan and Eric Mottram, Letters and Essays* (Bern: Peter Lang, 1995), 124.

9 Duncan's introduction to Welch's reading begins at 1:01:25 on http://www.kpfahistory.info/bpc/readings/Loewinsohn.mp3.

10 http://www.kpfahistory.info/bpc/readings/Loewinsohn.mp3, 1:05:02.

11 Interestingly, at the reading to commemorate the reissue of *Ring of Bone* at the Los Angeles Public Library in Spring 2011, Huey Lewis states that jazz great John Handy had helped Welch write the music for "Graffiti."

12 William Hjortsberg, *Jubilee Hitchhiker: The Life and Times of Richard Brautigan* (Berkeley, CA: Counterpoint Press, 2012), 258.

13 Typescript of "A Preface to Leather Prunes," by Lew Welch, n.d., box 3, folder 7, Lew Welch Papers, MSS 13, Mandeville Special Collections Library, UCSD.

14 Typescript of "A Preface to Leather Prunes," by Lew Welch, n.d.

15 Typescript of "Abner Won't Be Home for Dinner: A Play," by Lew Welch, n.d., box 3, folder 7, Lew Welch Papers, MSS 13, Mandeville Special Collections Library, UCSD.

16 Typescript of poem "A perception, by definition, is discontinuous—a surprise, a bafflement:"

17 Typescript of untitled essay by Lew Welch, n.d., box 3, folder 4, Lew Welch Papers, MSS 13, Mandeville Special Collections Library, UCSD.

18 In a written interpretation of his 1808 painting, "A Vision of the Last Judgment," Blake wrote "I question not my Corporeal or Vegetative Eye any more than I would Question a Window concerning a Sight I look thro it & not with it." See David Erdman's *The Complete Poetry and Prose of William Blake* (New York: Anchor Books, 1988), 565–566.

19 In the same essay, Welch quotes Edwin N. Garlan, a professor of philosophy at Reed College, who said, "If Van Gogh had not painted that curious yellow world of his, it is very improbable that any of us could see it."

CHAPTER 23

1 Flyer for the UC Letters and Science Extension Berkeley, September 26–November 28, 1965, box 87, folder 7, Donald Allen Collection, MSS 3, Special Collections & Archives, UC San Diego.

2 Handwritten notebook entry with notes for teaching by Lew Welch, n.d., box 5, folder 9, Lew Welch Papers, MSS 13, Mandeville Special Collections Library, UCSD.

3 Richard Hughey, "Whatever Happened to Lew Welch," in *Hey Lew*, ed. Magda Cregg (Bolinas, CA: Magda Cregg, 1997), 60.

4 Hughey, "Whatever Happened to Lew Welch," 61.

5 It should be noted here that Burns never actually wrote the lyrics to the song. The composer of the ballad remains a mystery, although it is widely attributed to a Jacobite prisoner in 1746 awaiting his fate in prison after the Jacobite rebellion.

6 Several letters concerning this complainant are housed in the UCSD archives, including his original letter plus the subsequent correspondence between Welch and the course supervisor, Elspeth Smith.

7 Correspondence from Gene Harney to Dr. Morton Gordon, June 1969, box 6, folder 13, Lew Welch Papers, MSS 13, Mandeville Special Collections Library, UCSD.

8 Correspondence from Elspeth Smith to Lew Welch, June 1969, box 6, folder 13, Lew Welch Papers, MSS 13, Mandeville Special Collections Library, UCSD.

9 Correspondence from Lew Welch to Elspeth C. Smith, June 1969, box 6, folder 13, Lew Welch Papers, MSS 13, Mandeville Special Collections Library, UCSD.

10 Correspondence from Lew Welch to Elspeth C. Smith, June 1969.

11 Correspondence from Lew Welch to Elspeth C. Smith, June 1969.

12 Typescript of essay "How to Survive in the United States," by Lew Welch, n.d., box 4, folder 5, Lew Welch Papers, MSS 13, Mandeville Special Collections Library, UCSD.

13 In Lawrence Lipton's *The Holy Barbarians*, Rexroth is quoted as saying, "you can't hop up your young people with sadism in the movies and television and train them to commando tactics in the army camps, to say nothing of brutalizing them in wars, then expect them to "untense" them with Coca-Cola and Y.M.C.A. hymn sings," 150.

14 Typescript of essay "How to Survive in the United States" by Lew Welch, n.d., box 4, folder 5, Lew Welch Papers, MSS 13, Mandeville Special Collections Library, UCSD.

15 Typescript essay "How to Survive in the United States," by Lew Welch, n.d.

16 Typescript essay "How to Survive in the United States," by Lew Welch, n.d.

17 Typescript essay "How to Survive in the United States," by Lew Welch, n.d.

18 Typescript of untitled essay by Lew Welch, December 10, 1966, box 3, folder 4, Lew Welch Papers, MSS 13, Mandeville Special Collections Library, UCSD.

19 Typescript of untitled essay by Lew Welch, December 10, 1966.

20 Typescript of untitled essay by Lew Welch, December 10, 1966.

21 Ann Charters, "The Best Minds of a Generation," in *The Portable Beat Reader*, ed. Ann Charters (London: Penguin Books, 1992), 1.

22 Bertolt Brecht, "On Teaching without Pupils," in *Bertolt Brecht: Poems 1913–1956* (London: Methuen, 1976), 255–256.

CHAPTER 24

1 Stephen Vincent, "Poems in Street, Coffeehouse, and Print—The Mid-1960s," FoundSF, http://www.foundsf.org/index.php?title=Poems_in_Street,_Coffeehouse, _and_Print%E2%80%94The_Mid-1960s.

2 Doug Palmer, "My Experience as a Poet in 60s," *East Bay Review of Books*, http://www. bigbridge.org/DougPalmerMemoir.pdf.

3 Correspondence from Lew Welch to Doug Palmer, December 5, 1965, box 247c, folder 2:7. Doug Palmer Papers, 1952–1974, MSS 72, The Bancroft Library, University of California, Berkeley.

4 Correspondence from Lew Welch to Doug Palmer, December 5, 1965.

5 Correspondence from Lew Welch to Doug Palmer, December 5, 1965.

6 Doug Palmer, *Poems for the People* (Berkeley: Peace & Gladness Press, 1965), 1.

7 Magda Cregg, ed., *Hey Lew* (Bolinas, CA: Magda Cregg, 1997), 88.

8 Dennis Hevesi, "W. J. Wilkins, 98; Was Judge at Trial Of Nazi Industrialists," *New York Times*, September 14, 1995, Section B, 15, https://www.nytimes.com/1995/09/14/ obituaries/w-j-wilkins-98-was-judge-at-trial-of-nazi-industrialists.html.

9 Correspondence from George Redman to Donald Allen, May 10, 1972, box 87, folder 7, Donald Allen Collection, MSS 3, Special Collections & Archives, UC San Diego.

10 Correspondence from Jack Shoemaker to Lew Welch, January 22, 1967, box 58, folder 51-53, Sand Dollar Collection, PCMS-0015, the Poetry Collection of the University Libraries, University at Buffalo, The State University of New York.

11 Correspondence from Lew Welch to Jack Shoemaker, April 26, 1967, box 61, folder 1-7, Sand Dollar Collection, PCMS-0015, the Poetry Collection of the University Libraries, University at Buffalo, The State University of New York.

12 Correspondence from Lew Welch to Jack Shoemaker, April 26, 1967.

13 Handwritten entry in a spiral-bound notebook by Lew Welch, n.d., box 5, folder 9, Lew Welch Papers, MSS 13, Mandeville Special Collections Library, UCSD.

14 Handwritten entry in a spiral-bound notebook by Lew Welch, n.d.

15 Handwritten entry in a spiral-bound notebook by Lew Welch, n.d.

16 Handwritten entry in a spiral-bound notebook by Lew Welch, n.d.

17 Correspondence from Donald Allen to Richard Seaver, September 8, 1969, box 1, folder 1, Lew Welch Papers, MSS 13, Mandeville Special Collections Library, UCSD.

18 Correspondence from Marilyn Meeker to Lew Welch, April 20, 1970, box 2, folder 20, Lew Welch Papers, MSS 13, Mandeville Special Collections Library, UCSD.

19 In the original transcript for David Meltzer's interview with Welch that featured in *The San Francisco Poets*, Welch rages about Eliot having "nothing to do with english [sic] literature at all" and that he is merely "fumbling around with his imitations of 17th century sermons."

20 Clifford Burke, letter to the author, August 21, 2017.

21 Clifford Burke, letter to the author, August 21, 2017.

22 Clifford Burke, letter to the author, August 21, 2017.

23 Clifford Burke, *Printing Poetry: A Workbook in Typographic Reification* (San Francisco: Scarab Press, 1980), 111.

CHAPTER 25

1 See Chris Carlsson's essay on the subsequent uprising following the shooting, FoundSF, https://www.foundsf.org/index.php?title=Hunters_Point_Uprising.

2 Gene Anthony, *Summer of Love: Haight-Ashbury at Its Highest* (San Francisco: Last Gasp, 1980), 93.

3 Anthony, *Summer of Love*, 98.

4 Gary Snyder, interview with the author, September 23, 2019, Kitkitdizze, California.

5 Correspondence from Lew Welch to Dorothy Brownfield, February 1, 1950, box 1, folder 11, Lew Welch Papers, MSS 13, Mandeville Special Collections Library, UCSD.

6 Michael Schumacher, *Dharma Lion: A Biography of Allen Ginsberg* (New York: St. Martin's Press, 1992), 480.

7 David Schneider, *Crowded by Beauty* (Oakland: University of California Press, 2005), 286.

8 The Gestetner was a stencil duplicator, invented by David Gestetner in the early 1920s. The machine revolutionized the printing industry with its ability to produce numerous copies of printed matter at low cost. Claude Hayward and Chester Anderson had two machines, which they used to print pamphlets for the Communications Company. By that time, "gestetner" was also a commonly used verb, hence Welch's use of "gestetnered" here. For more information on the Communications Company, see https://www.diggers.org/Outrageous_Pamphleteers-A_History_Of_The_Communication_Company.pdf

9 Schumacher, *Dharma Lion*, 480.

10 Handwritten essay, "Co-Lively" by Lew Welch, n.d., box 4, folder 1, Lew Welch Papers, MSS 13, Mandeville Special Collections Library, UCSD.

11 Typescript of essay, "The Mindless Conspiracy," by Lew Welch, n.d., box 3, folder 8, Lew Welch Papers, MSS 13, Mandeville Special Collections Library, UCSD.

12 Peter Coyote, *Sleeping Where I Fall* (New York: Counterpoint Press, 1998), 35.

13 Coyote, *Sleeping Where I Fall*.

14 Peter Berg in an interview with Judy Goldhalf, in *San Francisco Chronicle*, May 20, 2007, https://arthurmag.com/2007/06/04/diggers-peter-berg-and-judy-goldhaft-on-1967/

15 John Anthony Moretta, *The Hippies: A 1960s History* (Jefferson, NC: MacFarland & Company, 2017), 189.

16 Peter Coyote in an interview with Etan Ben-Ami, Mill Valley, California, January 12, 1989, http://www.diggers.org/oralhistory/peter_interview.html.

17 Typescript of "Co-Lively," by Lew Welch, n.d., box 4, folder 1, Lew Welch Papers, MSS 13, Mandeville Special Collections Library, UCSD.

CHAPTER 26

1 Correspondence from Philip Whalen to Lew Welch, September 30, 1969, box 2, folder 34, Lew Welch Papers, MSS 13, Mandeville Special Collections Library, UCSD.

2 Correspondence from Philip Whalen to Lew Welch, September 30, 1969.

3 Correspondence from Lew Welch to Philip Whalen, October 14, 1969, box 27, folder 7, Philip Whalen papers, BANC MSS 2000/93 p, The Bancroft Library, University of California, Berkeley.

4 A review of the visit was published in *The San Quentin News* on August 22, 1969.

5 Correspondence from Terry Cuddy to Lew Welch, August 31, 1969, box 2, folder 6, Lew Welch Papers, MSS 13, Mandeville Special Collections Library, UCSD.

6 Correspondence from Magda Cregg to Donald Allen, June 30, 1970, box 90, folder 14, Donald Allen Collection, MSS 3, Special Collections & Archives, UC San Diego.

7 Correspondence from Magda Cregg to Donald Allen, June 30, 1970.

8 Correspondence from Neal Cross to Lew Welch, February 23, 1970, box 6, folder 9, Lew Welch Papers, MSS 13, Mandeville Special Collections Library, UCSD.

9 Correspondence from Neal Cross to Lew Welch, March 6, 1970, box 6, folder 9, Lew Welch Papers, MSS 13, Mandeville Special Collections Library, UCSD.

10 "Greeley's Poet in Residence Slates First Reading June 14," *Greeley Tribune*, June 5, 1970, 23.

11 Carol Carmody, "Welch Satisfied with Stay," in *UNC Summer Mirror* 1, no. 6 (July 24, 1970): 7.

12 Carmody, "Welch Satisfied with Stay."

13 Handwritten list of poems to write, n.d., box 3, folder 18, Lew Welch Papers, MSS 13, Mandeville Special Collections Library, UCSD.

14 The event was planned to last seventy-two hours but was cancelled by the church after only eight because of what Pat Nolan later described as "inevitable anarchy." In the end, Welch read the sermon in poet Steve Carey's living room. See https://thenew-blackbartpoetrysociety.wordpress.com/2014/12/19/steve-carey-smith-going-backward/.

15 Whalen's lectures on Welch were part of a series called "In the Pressure Tank." They can be found at https://archive.org/details/naropa?and[]=creator%3A%22whalen%2C+philip%22.

CHAPTER 27

1 Finis Dunaway, *Seeing Green: The Use and Abuse of American Environmental Images* (Chicago: University of Chicago Press, 2015), 105.

2 "Slick Hits Bolinas—East Bay," *San Francisco Examiner*, January 20, 1971, p. 4.

3 "Gloomy Marin City Poet Vanishes in Mother Lode," *Daily Independent Journal*, San Rafael, California, May 28, 1971.

4 Donna L. Potts, *The Wearing of the Deep Green* (Cham: Palgrave Macmillan, 2018), 13.

5 Handwritten journal entry by Gary Snyder, January 10, 1971, box i 86, folder 7, Gary Snyder Papers, D-050, Special Collections, UC Davis Library, University of California, Davis.

6 Jeff Cregg, interview with the author, October 16, 2017, San Rafael, California.

7 Welch (IR2), 179.

8 Magda Cregg, *Hey Lew* (Bolinas, CA: Magda Cregg, 1997), 91.

9 This poem, which was apparently written for Lloyd Reynolds, was sent to Donald Allen after Welch's disappearance by the poet Gene Detro. Allen then sent it on to Dorothy Brownfield.

10 A stupa is a monument built in memory of the Buddha. They were originally built to commemorate the eight great deeds that Buddha accomplished during his life and to enshrine relics after his death.

11 Handwritten poem, "What the Turkey Buzzard Said," by Lew Welch, box 3, folder 8, Lew Welch Papers, MSS 13, Mandeville Special Collections Library, UCSD.

12 Correspondence from Lew Welch to Dorothy Brownfield, May 14, 1958, box 1, folder 16, Lew Welch Papers, MSS 13, Mandeville Special Collections Library, UCSD.

13 Tom Killion and Gary Snyder, *Tamalpais Walking* (Berkeley, CA: Heyday, 2010), 2.

14 Killion and Snyder, *Tamalpais Walking*, 13.
15 Britannica, eds., "Pradakshina," *Encyclopaedia Britannica*, July 16, 2012, https://www.britannica.com/topic/pradakshina.
16 Handwritten journal entry by Gary Snyder, October 30, 1966, box I 85, folder 4, Gary Snyder Papers, D-050, Special Collections, UC Davis Library, University of California, Davis.
17 Jeff Cregg was more forthright in his condemnation of Brownfield, suggesting that as a "co-dependent" she had intentionally exacerbated her son's alcoholism as a means of manipulating him.
18 Philip Whalen, *In the Pressure Tank*, class 12, part 1, Jack Kerouac School of Disembodied Poetics, 58:10, https://archive.org/details/80P112.
19 Katsunori Yamazato, "Kitkitdizze, Zendo, and Place: Gary Snyder as a Reinhabitory Poet," *ISLE* 1, no. 1 (Spring 1993): 51–63, http://www.jstor.org/stable/44085343; Gary Snyder, "What Happened Here Before," *Turtle Island* (New York: New Directions, 1974), 78.
20 Carl Nolte, "Poet Steve Sanfield, Sierra Storytelling Festival Founder, Dies," *SFGate*, February 12, 2015, http://www.sfgate.com/bayarea/article/Poet-Steve-Sanfield-Sierra-Storytelling-Festival-6078140.php.
21 James Campbell, *Syncopations: Beats, New Yorkers, and Writers in the Dark* (Berkeley: University of California Press, 2008), 149.
22 Dana Goodyear, "Zen Master: Gary Snyder and the Art of Life," *New Yorker*, October 20, 2008.
23 Bill Morgan, ed., *The Selected Letters of Allen Ginsberg and Gary Snyder* (Berkeley: Counterpoint, 2009), 130.
24 Paul Messersmit-Glavin, "Between Social Ecology and Deep Ecology: Gary Snyder's Ecological Philosophy," in *The Philosophy of the Beats* (Lexington: University Press of Kentucky, 2011), 243.
25 David Schneider, *Crowded by Beauty* (Oakland: University of California Press, 2005), 182.
26 Handwritten journal entry by Gary Snyder, May 26, 1971, box i 86, folder 7, Gary Snyder Papers, D-050, Special Collections, UC Davis Library, University of California, Davis.
27 Handwritten journal entry by Philip Whalen, Kyoto, May 1971, box 3, folder 5, Philip Whalen papers, BANC MSS 2000/93 p, The Bancroft Library, University of California, Berkeley.
28 Handwritten journal entry by Philip Whalen, Kyoto, May 1971.
29 Handwritten journal entry by Philip Whalen, Kyoto, May 1971.
30 Handwritten journal entry by Philip Whalen, Kyoto, May 1971.
31 "The Vulture" was published in the London weekly *Universal Chronicle* as part of a series of 103 essays called "The Idler" between 1758 and 1760; it was no. 22.

EPILOGUE

1 Clifford Burke, letter to the author, August 21, 2017.
2 Correspondence from Dorothy Brownfield to Donald Allen, January 31, 1973, box 74, folder 3, Donald Allen Collection, MSS 3, Special Collections & Archives, UC San Diego.

Index

Italicized numbers refer to figures. An "n" designates notes.